Solar Energy Storage

Solar Energy Storage

Edited by

Bent Sørensen
Roskilde University, Denmark,
Department of Environmental,
Social and Spatial Change,
http://energy.ruc.dk
boson@ruc.dk

AMSTERDAM • BOSTON • HEIDELBERG • LONDON
NEW YORK • OXFORD • PARIS • SAN DIEGO
SAN FRANCISCO • SINGAPORE • SYDNEY • TOKYO
Academic Press is an imprint of Elsevier

Academic Press is an imprint of Elsevier
125 London Wall, London, EC2Y 5AS, UK
525 B Street, Suite 1800, San Diego, CA 92101–4495, USA
225 Wyman Street, Waltham, MA 02451, USA
The Boulevard, Langford Lane, Kidlington, Oxford OX5 1GB, UK

Notices
Knowledge and best practice in this field are constantly changing. As new research and experience
broaden our understanding, changes in research methods, professional practices, or medical
treatment may become necessary.

Practitioners and researchers must always rely on their own experience and knowledge in evaluating
and using any information, methods, compounds, or experiments described herein. In using such
information or methods they should be mindful of their own safety and the safety of others, including
parties for whom they have a professional responsibility.

To the fullest extent of the law, neither the Publisher nor the authors, contributors, or editors, assume
any liability for any injury and/or damage to persons or property as a matter of products liability,
negligence or otherwise, or from any use or operation of any methods, products, instructions, or
ideas contained in the material herein.

British Library Cataloguing in Publication Data
A catalogue record for this book is available from the British Library

Library of Congress Cataloging-in-Publication Data
A catalog record for this book is available from the Library of Congress

ISBN: 978-0-12-409540-3

For information on all Academic Press publications
visit our website at http://store.elsevier.com/

Working together
to grow libraries in
developing countries

www.elsevier.com • www.bookaid.org

Contents

Part II
Economic Assessment of Solar Storage

Part III
Environmental and Social Impacts

Part IV
Case Studies

Contributors

Naoya Abe, Department of International Development Engineering, Tokyo Institute of Technology, Tokyo, Japan

M.A. Baseer, Mechanical and Aeronautical Engineering Department, University of Pretoria, Pretoria, South Africa, and Jubail Industrial College, Jubail, Saudi Arabia

B. Böcker, Chair for Management Sciences and Energy Economics, University Duisburg-Essen, Essen, Germany

Alexandre Beluco, Instituto de Pesquisas Hidráulicas (IPH), Universidade Federal do Rio Grande do Sul (UFRGS), Porto Alegre, Brazil

Johann Caux, École d'Ingénieurs de l'Energie, l'Eau et l'Environnement (Ense3), Institut Polytechnique de Grenoble (INP), Grenoble, France

Paulo Kroeff de Souza, Instituto de Pesquisas Hidráulicas (IPH), Universidade Federal do Rio Grande do Sul (UFRGS), Porto Alegre, Brazil

R. Dhital, Alternative Energy Promotion Center, Ministry of Science, Technology and Environment, Lalitpur, Nepal

Yulong Ding, School of Chemical Engineering & Birmingham Centre of Energy Storage, University of Birmingham, Birmingham, UK

Z. Glasnovic, University of Zagreb, Zagreb, Croatia

Junichiro Ishio, Department of International Development Engineering, Tokyo Institute of Technology, Tokyo, Japan

J.K. Kaldellis, Lab of Soft Energy Applications & Environmental Protection, TEI of Piraeus, Athens, Greece

Teppei Katatani, Department of International Development Engineering, Tokyo Institute of Technology, Tokyo, Japan

Nolwenn Le Pierrès, LOCIE, CNRS UMR 5271-Université de Savoie, Polytech Annecy-Chambéry, Campus Scientifique, Savoie Technolac, Le Bourget-Du-Lac Cedex, France

Yongliang Li, School of Chemical Engineering & Birmingham Centre of Energy Storage, University of Birmingham, Birmingham, UK

Chuanping Liu, Department of Thermal Engineering, University of Science and Technology Beijing, Beijing, China

Flávio Pohlmann Livi, Instituto Mensura, Porto Alegre, Brazil

Lingai Luo, Laboratoire de Thermocinétique de Nantes (LTN-CNRS UMR 6607), La Chantrerie, Nantes Cedex, France

Iain MacGill, School of Electrical Engineering and Telecommunications, University of NSW, Sydney, NSW, Australia

B. Mainali, Energy and Climate Studies, Royal Institute of Technology, KTH, Stockholm, Sweden

K. Margeta, University of Zagreb, Zagreb, Croatia

Roman Marx, Institute of Thermodynamics and Thermal Engineering (ITW), Research and Testing Centre for Thermal Solar Systems (TZS), University of Stuttgart, Stuttgart, Germany

Ashmore Mawire, Department of Physics and Electronics, Northwest University, Mmabatho, South Africa

Toshihiro Mukai, Central Research Institute of Electric Power Industry, Tokyo, Japan

F. Rahman, Center for Refining & Petrochemicals, Research Institute, and Center of Research Excellence in Renewable Energy, King Fahd University of Petroleum & Minerals, Dhahran, Saudi Arabia

S. Rehman, Center for Engineering Research, Research Institute, King Fahd University of Petroleum and Minerals, Dhahran, Saudi Arabia

Bent Sørensen, Department of Environmental, Social and Spatial Change, Roskilde University, Roskilde, Denmark

B. Steffen, Chair for Management Sciences and Energy Economics, University Duisburg-Essen, Essen, Germany

Ze Sun, National Engineering Research Centre for Integrated Utilization of Salt Lake Resources, East China University of Science and Technology, Shanghai, China

Muriel Watt, School of PV and Renewable Energy Engineering, University of NSW, Sydney, NSW, Australia

C. Weber, Chair for Management Sciences and Energy Economics, University Duisburg-Essen, Essen, Germany

H.A. Zondag, Department of Mechanical Engineering, Eindhoven University of Technology, Eindhoven, The Netherlands

Preface

The aim of this book is to provide a state-of-the-art description and discussion of the energy storage issues relevant for most solar energy systems, particularly for situations in which solar energy constitutes more than a marginal fraction of the total energy supply. The disposition of the material is as follows. First, an overview of solar systems with energy storage is given, separately for solar electricity and solar thermal systems. Then, two more detailed presentations of solar storage options are given, with emphasis on suitability for particular geographical and thus climatic situations, again divided between electric power and heat applications. After that come three specific discussions of sorption heat stores and hydroelectric storage systems, including comparison with battery storage.

Following these technology chapters, two chapters describe the economic status of solar storage facilities, from both a theoretical and a practical point of view. Advantages of solar and other renewable energy systems may be modest when looking only at direct economic viability, but the two next chapters add a discussion of indirect economic factors that invite employment of a full life-cycle analysis and assessment. These factors include environmental concerns as well as social concerns influencing the choices made by societies and individual consumers.

Rounding up the book's survey, three chapters present case studies of the particular challenges that solar systems meet on remote islands and locations far from existing power grids, looking at cooking as well as basic electricity demands for communication, education, and local machinery. The alternative for economic comparison in these cases would be grid extension, over land or by sea cables. The functioning of local solar systems with storage is analyzed both for individual systems and for units connected into a mini-grid, exploring the mini-grid's possible advantage by reducing storage requirements.

The subjects treated in the book are timely because solar technologies have recently experienced a substantial cost reduction that makes it worthwhile to explore barriers to reaching a sizable fraction of energy supply under the different conditions prevailing in different parts of the world, and when possible to devise solutions to overcome these barriers, even in cases where increased solar energy penetration implies augmented need for energy storage solutions.

Bent Sørensen

Chapter 1

Introduction and Overview

Bent Sørensen
Department of Environmental, Social and Spatial Change, Roskilde University, Roskilde, Denmark

For some five decades, the cost of solar photovoltaic systems has declined from space industry levels toward affordability for individual consumers, helped by the occasional government subsidy programs in effect in one country or another for a limited time. The current decade is seeing solar panel prices being increasingly accepted by customers without subsidies; consequently, a rapid augmentation of installed capacity has taken place (IEA, 2014). A similar but less dramatic development has happened for solar thermal systems, having become common first as hot water providers, and now also for space heating in the intermediate latitude regions that need winter space heating but are still close enough to the equator to have a decent number of sunshine hours in winter. These developments make it timely to think of the obstacles for solar energy techniques to eventually become a major component in our energy systems. Among these problems, intermittency seems to be the most important. Either solar energy should be combined with other energy sources capable of complementing supply during dark hours, or devices for storing and regenerating energy must be part of the solar energy system. The best solution is probably a combination of including energy sources with a different delivery profile, having energy storage available, and taking advantage of energy transmission and trade possibilities, as well as any load management options that would be acceptable to the energy users.

This book focuses on energy stores suitable for integration into solar energy systems for delivering electric or thermal power to the end users. The combination of all the methods hinted at above is the subject of another recent book on energy intermittency, being the first to chart the global intermittency problem and to look at all the options for a sustainable energy system based on several renewable energy sources, storage, trade, and demand management, including the possible synergies offered by combining the solutions for dealing with intermittency (Sørensen, 2014).

In looking specifically at energy storage, two distinct areas of application for solar energy can be discerned. One is that of storage systems aimed at handling the obvious day-to-night storage requirement, caused by human activities not being restricted to hours of high solar radiation input. The second area is the seasonal storage that will store energy generated in summer or adjacent periods for use in winter at mid or high latitudes. Electricity demand varies over the year, with differences between different types of societies, but it is still fairly evenly distributed over the year. This is at least true compared with the situation for solar space heating, where the demand is obviously inversely correlated with solar supply, as soon as one departs substantially from the equator. Between these two extremes are the intermittency problems caused by passage of weather systems, with clouds and other atmospheric obstacles to receiving solar radiation at the Earth's surface. If solar intermittency is to be handled by energy stores, the storage systems must comprise units operating on diurnal, weekly, and seasonal time scales.

Several of the energy stores currently developed and ready for use are directed at shorter periods of storage. The shortest storage periods, related to fluctuations of solar energy production over seconds or minutes, are usually unimportant for heating purposes; for electric power production, any grid-connected system with some kilometer spacing between solar panels will smooth out such small-scale fluctuations. For isolated systems, short-term fluctuations may be handled by adding capacitor stores or flywheels, but even in this case the importance is declining, because the present generation of electricity-using equipment is either not very sensitive to modest variations in voltage or frequency (if AC), or it has built-in short-term storage, usually in the form of supercapacitors.

Moving on to storage levels capable of handling the variations in solar energy production during the day, or the nighttime nonproduction, the relevant storage periods would be 12-18 h, considering that the reduced solar radiation away from noontime may not correlate with the variations in demand. Stores of interest for this length of drawing time (in most cases also with a prolonged store-filling time) include various types of batteries, compressed air underground stores, pumped hydro, as well as fuel containers, which in a renewable energy scenario would have to contain fuels that can be created from biomass or electricity (such as certain biofuels and hydrogen). A range of such stores will be described and evaluated in the following chapters.

Finally, for storage lasting more than a week (as needed during episodes of heavy cloud cover or extended rainy periods) and seasonal storage (needed at latitudes with a substantial difference between summer and winter levels of incoming solar radiation, and for heat also because of temperature-related variations in space-heating demands), there are fewer storage options available. Economic considerations rule out storage in containers as compressed air and probably also in batteries, due to the cost being proportional to the amount of energy stored. Pumped hydro will remain an option as long as the upper reservoir is large enough, and so will fuel storage if containers can be replaced by inexpensive underground cavity options. Such cavities can be found in abandoned oil or gas wells, or they can be formed in certain aquifers and geological

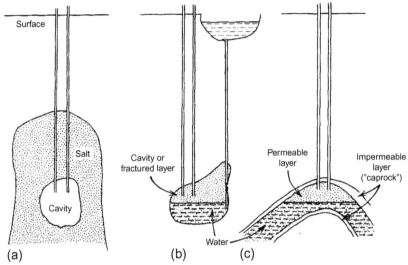

FIGURE 1.1 Types of underground gas stores (for hydrogen, natural gas, or air at various levels of compression): (a) storage in salt domes, optionally using canisters if cavity walls are too porous or do not accept liners; (b) drilled rock cavities or abandoned wells, possibly with a compensating surface reservoir; and (c) storage in an upward bend of an aquifer running between nonporous layers (e.g., of clay). *(From Sørensen (2010), used with permission from Elsevier.)*

salt deposits (Figure 1.1). Both aquifer bends offering sections suitable for displacing water by a gas and slowly washed out cavities in salt domes are in substantial use today for natural gas storage, such as the strategic stores common in some European countries, holding 4-6 months of gas demand as a buffer against disruptions caused by major rupture of undersea gas supply pipelines or by political delivery blockades (Russian gas).

In connection with solar power and other sources of power generated by variable renewable energy flows, the relevant gas to store would be hydrogen. This is independent of the fate of the current efforts to develop hydrogen fuel cells for the automotive sector, because the solar-produced hydrogen stored could be used to regenerate electricity by conventional gas turbines, a low-cost technology available today. By the way, one should remember that fuel cells and electrolyzers are the same thing: all fuel cells can be used in reversed mode to generate hydrogen from electricity. Currently, common high-efficiency electrolyzers are alkaline fuel cells, but also the proton-exchange membrane fuel cells relevant for the transportation sector or the solid-oxide fuel cells aimed at electric power sector applications may be operated in reverse mode as electrolyzers (electricity to hydrogen) (Sørensen, 2011).

The underground hydrogen stores seeming to offer the most viable solution for longer time storage of solar electricity are often missing from the scientific literature on stores for renewable energy systems, but have in recent years taken the foreground in national and regional planning studies both in the United States (Lord et al., 2011) and in the European Union (Simón et al., 2014). Here, they are dealt with in the chapter on environmental issues.

As regards thermal energy storage, the most common technology is heat capacity storage using water, either in connection with individual building solar installations or in a central location for district heating based on solar collector arrays. Alternatives using gravel or other materials have been tested, as have phase change materials, but without pointing to any clear winners. In many locations, the best option for covering winter space loads in the absence of sufficient solar energy collection is to use renewable electricity or electricity from stores through heat pumps (Sørensen, 2014). A selection of such systems is discussed in the chapters below.

REFERENCES

IEA, 2014. Trends 2014 in photovoltaic applications. PVPS report T1-25:2014. International Energy Agency, Paris.

Lord, A., Kobos, P., Klise, G., Borns, D., 2011. A life cycle cost analysis framework for geological storage of hydrogen: a user's tool. Report SAND2011-6221. Sandia National Laboratories, Albuquerque/Livermore.

Simón, J., et al., 2014. Assessment of the potential, the actors and relevant business cases for large scale and long term storage of renewable electricity by hydrogen underground storage in Europe. Final report D 6.3 from the European Commission funded project "HyUnder", Zaragosa.

Sørensen, B., 2010. Renewable Energy—Its Physics, Engineering, Environmental Impacts, Economics & Planning. fourth ed. Academic Press/Elsevier, Burlington (5th edition currently in preparation).

Sørensen, B., 2011. Hydrogen and Fuel Cells, second ed. Academic Press/Elsevier, Burlington.

Sørensen, B., 2014. Energy Intermittency. CRC Press, Taylor & Francis, Baton Rouge.

Part I

Solar Energy Storage Options

Chapter 2

Solar Electrical Energy Storage

Yulong Ding[1], Yongliang Li[1], Chuanping Liu[2] and Ze Sun[3]
[1]*School of Chemical Engineering & Birmingham Centre of Energy Storage, University of Birmingham, Birmingham, UK*
[2]*Department of Thermal Engineering, University of Science and Technology Beijing, Beijing, China*
[3]*National Engineering Research Centre for Integrated Utilization of Salt Lake Resources, East China University of Science and Technology, Shanghai, China*

Chapter Outline

2.1 BACKGROUND

Modern societies become increasingly dependent on reliable and secure supplies of electricity to underpin economic growth and community prosperity. This makes electricity an important vector in current and future energy systems, with the latter particularly related to electrification of heat and transportation. In the United Kingdom, the current end user demand on electricity accounts for around 18%. Under the Carbon Plan scenarios, this share will increase to

25-31% by 2030 and 33-44% by 2050 (Taylor et al., 2013). Globally, the net electricity generation will increase by 93% in the IEO2013 reference case, from 20.2 trillion kWh in 2010 to 39.0 trillion kWh in 2040 (EIA, 2013). This implies that the world electricity generation will have to rise by 2.2% per year from 2010 to 2040, compared with an average growth of 1.4% per year for all delivered energy sources. Electricity will supply an increased share of the world's total energy demand, and hence become the world's fastest growing form of delivered energy.

Currently, electricity is produced mainly from fossil fuels. However, due to the long-term pernicious effects of greenhouse gas emissions on the environment, the decreased availability of fossil fuel resources, and the growing sense of urgency toward energy security, the use of more and more renewable and environmentally sustainable energy resources is inevitably happening and is expected to be dominant in the foreseeable future. Solar energy is regarded as a leading contender for green energy production. In fact, solar power installations are currently increasing by 40% per year worldwide (Ginley et al., 2008). According to the estimation of Energy Technology Perspectives 2014, solar power could be the dominant source of green energy by 2050 (IEA, 2014).

Solar energy can be converted to electrical energy in two main ways (Li et al., 2012). One is through solar cells (photovoltaic technology), which directly convert the short wave range of solar radiation energy into electrical energy. The other is via an indirect solar thermal route, which converts the solar radiation energy into thermal energy by means of solar collectors or concentrators, which then generates electricity through a conventional thermal cycle. However, sunlight is diffuse and intermittent. Weather conditions also determine the availability; power generation using both the technologies is unpredictable and unreliable. Therefore, substantial use of solar power to meet humanity's needs requires electrical energy storage to ensure a reliable power supply.

2.2 TECHNICAL REQUIREMENTS OF A SOLAR ELECTRICAL ENERGY STORAGE FACILITY

Currently, solar cells and solar thermal power systems cover a wide range of applications, from less than 1 W to 100s MW, as shown in Figure 2.1 (Quaschning and Muriel, 2001). It should be noted that solar thermal power plants can only use direct solar irradiance for power generation, while solar cells can convert both direct irradiance and diffuse irradiance. Therefore, solar cells can produce some electricity even with cloud-covered skies, making them applicable even with very low solar irradiation. Generally speaking, solar cells are most suitable for small-scale low-power demands, which are able to operate as standalone systems as well as grid-connected systems, whereas a solar thermal power plant is often a better option for large-scale and grid-connected

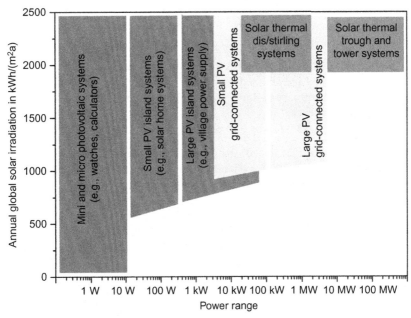

FIGURE 2.1 Operational regime for solar power generation.

systems. Due to different applications, there are different configurations to create a solar electricity installation with a solar cell facility, a solar thermal power plant, or both. However, from the role in the electricity supply chain, the installation can be classified into two categories: utility-scale solar electrical facility and distributed solar electrical facility.

Utility-scale solar power generation refers to medium- to large-scale solar energy installations, which can either be thermal power plants or solar cells. These units are designed to generate large amounts of electricity, which require large vacant lands and therefore are located in rural or semi-wild regions. As a result, they are far from end users and have to be physically connected to existing grids at discrete points. With increasing solar electricity penetration, utility-scale energy storage systems are required to provide utility-controlled functions, including long-duration electricity shift and capacity firming. Although there are no recognized standards at present, it is expected that the storage systems should have a maximum power rating of 1-20 MW (charging and discharging) and the ability to store 2-6 h of energy for on-demand delivery to the electric grid (EPRI, 2011). With such capacity, the storage system can provide a tremendous advantage to solar power generation efficiency and production, while lessening the negative effects of solar power generation on the grid. The energy storage systems are also expected to be used as a spinning reserve to delay committing additional fossil fuel power generation units. These imply that the following are required for energy storage systems for utility-scale solar power generation:

- Storage properties—high storage capacity, long charge/discharge times, good partial-load feature, and acceptable round-trip efficiency
- Financial performance—low capital cost, easy to maintain, and environment-friendly
- Other aspects—fast start-up and response for load following

On the other hand, distributed solar power generation refers to small- to medium-scale systems. Such systems are most commonly solar cell based, except for dish/Stirling solar thermal power systems. They are designed to generate moderate amounts of electricity, which require a small amount of land; hence, they can be placed in local electrical distribution systems at both the generation and use points. They could either be stand-alone systems, or they could be used to generate more electrical energy in conjunction with nearby installations. Compared to utility-scale systems, the generated electricity from distributed systems can only be used locally in most cases, rather than sold to electricity grids. As a result, a robust energy storage system that can charge/discharge more frequently is a necessity in order to offer inherently high service reliability to local electrical systems. The main characteristics required for energy storage technologies in distributed solar electricity systems include load response, round-trip efficiency, lifetime, and reliability.

2.3 OPTIONS FOR SOLAR ELECTRICAL ENERGY STORAGE TECHNOLOGIES

Except for thermal energy storage (TES) in concentrated solar power and solar fuels, electricity is generated by solar radiation first before charging into storage units. As a result, current available electrical energy storage technologies are potential options for solar electrical energy storage. These technologies can be categorized according to the following forms of stored energy (Chen et al., 2009):

- Electrical and magnetic forms: (i) Electrostatic energy storage (capacitors and supercapacitors); (ii) Magnetic/current energy storage (superconducting magnetic energy storage)
- Mechanical form: (i) Kinetic energy storage (flywheels); (ii) Potential energy storage (pumped-hydroelectric storage and compressed air energy storage (CAES))
- Chemical form: (i) Electrochemical energy storage (conventional batteries such as lead-acid, nickel metal hydride, lithium-ion, and flow batteries such as zinc bromine and vanadium redox); (ii) Chemical energy storage (hydrogen, solar fuel)

- Thermal form: (i) Cold and cryogenic energy storage (CES) (e.g., in solid and liquid materials); (ii) Heat storage (sensible heat storage using, for example, solid and liquid materials; latent heat storage using phase change materials); (iii) Thermochemical energy storage using heat of a reversible reaction or sorption process (thermochemical energy storage may also be regarded as energy stored in the chemical form); (iv) Combined TES (pumped thermal electricity storage (PTES))

The previous categorization is widely applied; however, more fundamentally, all energy storage technologies can be categorized into two forms of kinetic and/or potential energy at different spatial and time scales. Another method of classification is as follows, depending on the relationship between storage medium and charging/discharging devices:

- Coupled energy storage: If the charging/discharging devices are fully integrated with the storage medium, the technology is regarded as coupled. Examples of such storage technologies include rechargeable batteries, flywheels, capacitors, and so on. The advantage of coupled energy storage is the fast response rate and high efficiency, which allow the technology to manage short timescale fluctuations on the electrical supply chain.
- Decoupled energy storage: If the storage medium can be fully separated from the charging/discharging device, namely if the storage is in an independent container, the system is considered to be decoupled energy storage technology. Examples of decoupled energy storage technologies include pumped-hydro storage (PHS), CAES, flow batteries, TES, and hydrogen/fuel storage. In decoupled energy storage systems, the energy content is only limited by the storage vessel and is therefore potentially cheaper for long charging/discharging period applications.

Coupled energy storage technologies deal with "power" applications, while decoupled energy storage technologies are suitable for "energy" applications. Figure 2.2 (Li, 2011; Li et al., 2011) shows a map of the application regime of most storage technologies. Decoupled storage technologies with a large energy storage capacity, such as PHS and CAES, TES, flow batteries, and solar fuels/hydrogen, can provide enough capacity to smooth diurnal fluctuations in solar power supply and are therefore suitable for utility-scale solar electrical storage. On the other hand, as both charging and discharging processes have energy loss, the round-trip efficiency of decoupled electrical energy storage technologies is lower. For small-scale distributed solar electrical facilities, rechargeable batteries are better options because they offer faster responses and higher round-trip efficiencies, although the capital cost is higher. Flywheels and supercapacitors can smooth very short-term fluctuations (at millisecond levels), such as those caused by line faults. However, they are very expensive for use in a solar electricity system for "energy" management, and therefore will not be discussed in this chapter.

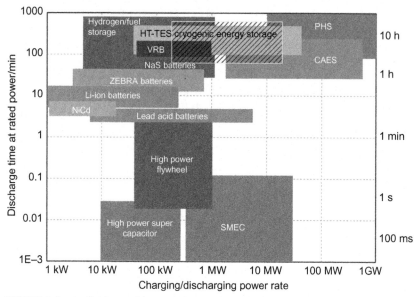

FIGURE 2.2   Application regime map of energy storage technologies (Note that the figure is intended for conceptual purposes only, and many of the options could have broader duration and power ranges than shown.).

2.4   UTILITY-SCALE STORAGE TECHNOLOGIES

2.4.1   Pumped-Hydro Storage

PHS is the most widely used and technically matured bulk electrical energy storage technology. It currently accounts for approximately 98% of all global energy storage installations. PHS is based on the hydroelectric principle and hence does not have issues associated with SOx, NOx, greenhouse gases, and particulate matter emissions. PHS stores excessive electricity as the potential energy of raised water against gravity. Due to the use of efficient water wheels or hydraulic turbines to extract the kinetic energy of moving water and convert the energy into electricity through generators, it can provide ancillary services with better part-load efficiency, better controllability, lower maintenance costs, and minimal start-up costs (Succar and Williams, 2008). These advantages enable PHS (as well as hydroelectric generation) to play a central role in the operation of large-scale electricity grids, in particular when there is a large share of renewable electricity such as solar power. As a stabilizer for electricity grids, PHS systems could ensure that solar electricity supply is both regular and steady.

The major drawback of PHS technology lies in the fact that suitable and available sites for water reservoirs, essential for a PHS plant, are scarce, and in most cases are far from solar power plants. As a result, some new ideas have been developed in the past decade to address the issues. The following are two examples:

- Underground PHS uses underground caverns or water structures as lower reservoirs (Gonzalez et al., 2004). Conceptually, this seems a logical and sound solution for energy storage. However, the practical design and actual construction of large underground PHS systems are highly challenging. Consequently, this is still in the stage of feasibility study, and no project has ever been built (Crotogino et al., 2001; Schulte, 2001).
- Ocean PHS uses the high-elevation coastal region as the upper reservoir to retain ocean water, and the ocean itself as the lower reservoir. Challenges associated with this technology include reservoir sealing and corrosion of the pump turbine (DIT, 2004). Despite the challenges, there is already an ocean PHS in Okinawa, Japan. This plant, the first in the world, has been in operation for more than 10 years (EPRI, 2012).

Other recent developments in the area of PHS technology are the introduction of adjustable-speed motor/generators. Conventional PHS plants with a single-speed machine can only provide frequency regulation in the generation mode. With adjustable speed, it is possible to operate in pump mode as well in the range of 60-100% of rated capacity (Pimm and Garvey, 2009). This means an adjustable speed unit can change pumping power and provide load following and frequency regulation in both the generation mode and the pumping mode, making the technology more attractive in absorbing fluctuating solar power. Furthermore, with conventional single-speed machines, the pump turbines can only achieve their peak efficiency in one of the two modes, not both, whereas the adjustable technology can achieve both. Two Japanese companies, Toshiba and Hitachi, are leading the design and manufacture of high-power adjustable-speed motor/generators. In fact, eleven adjustable-speed pumped storage units have entered commercial service, with nine in Japan and two in Germany (Havel, 2011; Lima et al., 2004).

2.4.2   Compressed Air Energy Storage

CAES is a commercially available technology capable of providing very large energy storage capacity (above 100 MW in single units). Currently, CAES accounts for approximately 20% of global energy storage installations, except for pumped hydro. CAES is based on conventional gas turbine technology, with the compression and expansion processes decoupled. In a traditional gas turbine power station, around two-thirds of the turbine output is needed for compressing the combustion air (Ibrahim et al., 2008). In a CAES power station, however, no compression is needed during turbine operation because the required enthalpy is already included in the compressed air. As a result, the net power output for a given sized turbine expander can be 2.5 times that for the compressor load (Septimus, 2006).

Large CAES systems are designed as central storage facilities to cycle on a daily basis and to operate efficiently during partial-load conditions. This design approach allows CAES units to swing quickly from generation to compression modes, which can efficiently absorb or compensate solar power. CAES plants can also respond to load changes to provide load following, because they are designed to sustain frequent start-up/shut-down cycles.

The technical feasibility of CAES has been proven by two large-scale installations. These installations have been successfully running for decades, and as a result, CAES can be regarded as a mature technology. However, the fact that there have been no installations in recent years raises some concerns about the technology. (In fact, some planned large-scale CAES projects have been either terminated/canceled or postponed.) Technically, this may be associated with low round-trip efficiency and economics. The low round-trip efficiency is due to the following aspects:

- The air compression process is highly irreversible, particularly for high-pressure ratios.
- Storage of compressed air is commonly under constant volume conditions. This leads to a great loss of energy due to the throttling process of higher pressure charging and lower pressure discharging (Jovan et al., 2011).
- The expansion process is irreversible, though it is less serious compared to the compression process.

The combination of the previous points gives relatively low round-trip efficiency of around 40%. Even for the new planned facilities in the United States, the round-trip efficiency levels are expected to be between 42% and 54%, depending on the way the waste heat is used (Safaei et al., 2013).

To resolve the issues, recent developments of CAES technology have been in the storage vessel and energy conversion processes, which could ultimately contribute to round-trip efficiency improvement and cost reduction. These include constant-pressure storage vessel (e.g., water-compensated container (Kim et al., 2011), underwater container (Pimm and Garvey, 2009), adsorption-enhanced CAES) and novel energy conversion processes (e.g., near-isothermal CAES (Li, 2011; Li et al., 2011), advanced adiabatic CAES (Bullough et al., 2004)). However, most of these are still in their early research stage, and no reports have been seen on demonstrative scale systems.

2.4.3 Thermal Energy Storage

Thermal energy storage (TES) refers to technologies that store a thermal form of energy, which can be heat or cold. TES technologies can be broadly divided into two categories of thermophysical and thermochemical, with the principle of the former based on the temperature difference and/or latent heat due to phase change, and that of the latter on reaction or sorption enthalpy. Thermophysical-based TES is much more technically developed than thermochemical-based TES. Globally,

TES technologies account for approximately 54% of global energy storage installations, except for pumped hydro. TES technologies can be used in both supply side and demand side management. In the demand side, TES is largely distributed and mostly at an individual building scale for domestic heating or cooling. For example, almost 14 million households in the United Kingdom have a hot water cylinder, giving a maximum combined storage capacity of around 80 GWh (Taylor et al., 2013). However, only recently has it been recognized that TES can also be an effective means of electricity management at the supply side.

Potential applications of TES technologies are ample. Examples include solar electrical energy management, particularly solar thermal power generation such as concentrating solar power (CSP), cryogenic energy storage (CES), Pumped Thermal Electricity Storage (PTES), and advanced adiabatic CAES (Bullough et al., 2004).

CSP is unique compared with solar photovoltaics and other renewable energy generation methods because it can be effectively integrated with TES. In addition, TES can decouple the thermal energy production process from the thermal energy utilization (power generation) process in CSP plants. This makes TES highly attractive due to fewer energy conversion steps and hence lower energy losses. At the heart of TES technologies is the storage medium (material). Currently, solar thermal power generation uses synthetic oil (e.g., VP1) and molten salt (e.g., solar salt) as the storage materials. Because these materials also act as heat transfer fluid, they are in the liquid form. As a result, they only store sensible heat with limited energy density. The use of phase change materials and the thermochemical method (with high energy density) is still in the development stage (Gil et al., 2010).

CES is a newly developed electrical energy storage technology that uses cryogen (e.g., liquid air/nitrogen) as the storage medium (Li, 2011; Li et al., 2011). The charging process is through air liquefaction, which consumes excessive electricity, whereas the discharging process is liquid air/nitrogen fuelled power generation. CES is potentially attractive for utility-scale electrical energy storage, as cryogen can be stored at low pressures (close to ambient) and has a much higher volumetric storage density (>10 times that of CAES and >100 times that of PHS) (Li et al., 2010). Additionally, cryogen is not only the energy carrier, but also the working fluid in the energy release process. Because cryogen has a relatively low critical point compared to steam, waste heat (e.g., from the exhaust of a new or existing simple-cycle gas turbine) can be recovered efficiently in the CES system (around four times more efficiently than in the Organic Rankine Cycle system) to achieve high energy-storage efficiencies (Strahan, 2013). However, without an external heat source, an independent CES system has relatively low efficiency (below ~60%) due to the high energy consumption of the air liquefaction process (Chen et al., 2009). Another potential issue is that current liquid nitrogen/oxygen production facilities are designed for continuous operation. This is in contradiction to the main function of the storage system to absorb excessive electricity. CES is a pre-commercial technology, with a 350 kW/2.5 MWh scale demonstration in operation since 2011 and a 5 MW/15 MWh demonstration plant to be completed in early 2015 in the United Kingdom (HPS, 2014). Current research and development efforts are mainly on round-trip efficiency improvement through process integration, novel liquefaction process for better flexibility, and system scale-up.

PTES is a recently proposed concept based on reversible heat pump cycles (Howes, 2012). In the charging process, the system works as a heat pump converting electricity into thermal energy (both heat and cold); in the discharging process, the system works as a heat engine to convert cold and heat into electricity. The greatest challenge of the PTES technology is associated with the mechanical components (compressor-expander coupled machinery), which need to have high efficiencies in both forward (charging process) and reverse (discharging process) operations (Desrues et al., 2010). In particular, a PTES system has to use an inert gas such as helium or argon instead of commonly used refrigerants to produce high-grade thermal energy (below $-150\ °C$ at the cold end and above 600 °C at the hot end). The high-grade heat and cold are stored as "sensible heat" in insulated cylindrical vessels containing an appropriate thermal storage medium (e.g., a packed bed of pebbles or gravel, or a matrix of ceramic material). Heat exchange is through direct contact between the working fluid and storage medium, enabling a quick switch between charging/discharging modes and efficient integration with the thermodynamic cycle, and avoiding the "pinch-point" difficulties associated with phase-change storage methods. However, because a complete charging and discharging cycle involves twice as many compression, expansion, and heat transfer processes, the round-trip efficiency of PTES is limited due to the irreversibility of these processes. PTES technology is still in the early stage of development, and according to a UK company's plan, a 1.5 MW/6 MWh demonstrative storage unit will be deployed on a UK grid-connected primary substation in the near future (ETI, 2013).

2.4.4 Flow Battery

Modern redox flow batteries (RFBs) were invented in 1976 by Lawrence Thaller at the National Aeronautics and Space Administration (NASA) (Thaller, 1976). RFBs convert electrical energy into chemical potential energy by means of a reversible electrochemical reaction between two liquid electrolyte solutions contained in external electrolyte tanks. The conversion between electrical energy and chemical energy occurs as the liquid electrolytes from storage tanks flow through electrodes in a cell stack. In contrast to conventional batteries, they store energy in electrolyte solutions and the energy is proportional to the amount of electrolytes in the tanks (Ferreira et al., 2013). The cell has two compartments (positive half-cell and negative half-cell) separated by a membrane to prevent mixing of the electrolytes (Wang et al., 2013; Tan et al., 2013). Compared with other decoupled electrical energy storage technologies, RFBs have relatively high round-trip efficiencies, short response times, a symmetrical charge and discharge process, and quick cycle inversion.

RFB energy storage systems are being developed for use in small-scale applications, such as stand-alone power systems with solar photovoltaic arrays or wind turbines, and distributed energy installations for electric utility services. The first true RFB used a ferric/ferrous (Fe^{2+}/Fe^{3+}) halide solution electrolyte in the positive half-cell and a chromic/chromos (Cr^{2+}/Cr^{3+}) halide solution electrolyte in the negative half-cell. The battery soon encountered severe cross-contamination that resulted in dramatic capacity decay. The effort made to mitigate the cross-contamination led to the invention of the all-vanadium redox flow battery (VRB), which concerns only one active element in both positive and negative electrolytes, utilizing four different oxidation states of vanadium ions to form the redox couples. VRBs could have a round-trip efficiency of >70% (Skyllas-Kazacos and Robins, 1988). They have an expected lifespan of 15 years with a low environmental impact (Rydh, 1999), and as a result, a considerable amount of fundamental and applied research has been carried out (Zhao et al., 2006).

In addition to VRBs, there has been work on other flow batteries (Chakrabarti et al., 2011; Liu et al., 2013; Yan et al., 2013; Ponce et al., 2006). Some of them have been studied for solar energy storage (Chakrabarti et al., 2011; Liu et al., 2013; Yan et al., 2013). Examples include ruthenium-based RFBs (Chakrabarti et al., 2011) and a combination of RFBs with dye-sensitized solar cells (Liu et al., 2013; Yan et al., 2013).

2.4.5 Solar Fuels

Solar fuels store solar energy in a chemical form for use when sunlight is not available. The most widely investigated solar fuels are hydrogen and hydrocarbon (Licht et al., 2001; Yamada et al., 2003). There are a number of methods for solar fuel production. For example, hydrogen production can be done by water electrolysis, thermolysis, and photolysis. These production methods are either indirect or direct. Indirect hydrogen production occurs through converting solar radiation into another form of energy, such as heat and electricity, followed by fuel production. Direct methods harness solar energy to produce fuels without any intermediary conversion steps. In principle, the indirect methods have the disadvantage of low efficiency due to losses in the intermediary conversion steps, but practically, they are easier to implement. Today, many advanced large-scale water electrolysis units are in operation with an electricity-to-hydrogen efficiency of over 75%, while most other solar hydrogen production processes are still at the stage of laboratory research (Licht et al., 2001; Yamada et al., 2003). Production of solar fuels is not the only challenge, particularly for solar hydrogen; challenges are also found in hydrogen storage and electricity regeneration steps. Although hydrogen can be stored as compressed gas, liquefied gas, or in a solid hydrogen carrier, none of these are currently energy-efficient or cost-effective. Fuel cells are currently regarded as a promising energy extraction technology for hydrogen. They are about four times more expensive than combustion engines/turbines, and two to three times shorter in terms of lifespan (Li et al., 2010). Significant efforts are therefore needed to address these issues.

Instead of hydrogen production, combining the water-splitting process with CO_2 can generate liquid fuels (Rakowski and Dubois, 2009; Benson et al., 2009; Centi et al., 2007). These fuels have higher volumetric energy densities compared to hydrogen, and could potentially alleviate challenges associated with hydrogen storage. The liquid solar fuel routes have been shown to be achievable in research labs, but scaling up to commercial level faces a number of challenges, including system efficiency and lifetime and capital costs (Cook et al., 2010).

2.5 DISTRIBUTED SCALE STORAGE TECHNOLOGIES— RECHARGEABLE BATTERIES

A rechargeable battery stores and releases energy through reversible chemical reactions (Ribeiro et al., 2001). Battery energy storage systems have been shown to play an important role in household demand smoothening (Purvins et al., 2013). Integrated renewable electrical energy systems often use rechargeable batteries, particularly in regions with distributed power systems and in remote areas. To meet the actual demands, rechargeable battery systems are sometimes integrated with weather forecasts and market signals, and are always colocated with renewable energy resources to improve system stability through frequency response and ramp control and to improve the economics of the renewable generator through leveling the output of solar generators (Hill et al., 2012). Several types of batteries could be used for distributed solar electrical energy system, and they are described in the following subsections. As will be seen, the main difference between these batteries lies in the electrodes and electrolytes, which also determine their specific characteristics.

2.5.1 Lead-Acid Battery

The lead-acid battery, invented in 1859 by Gaston Plante, is the oldest and most widely used rechargeable electrochemical device in automobile, uninterrupted power supply (UPS), and backup systems for telecom and many other applications. Such a device operates through chemical reactions involving lead dioxide (cathode electrode), lead (anode electrode), and sulfuric acid (electrolyte) (Parker and Garche, 2004). Lead-acid batteries have a high round-trip efficiency, and are cheap and easy to install. It is the affordability and availability that make this type of battery dominant in the renewable energy sector. It is also well known that lead-acid batteries have low energy density and short cycle life, and are toxic due to the use of sulfuric acid and are potentially environmentally hazardous. These disadvantages imply some limitations to this type of battery.

Indeed, a recent study on economic and environmental impact suggests that lead-acid batteries are unsuitable for domestic grid-connected photovoltaic systems (McKenna et al., 2013).

2.5.2   Lithium-Ion Battery (Li-Ion)

The first lithium battery was built in 1979 (Mizushima et al., 1980). Lithium batteries in the early days suffered from poor cycle life and safety problems. It was the development of lithium-cobalt batteries with carbon as negative electrodes that led to the successful commercialization of lithium-ion batteries by Sony in 1990. The term "lithium batteries" actually means a family of dozens of different battery technologies based on moving lithium ions between a positive electrode consisting of a lithium and transition metal compound and a negative electrode material. Due to the nature of the reactions involved and the structure of electrodes, lithium-ion batteries have a much longer cycle life than lead acid batteries do (McDowall, 2008). Lithium-based batteries have high round-trip efficiency, high energy and power density, and a low self-discharge rate, and have been widely used in portable electronics. High power density also makes lithium-ion batteries an option for powering electrical vehicles. Efforts have also been made to extend the application range to solar energy storage applications (Guo et al., 2012).

2.5.3   Nickel-Based Battery

Nickel-based batteries mainly refer to nickel-cadmium (Ni-Cd), nickel-metal hydride (Ni-MH), and nickel-zinc (Ni-Zn) batteries. Ni-Cd batteries consist of a positive electrode with nickel oxyhydroxide as active material, and a metallic cadmium-based negative electrode with aqueous potassium hydroxide as electrolyte (Shukla et al., 2001). The advantages of such batteries include robustness to deep discharges, long cycle life, temperature tolerance, and high energy density (compared with lead-acid batteries). However, the cadmium used in Ni-Cd batteries is highly toxic. This leads to the development of Ni-MH batteries, which are much more environmentally friendly, though they have a high self-discharge rate. Despite various developments, only Ni-Cd batteries have found commercial application in UPS systems. Recently, Ni-Cd batteries became one of the popular storage technologies for solar energy generation because they can withstand high temperatures, though the high initial investment of Ni-Cd battery systems may hinder widespread application in the sector (Nair and Garimella, 2010).

2.5.4   Sodium-Sulfur Battery

The development of sodium-sulfur (NaS) battery technology has been ongoing for more than 50 years. NGK of Japan was the first to commercialize the technology successfully. NaS batteries consist of liquid (molten) sulfur at the positive electrode and liquid (molten) sodium at the negative electrode as active material, separated by a solid beta alumina ceramic electrolyte (Hadjipaschalis et al., 2009). They have high energy density, long cycle life, and high round-trip energy efficiency. These features make the technology suitable for load leveling, power quality, and peak shaving in distributed energy systems, including solar power (Baker, 2008). In fact, current global installations of NaS batteries exceed 300 MW, despite recent concerns about the safety aspects.

2.5.5   Other Battery Technologies

Intensive research efforts have been made in recent years, seeking new storage technologies that can economically provide the power, cycle life, and energy efficiency needed to respond to the short-term transients arising from renewables and other aspects of grid operation. The following are some examples of recent developments.

A new type of aqueous electrolyte battery has been proposed and demonstrated (Pasta et al., 2012). Such a technology relies on the insertion of potassium ions into a copper hexacyanoferrate cathode and a novel activated carbon/polypyrrole hybrid anode. The cathode reacts rapidly, with very little hysteresis. The hybrid anode uses an electrochemically active additive to tune its potential. It has been demonstrated that a 95% round-trip energy efficiency can be achieved when cycled at a 5C rate, and the efficiency reduces to 79% when cycled at 50C. The results also show a zero-capacity loss after 1000 deep-discharge cycles.

An aqueous rechargeable sodium-ion battery has recently been proposed. This technology uses NaCuHCF as cathode, $NaTi_2(PO_4)_3$ as anode, and aqueous Na_2SO_4 solution as electrolyte. Lab experiments have shown a voltage of 1.4 V and a specific energy of 48 Wh/kg based on the total weight of the active electrode materials (Wu et al., 2014). In addition, the aqueous Na-ion battery has also shown an excellent high-rate discharge capability and cycle capability, with 88% capacity retention over 1000 cycles at a 10C rate.

A novel symmetric open-framework electrode battery is proposed and demonstrated experimentally (Pasta et al., 2014). The battery is shown to provide a maximum specific energy of 27 Wh/kg at a 1C rate based on the mass of the active material (Pasta et al., 2014). At a 50C rate, the battery has a specific energy of 15 Wh/kg, a specific power of 693 W/kg, and an energy efficiency of 84.2%. These performance data suggest that this type of battery is suitable for grid-related applications, including smoothing of intermittent variations in power production associated with the integration of renewable energy.

2.6   ECONOMICS OF SOLAR ELECTRICAL ENERGY STORAGE TECHNOLOGIES

Apart from technical performance, capital and operating costs also determine whether an electrical energy storage technology is viable for solar power generation. The principal components for comparing different electrical energy storage technologies are the costs per unit charged/discharged power that storage can deliver ($/kW), and costs per unit energy capacity ($/kWh) stored in the storage system. Although it is difficult to evaluate a specific technology because the costs are influenced by a wide range of factors, including system size, location, local labor rate, market variability, environmental considerations, and transport/access issues, Figure 2.3 (Chen et al., 2009; IME, 2014) shows a comparison that provides a high-level understanding of the issues. From the view of capital cost, the lead-acid battery is the best option for distributed-scale solar power storage, while TES is preferred for utility-scale solar power storage.

In practical applications, however, the lifetime cost has to be considered. This is affected mainly by two additional factors: round-trip efficiency and cycle life. Cycle life refers to the number of charge and discharge cycles that a storage device can provide before performance decreases to an extent that it cannot perform the required functions. This is extremely critical for distributed-scale applications. Figure 2.4 shows the round-trip efficiency and cycle life of different energy storage technologies (Chen et al., 2009; IME, 2014). One can see that the lifetime of lead-acid batteries is much shorter than that of other technologies.

From Figures 2.3 and 2.4, one can also find that hydrogen/fuel storage is very expensive and of low round-trip efficiency. This makes it an unfavorable option for stationary applications. However, it should be noted that

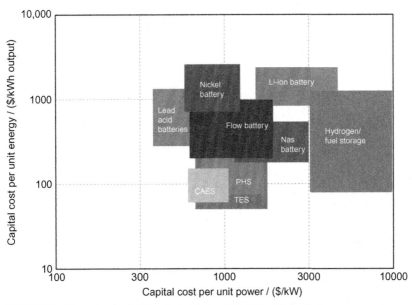

FIGURE 2.3 Capital costs of different electrical energy storage technologies.

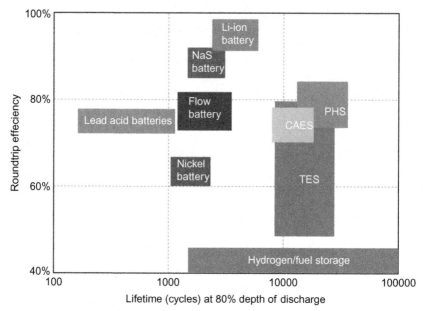

FIGURE 2.4 Round-trip efficiency and lifetime of different electrical energy storage technologies.

hydrogen/fuel is the only dispatchable option that is a potential replacement for transportation fuels.

2.7 FINAL REMARKS

Solar electrical energy storage is an emerging area that is still under extensive research and development. Many challenges remain to be overcome. Solar electrical power is rarely used as the only energy resource in an energy system or a grid. In most cases, solar power is integrated with other renewable energy sources or fossil-fuel-based power generation to provide a reliable electricity supply to end users. As a result, the electrical energy storage units are used to provide services to the integrated system/grid instead of a solo solar power facility (with the exception of TES in a CSP plant). The selection of specific storage technology must therefore be considered at a system/grid level, including not only the characteristics of solar power generation, but also features of other energy resources as well as the characteristics of transmission and end users.

REFERENCES

Baker, J., 2008. New technology and possible advances in energy storage. Energy Policy 36, 4368–4373.

Benson, E.E., Kubiak, C.P., Sathrum, A.J., Smieja, J.M., 2009. Electrocatalytic and homogeneous approaches to conversion of CO_2 to liquid fuels. Chem. Soc. Rev. 38, 89–99.

Bullough, C., Gatzwn, C., Jakiel, C., Koller, M., Nowi, A., Zunft, S., 2004. Advanced adiabatic compressed air energy storage for the integration of wind energy. In: European Wind Energy Conference, London, UK.

Centi, G., Perathoner, S., Wine, G., Gangeri, M., 2007. Electrocatalytic conversion of CO_2 to long carbon-chain hydrocarbons. Green Chem. 9, 671–678.

Chakrabarti, M.H., Roberts, E.P.L., Bae, C., Saleem, M., 2011. Ruthenium based redox flow battery for solar energy storage. Energy Convers. Manag. 52, 2501–2508.

Chen, H., Cong, T.N., Yang, W., Tan, C., Li, Y., Ding, Y., 2009. Progress in electrical energy storage system: a critical review. Prog. Nat. Sci. 19, 291–312.

Cook, T.R., Dogutan, D.K., Reece, S.Y., Surendranath, Y., Teets, T.S., Nocera, D.G., 2010. solar energy supply and storage for the legacy and nonlegacy worlds. Chem. Rev. 110, 6474–6502.

Crotogino, F., Mohmeyer, K.U., Scharf, R., 2001. In: Huntorf CAES: More Than 20 Years of Successful Operation, Orlando, Florida, USA.

Desrues, T., Ruer, J., Marty, P., Fourmigué, J.F., 2010. A thermal energy storage process for large scale electric applications. Appl. Therm. Eng. 30, 425–432.

DIT, 2004. Review of Electrical Energy Storage Technologies and Systems and of their Potential for the UK. Department of Trade and Industry, London.

EIA, 2013. International energy outlook 2013, Washington, DC.

EPRI, 2011. Functional requirements for electric energy storage application on the power system grid, Palo Alto, California, Product ID 1021936.

EPRI, 2012. U.S. energy storage project activities—demonstrations & commercial installations.

ETI, 2013. Delivering the UK's future energy technologies: affordable, clean and secure energy solutions for 2050.

Ferreira, H.L., Garde, R., Fulli, G., Kling, W., Lopes, J.P., 2013. Characterisation of electrical energy storage technologies. Energy 53, 288–298.

Gil, A., Medrano, M., Martorell, I., Lázaro, A., Dolado, P., Zalba, B., et al., 2010. State of the art on high temperature thermal energy storage for power generation. Part 1—concepts, materials and modellization. Renew. Sust. Energ. Rev. 14, 31–55.

Ginley, D., Green, M.A., Collins, R., 2008. Solar energy conversion toward 1 terawatt. MRS Bull. 33, 355–364.

Gonzalez, A., Gallachoir, B., McKeogh, E., 2004. Study of electricity storage technologies and their potential to address wind energy intermittency in Ireland.

Guo, W., Xue, X., Wang, S., Lin, C., Wang, Z.L., 2012. An integrated power pack of dye-sensitized solar cell and Li battery based on double-sided TiO_2 nanotube arrays. Nano Lett. 12, 2520–2523.

Hadjipaschalis, I., Poullikkas, A., Efthimiou, V., 2009. Overview of current and future energy storage technologies for electric power applications. Renew. Sust. Energ. Rev. 13, 1513–1522.

Havel, T.F., 2011. Adsorption-enhanced compressed air energy storage. In: Proceeding for Clean Technology Conference & Expo 2011, Boston, MA, June 13–16.

Hill, C.A., Such, M.C., Dongmei, C., Gonzalez, J., Grady, W.M., 2012. Battery energy storage for enabling integration of distributed solar power generation. IEEE Trans. Smart Grid 3, 850–857.

Howes, J., 2012. Concept and development of a pumped heat electricity storage device. Proc. IEEE 100, 493–503.

Ibrahim, H., Ilinca, A., Perron, J., 2008. Energy storage systems—characteristics and comparisons. Renew. Sust. Energ. Rev. 12, 1221–1250.

IEA, 2014. Energy Technology Perspectives 2014. IEA, France.

IME, 2014. Energy storage: the missing link in the UK's energy commitments.

Jovan, I., Marija, P., Shawn, R., Jesse, G., David, W., Chungyan, S., 2011. In: NETL, (Ed.), Technical and Economic Analysis of Various Power Generation Resources Coupled with CAES Systems. National Energy Technology Laboratory, USA.

Kim, Y.M., Shin, D.G., Favrat, D., 2011. Operating characteristics of constant-pressure compressed air energy storage (CAES) system combined with pumped hydro storage based on energy and exergy analysis. Energy 36, 6220–6233.

Li, Y., 2011. Cryogen Based Energy Storage: Process Modelling and Optimisation. PhD Thesis University of Leeds, Leeds.

Li, Y., Chen, H., Zhang, X., Tan, C., Ding, Y., 2010. Renewable energy carriers: hydrogen or liquid air/nitrogen? Appl. Therm. Eng. 30, 1985–1990.

Li, P.Y., Loth, E., Simon, T.W., Van, V.D., Crane, S.E., 2011. Compressed air energy storage for offshore wind turbines.

Li, Y., Wang, X., Jin, Y., Ding, Y., 2012. An integrated solar-cryogen hybrid power system. Renew. Energy 37, 76–81.

Licht, S., Wang, B., Mukerji, S., Soga, T., Umeno, M., Tributsch, H., 2001. Over 18% solar energy conversion to generation of hydrogen fuel; theory and experiment for efficient solar water splitting. Int. J. Hydrogen Energy 26, 653–659.

Lima, A.C., Guimaraes, S.C., Camacho, J.R., Bispo, D., 2004. Electric energy demand analysis using fuzzy decision-making system. In: Proceedings of the 12th IEEE Mediterranean Electrotechnical Conference, vols. 1–3, pp. 811–814.

Liu, P., Cao, Y. l, Li, G.R., Gao, X.P., Ai, X.P., Yang, H.X., 2013. A solar rechargeable flow battery based on photoregeneration of two soluble redox couples. ChemSusChem 6, 802–806.

McDowall, J., 2008. Understanding Lithium-Ion Technology. Battcon, Marco Island.

McKenna, E., McManus, M., Cooper, S., Thomson, M., 2013. Economic and environmental impact of lead-acid batteries in grid-connected domestic PV systems. Appl. Energy 104, 239–249.

Mizushima, K., Jones, P.C., Wiseman, P.J., Goodenough, J.B., 1980. Li_xCoO_2 $(0 < x < 1)$: a new cathode material for batteries of high energy density. Mater. Res. Bull. 15, 783–789.

Nair, N.C., Garimella, N., 2010. Battery energy storage systems: assessment for small-scale renewable energy integration. Energy Build. 42, 2124–2130.

Parker, C.D., Garche, J., 2004. Battery energy-storage systems for power-supply networks. In: Rand, D.A.J., Garche, J., Moseley, P.T., Parker, C.D. (Eds.), Valve-Regulated Lead-Acid Batteries. Elsevier, Amsterdam, pp. 295–326 (Chapter 10).

Pasta, M., Wessells, C.D., Huggins, R.A., Cui, Y., 2012. A high-rate and long cycle life aqueous electrolyte battery for grid-scale energy storage. Nat. Commun. 3 (2139), 1149.

Pasta, M., Wessells, C.D., Liu, N., Nelson, J., McDowell, M.T., Huggins, R.A., et al., 2014. Full open-framework batteries for stationary energy storage. Nat. Commun. 5, 3007.

Pimm, A., Garvey, S., 2009. Analysis of flexible fabric structures for large-scale subsea compressed air energy storage. J. Phys. Conf. Ser. 181, 012049.

Ponce, D.C., Frías-Ferrer, A., González-García, J., Szánto, D.A., Walsh, F.C., 2006. Redox flow cells for energy conversion. J. Power Sources 160, 716–732.

Purvins, A., Papaioannou, I.T., Debarberis, L., 2013. Application of battery-based storage systems in household-demand smoothening in electricity-distribution grids. Energy Convers. Manag. 65, 272–284.

Quaschning, V., Muriel, M.B., 2001. Solar power—photovoltaics or solar thermal power plants. In: Proceeding VGB Congress Power Plants 2001, Brussels, pp. 1–8.

Rakowski, D.M., Dubois, D.L., 2009. Development of molecular electrocatalysts for CO_2 reduction and H_2 production/oxidation. Acc. Chem. Res. 42, 1974–1982.

Ribeiro, P.F., Johnson, B.K., Crow, M.L., Arsoy, A., Liu, Y., 2001. Energy storage systems for advanced power applications. Proc. IEEE 89, 1744–1756.

Rydh, C.J., 1999. Environmental assessment of vanadium redox and lead-acid batteries for stationary energy storage. J. Power Sources 80, 21–29.

Safaei, H., Keith, D.W., Hugo, R.J., 2013. Compressed air energy storage (CAES) with compressors distributed at heat loads to enable waste heat utilization. Appl. Energy 103, 165–179.

Schulte, R., 2001. Iowa stored energy park project terminated, Des Moines.

Septimus, V.D.L., 2006. Bulk energy storage potential in the USA, current developments and future prospects. Energy 31, 3446–3457.

Shukla, A.K., Venugopalan, S., Hariprakash, B., 2001. Nickel-based rechargeable batteries. J. Power Sources 100, 125–148.

Skyllas-Kazacos, M.R., Robins, R., 1988. All-vanadium redox battery. US patent 4,786,567.

Strahan, D., 2013. Liquid air in the energy and transport systems: opportunities for industry and innovation in the UK. CLCF.

Succar, S., Williams, R.H., 2008. Compressed Air Energy Storage: Theory, Resources, and Applications for Wind Power. Princeton Environmental Institute, Princeton University, Princeton.

Tan, X., Li, Q., Wang, H., 2013. Advances and trends of energy storage technology in microgrid. Int. J. Electr. Power Energy Syst. 44, 179–191.

Taylor, P.G., Bolton, R., Stone, D., Upham, P., 2013. Developing pathways for energy storage in the UK using a coevolutionary framework. Energy Policy 63, 230–243.

Thaller, L.H., 1976. Electrically rechargeable redox flow cells. US patent 3,996,046.

Wang, W., Luo, Q., Li, B., Wei, X., Li, L., Yang, Z., 2013. Recent progress in redox flow battery research and development. Adv. Funct. Mater. 23, 970–986.

Wu, X.Y., Sun, M.Y., Shen, Y.F., Qian, J.F., Cao, Y.l., Ai, X.P., et al., 2014. Energetic aqueous rechargeable sodium-ion battery based on $Na_2CuFe(CN)_6$–$NaTi_2(PO_4)_3$ intercalation chemistry. ChemSusChem 7, 407–411.

Yamada, Y., Matsuki, N., Ohmori, T., Mametsuka, H., Kondo, M., Matsuda, A., et al., 2003. One chip photovoltaic water electrolysis device. Int. J. Hydrogen Energy 28, 1167–1169.

Yan, N.F., Li, G.R., Gao, X.P., 2013. Solar rechargeable redox flow battery based on Li_2WO_4/LiI couples in dual-phase electrolytes. J. Mater. Chem. A 1, 7012–7015.

Zhao, P., Zhang, H., Zhou, H., Chen, J., Gao, S., Yi, B., 2006. Characteristics and performance of 10 kW class all-vanadium redox-flow battery stack. J. Power Sources 162, 1416–1420.

Chapter 3

Innovative Systems for Storage of Thermal Solar Energy in Buildings

Lingai Luo[1] and Nolwenn Le Pierrès[2]

[1]*Laboratoire de Thermocinétique de Nantes (LTN-CNRS UMR 6607), La Chantrerie, Nantes Cedex, France*

[2]*LOCIE, CNRS UMR 5271-Université de Savoie, Polytech Annecy-Chambéry, Campus Scientifique, Savoie Technolac, Le Bourget-Du-Lac Cedex, France*

Chapter Outline

Solar Energy Storage. http://dx.doi.org/10.1016/B978-0-12-409540-3.00003-7

NOMENCLATURE

Notations

A	area (m^2)
C_p	specific heat ($J\ kg^{-1}\ °C^{-1}$)
h	specific enthalpy ($J\ kg^{-1}$)
Δh_{cr}	specific enthalpy of solution of LiBr·2H$_2$O ($J\ kg^{-1}$)
k	mass fraction of anhydrous LiBr in the crystal hydrate
LiBr	lithium bromide
m	mass flow rate ($kg\ s^{-1}$)
m%	mass percent
M	mass (kg)
M_{LiBr}	mass of anhydrous LiBr (kg)
P	pressure (Pa)
q	heat (J)
Q	power (W)
SHX	solution heat exchanger
t	time (s)
T	temperature (°C)
T_{tank}	ambient temperature around the storage tanks (°C)
U	heat transfer coefficient ($W\ m^{-2}\ °C^{-1}$)
v	volume flow ($m^3\ s^{-1}$)
w	work (W)
x	mass fraction of lithium bromide in the solution (m%)

Greek Symbols

α	absorption percentage
η	efficiency
ρ	density ($kg\ m^{-3}$)

Indices

b	building
c	condenser external loop fluid
cr	crystal
end	end of the process
eq	equilibrium conditions
ext	exterior (outside)
g	generator external loop fluid
i	inlet
ini	beginning of the process
int	interior (building)
is	isentropic
liq	liquid
loss	loss

max	maximum
o	outlet
p	propylene glycol
ref	surrounding
sh	tank shell
shext	external side of the tank shell
shint	internal side of the tank shell
sol	solution
v	vapor phase
w	water
1	solution storage
2	generator
3	condenser or evaporator
4	water storage
1→2	tube connecting 1-2
'	value from previous time step

3.1   INTRODUCTION

With the diminishing reserves of fossil fuels, the increasing energy demand, and the greenhouse gas emissions rise, energy consumption is at a critical stake. Developing more energy-efficient and environmentally friendly devices is essential to reach the "3E" objectives, conciliating clean environment, sustainable energy policy, solid economy, and social development.

The main role of storage systems is to reduce the time or rate mismatch between supply and demand (Figure 3.1). Solar energy seems to be the most promising renewable energy source (Goswami et al., 2000; Asif and Muneer, 2007). However, its intermittent and unstable nature is a major drawback, which leads to a disparity between supply and demand. Some of its fluctuations are predictable (day/night, season), but some are unpredictable (weather effect). Energy storage is an appropriate method of correcting these disparities. To enhance the fraction of energy use and make solar systems more efficient, practical, and attractive, heat storage systems are perceived as crucial components in solar energy applications.

Methods of solar thermal energy storage are mainly divided into three types: sensible, latent, and sorption/thermochemical (Hadorn, 2008; Odru, 2010).

FIGURE 3.1   Trend shift of energy chain between energy supply and demand sides.

Sensible and latent thermal storage have been the most studied technologies in the past decades. Most thermal storage devices used in practical solar systems involve sensible or latent storage methods. For sensible storage, heat is stored by the temperature difference of the storage medium; the value of its storage density depends on its specific heat and the temperature change.

In the latent thermal storage method, heat is stored through the phase change of a material at a relatively constant temperature. The materials are often referred to as phase change materials (PCMs). Compared to sensible thermal storage, latent heat storage has higher storage density and a much smaller temperature change; but it still involves important drawbacks, such as long-term stability of storage properties, low thermal conductivity, phase segregation, and subcooling during the phase change process (Farid et al., 2004).

Thermochemical storage can be divided into chemical reaction and sorption. Large amounts of heat can be stored in such processes (Figure 3.2), where heat is stored by breaking the binding force between the sorbent and the sorbate in terms of chemical potential.

The choice criteria for the type of storage to select are numerous, and often lead to a contradictory optimum system and material:

- High energy density of the storage (kWh m^{-3})
- High heat delivery power (good transfer between the heat transfer fluid (HTF) and the storage medium)
- Mechanical, thermal, and chemical stability of the storage material
- Chemical compatibility between storage material, heat exchanger, and storage medium
- Complete reversibility for a large number of charging/discharging cycles
- Easy integration in the user's system
- Low thermal losses
- Low cost (low price per kWh of heat stored)
- Low environmental impact

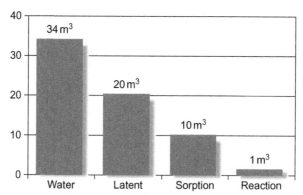

FIGURE 3.2 Volume required to store 1850 kWh (with consideration of 25% heat loss, based on a 70 °C temperature increase for water) (Hadorn, 2008).

In this chapter, the aforementioned three different solar heat storage systems in buildings will be presented. This chapter provides insight into recent developments about systems technologies and materials, their classification, limitations, and potential solutions for their application. According to the building needs, only the heat storage processes involving temperature levels between the ambient and 150 °C are considered. High temperature solar heat can also be stored, as discussed in Gil et al. (2010).

3.2   MAJOR TECHNOLOGIES FOR HEAT STORAGE IN BUILDINGS

3.2.1   Sensible Storage

One definition of sensible heat storage materials is that of materials that undergo no phase change in the temperature change of the heat storage process (Fernandez et al., 2010). The amount of energy q involved in a sensible heat storage process depends, as presented in Equation (3.1), on the specific heat of the material ($C_p(T)$), the temperature change (T_f, T_i), and the amount (M) of material.

$$q = \int_{T_{ini}}^{T_{end}} M \cdot C_p(T) \cdot dT \qquad (3.1)$$

Some disadvantages are inherent to sensible heat storage systems. The most important of them are their relatively low energy density and the self-discharge (heat losses) of the system, which can be important particularly when long periods of storage are the aim.

Sensible heat storage in buildings has been widely investigated throughout the literature and can be divided into two groups: liquid and solid storage mediums. Liquids are most often limited to water for this application, and solids are rocks, bricks, concrete, iron, dry and wet earth, and many others.

3.2.1.1   Liquid Storage

Water has been widely used for heat storage as well as for heat transport purposes in energy systems. It appears to be the best of sensible heat storage liquids for temperatures lower than 100 °C because of its availability, its inexpensiveness and, most important, its relatively high specific heat (Fath, 1998). For a 70 °C temperature change (from 20 to 90 °C), water can store 290 MJ m^{-3}. Today, it is also the most widely used storage medium for solar-based warm water and space heating applications. There exists a large amount of published data on the design criteria for systems with water as a heat storage material (Garg et al., 1985; Abhat, 1980; Duffie and Beckman, 1989; Wyman et al., 1980).

3.2.1.2 Solid Storage

Solid media are widely used for low temperature storage. They usually consist of rocks, concrete, sand, bricks, and so on. Materials most commonly in use in buildings for solar heat storage are indeed the ones involved in the building structure, rock beds, and in borehole thermal storage. For solar heat storage in building applications, solid materials are mostly used for space heating and cooling (air conditioning) purposes. Their operating temperatures cover a wide range, from 10 to over 70 °C. The major drawback of using solids as heat storage materials is their low specific heat capacity ($\sim$1200 kJ m^{-3} K^{-1} in average), which results in a relatively low energy density (more than three times less energy stored than water in the same volume for the same temperature lift, for example). However, compared to liquid materials, two main advantages are inherent to solids materials: their viability at higher temperatures, and no leakage problem with their containment (Hasnain, 1998). The compatibility of the material with the HTF used is of importance (Fath, 1998). Also, the efficiency and the viability of heat storage systems with solid materials is strongly dependent on the solid material size and shape, the packing density, the type of HTF, the flow pattern, and so on.

3.2.2 Latent Heat Storage

By definition, latent heat storage is based on the heat absorbed or released when a material undergoes a phase change from one physical state to another. Practically, the phase change can occur through the following forms: solid-solid, solid-liquid, solid-gas, liquid-gas, and vice versa.

During the solid-solid transitions, heat is stored as the material is transformed from one crystalline form to another (Sharma et al., 2009). Due to the fact that there is just a change of the crystalline structure, small latent heat and small volume changes are generally observed in comparison to solid-liquid transitions. In return, solid-solid PCMs have the advantages of less stringent container requirements and greater design flexibility (Pillai and Brinkwarth, 1976). During the past decades, relatively few solid-solid PCMs that might have a heat of fusion and transition temperatures suitable for low temperature solar applications have been identified. One of the most promising is pentaglycerine (melting temperature 81 °C, latent heat of fusion 263 MJ m^{-3}) (Garg et al., 1985).

Solid-gas and liquid-gas transitions present higher latent heat, but the large volume change involved during the transition leads to more stringent containment requirements; they cannot be suitable for solar heat storage applications (Hasnain, 1998).

Solid-liquid PCMs have benefited from many developments during the two last decades, although the amount of heat involved during their phase change is comparatively smaller than that of solid-gas or liquid-gas PCMs. Solid-liquid PCMs can store and release a relatively large quantity of heat over a narrow temperature range, without a large volume change. Solid-liquid transitions have also proved economically attractive (Hasnain, 1998).

Globally, latent heat storage systems have some advantages on sensible heat storage systems. They are volumetric heat storage density, which is high; and operating temperature, which is relatively constant for PCM systems but can vary widely for sensible systems, corresponding to the load. As shown in Table 3.1, for the same amount of heat stored, latent heat storage systems using paraffin wax will need 1.5 times (or 3 times) less volume than sensible heat storage systems using water (or rocks), with a 50 °C temperature change. But there are some disadvantages associated with latent heat storage materials. Low thermal conductivity, material stability over several cycles, phase segregation, subcooling, and cost are some of the limitations that are currently under investigation (Dincer and Rosen, 2002; Farid et al., 2004; Sharma et al., 2009).

PCMs can be classified into the following major categories: inorganic PCMs, organic PCMs, and eutectic PCMs. Each of these groups can be further categorized into more detailed subgroups, as shown in Figure 3.3. Each of these groups has its typical range of melting temperature and melting enthalpy.

3.2.2.1   Inorganic PCMs

Inorganic PCMs are constituted of salt hydrates and metals.

Salts hydrates have been the most investigated for heat storage purposes at low temperatures. Salt hydrates can be considered alloys of inorganic salts (AB)

TABLE 3.1 Comparison of Various Heat Storage Media (For Sensible Storage Materials, Energy is Stored in the Temperature Range 25-75 °C) (Tatsidjodoung et al., 2013)

	Heat Storage Material			
	Sensible Heat Storage		Phase Change Materials	
Property	Rock	Water	Paraffin Wax	$CaCl_2 \cdot 6H_2O$
Latent heat of fusion (kJ kg^{-1})	[a]	[a]	174.4	266
Specific heat capacity (kJ kg^{-1} K^{-1})	0.9	4.18	[b]	[b]
Density (kg m^{-3}) at 24 °C	2240	1000	1802	795
Storage volume for storing 1 GJ (m^3)	9.9	4.8	3.2	4.7
Relative volume[c]	3.1	1.5	1.0	1.5

[a]Latent heat of fusion is not of interest for sensible heat storage.
[b]Specific heat capacity is not of interest for latent heat storage.
[c]Equivalent storage volume; reference taken on paraffin.

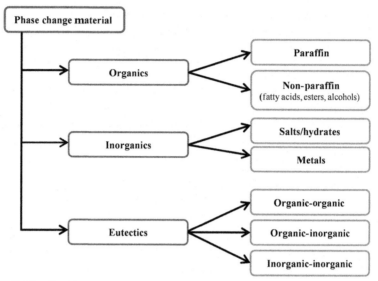

FIGURE 3.3 Classification of phase change materials (Tatsidjodoung et al., 2013).

and water (H_2O), resulting in a typical crystalline solid of general formula ($AB \cdot xH_2O$) (Sharma et al., 2009). Their phase change transition can actually be regarded as a dehydration or hydration of the salt, although this process can be assimilated to a melting or a freezing of the compound. Salt hydrates usually melt to either a salt hydrate with fewer moles of water, or to its anhydrous form. During the phase change transition, liquid water released from the hydrated salt dissolves the formed nonhydrated salt molecules.

Most of the selected metals with low melting temperature as members of the group of inorganic PCMs are fusible alloys of bismuth with other metals such as lead, tin, indium, and cadmium. As a general rule, the physical properties of these alloys will be close to those of bismuth, with some differences linked to the addition of the other compounds. Some important drawbacks of metals as energy storage media are their scarce availability and their high cost. However, when volume is a constraint, they are good candidates because of their high volumetric heat of fusion. Another advantage of these materials over other PCMs is their high thermal conductivity.

3.2.2.2 Organic PCMs

Organic heat storage materials are usually separated into two groups: paraffins and nonparaffin organic materials.

Paraffins can be defined as a mixture of pure alkanes, mostly straight chain configured. The crystallization/fusion of their molecular chain involves a large amount of latent heat. They are usually referred to as "paraffin waxes," with the chemical formula C_nH_{2n+2}, where $20 \leq n \leq 40$. The average heat of fusion of paraffins is 170 MJ m^{-3}, and is almost half the value of that of hydrated salts.

Nonparaffin organic PCMs are constituted of compounds such as fatty acids, esters, and alcohols (Hasnain, 1998; Sharma et al., 2009). They are highly flammable and therefore should not be exposed to intense temperature, flames, and oxidizing agents. It is the main drawback for their use as heat storage materials. Other features of these organic materials are reproducible melting and freezing behavior, freezing with no supercooling, low thermal conductivity, and varying level of toxicity. Their volumetric heat of fusion is comparable to that of paraffins.

3.2.2.3   Eutectics

Eutectics are alloys of inorganics (mostly hydrated salts) and/or organics. They have a single melting temperature, which is usually lower than that of any of the constitutive compounds. Eutectics form one single common crystal when crystallized (Hasnain, 1998). One of the most important characteristics of eutectics is their capability to melt/freeze congruently without phase segregation. A large number of eutectics have been reported throughout the literature and classified as inorganic, organic, and organic-inorganic eutectics. Organic eutectics have in general a lower melting temperature and a greater heat of fusion than inorganic eutectics do, making them adequate for low-temperature solar heat storage needs.

3.2.3   Sorption Heat Storage Systems

Sorption technologies, which used to be considered mainly for cooling and heat pumping, have gained a lot of interest for heat storage purpose in recent years, due to their high energy densities and long-term storage efficiency. Sorption heat storage is separated into four technologies: liquid absorption, solid adsorption, chemical reaction, and composite materials. After a general presentation of the different possible processes and the operating principle, we will illustrate an innovative technology: an absorption solar heat storage system.

3.2.3.1   Process Classification

Sorption technologies involve two chemical compounds: a sorbate and a sorbent. Sorbent/sorbate is called the "sorption couple." Depending on the link between the sorbent and the sorbate and the state of the sorbent, sorption can be divided into two types: solid/gas and liquid/gas (Figure 3.4). Moreover, based on the cohesive force between the two phases, adsorption is further divided into two types: physical adsorption (physisorption) and chemical adsorption (chemisorption) (N'Tsoukpoe et al., 2009).

Physical adsorption is a general phenomenon that occurs whenever an adsorbate is brought into contact with the surface of the adsorbent. The forces

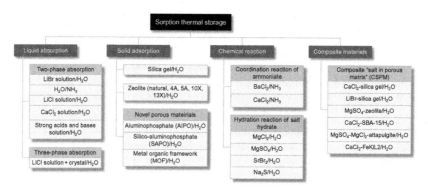

FIGURE 3.4 Sorption thermal storage classification (Yu et al., 2013).

involved are intermolecular forces (van der Waals forces). Chemical adsorption is due to covalent forces. The main difference between the two kinds of sorptions is the magnitude of the heat of adsorption (Yu et al., 2013). Because chemical forces are stronger than physical forces, heat of chemical sorption is usually larger than heat of physical sorption. Though this helps in recognizing physical and chemical adsorptions, in some cases it can be unclear which kind of adsorption (or both) is involved.

The word "absorption" is used when the molecules of the sorbate penetrate the surface layer and enter the structure of the bulk solid/liquid sorbent, causing the change of the composition of one or both bulk phases. Liquid absorption working pairs have been used in absorption chillers and heat pumps for decades.

3.2.3.2 Storage Operating Principle

The mechanism of heat storage through a sorption process can be represented by the following equation (Yu et al., 2013):

$$A \cdot (i+j)B + \Delta H_s \leftrightarrow A \cdot iB + jB \tag{3.2}$$

Here, A is the sorbent and B is the sorbate. For a chemical reaction process, $A \cdot (i+j)B$ and $A \cdot iB$ mean a compound of one mole of A with $(i+j)$ mole of B and i mole of B, respectively. For a liquid absorption process, $A \cdot (i+j)B$ represents a solution with a lower concentration of A than $A \cdot iB$. For a solid adsorption process, $A \cdot (i+j)B$ represents the enrichment of B on the surface of A, as j more mole of B is adsorbed.

During the endothermal charging process (direct sense of Equation (3.2)), when heat is added to $A \cdot (i+j)B$, the binding force between A and B is broken and some B is separated from A. Energy is stored in terms of the chemical potential ΔH_s (J mol^{-1}). During the exothermal discharging process (indirect sense of Equation (3.2)), $A \cdot iB$ is brought into contact with j mole of B to form $A \cdot (i+j)B$ again, and the heat ΔH_s is released.

3.3   FOCUS ON A SOLAR HEAT ABSORPTION STORAGE SYSTEM

To illustrate the most innovative storage technology, an example of one of the possible sorption processes will be detailed: an absorption solar heat storage system.

3.3.1   Basic Cycle Description

Liu et al. (2011) proposed a long-term solar heat storage absorption cycle for building heating. Figure 3.5 gives the operation principle of the system. The charging of solar energy takes place during summer and heat is released during winter. Comparative to heat sorption cooling systems, this process can be developed with only two heat exchangers that work reversely in the two functioning phases: a desorber in the charging phase that works as an absorber during the discharging phase, and a condenser in the charging phase that works as an evaporator during the discharging phase. The other originality of this cycle is the fact that it allows the solution to reach the crystallization point in the storage tank. Figure 3.6 describes the working principle of this cycle, using the $LiBr/H_2O$ sorption couple as an example. The high pressure (HP) level and low pressure (LP) level depend on the condensation temperature (point C) in summer and the evaporation temperature (point E) in winter, respectively. From point 2 to point 3 or 3′, solar heat is continually transferred from solar collectors to heat the solution (charging phase in Figure 3.5). The absorbate is driven out of the LiBr

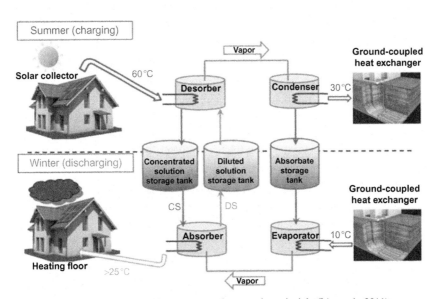

FIGURE 3.5   Long-term absorption storage cycle: operation principle (Liu et al., 2011).

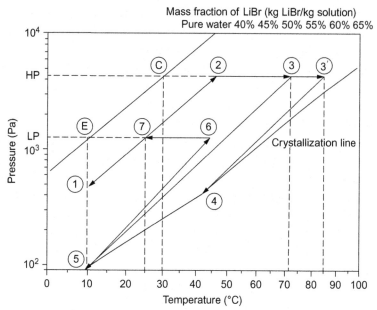

FIGURE 3.6 Long-term absorption storage cycle: working principle in pressure-temperature-concentration diagram for the LiBr-H₂O couple (Liu et al., 2011).

solution and transferred to the condenser. The strong solution at point 3 or 3′ flows back to the storage tank. Line 3-5 or 3′-5 represents the transition storage period during the summer to the winter, when the solution is kept in the tank and separated from other vessels. In winter, the absorbate is evaporated and then absorbed by the solution at point 5-7 (Figure 3.6 and discharging phase in Figure 3.5). Cycle 1-2-3-5-6-7-1 refers to a conventional long-term absorption storage cycle, in which point 3 should be carefully chosen to avoid the formation of crystal from point 3 to point 5. For the cycle 1-2-3′-4-5-6-7-1, the mass fraction of the solution at point 3′ can be higher than that at point 3. The solid crystal form of the salt is allowed to appear during the transition period in the solution tank. This improvement greatly enhances the storage density of the solution because of the increase of the interval of mass fraction between the weak and strong solutions, as will be explained in Section 3.3.2.3.

3.3.2   Process Modeling and Simulations

In order to evaluate the system performances (storage density, thermal and electrical efficiencies, required collector area, etc.) and identify the parameters that influence significantly these performances, a model has been developed for dynamic simulations of the process (N'Tsoukpoe et al., 2012). This model considers the transient characteristics of the system, inherent in the operating conditions of the process that change year-round: daily variations of the solar radiation, the environment (heat sink or source) and the building heating demand, variation of solution concentration in the storage, crystal formation, and so on. Thermal loss to the ambient, walls' thermal inertia, heat and mass transfer limitations in various components, and the energy consumption by auxiliary equipment (especially circulating pumps) are also considered.

3.3.2.1   System Modeling

The model is based on the energy and mass balance equations and the properties of the LiBr/H_2O couple (Saul and Wagner, 1987; Hellmann and Grossman, 1996). The main inputs of the model are the heat absorbed by solar collectors, the ambient temperature and the heating demand, the flow rate of the solution and water, and the overall heat transfer coefficients UA of the heat exchangers. The model calculates the system variables, such as the components' pressures and temperatures, the solution concentration in the generator and the solution tank, the mass of water in the water tank, the mass of solution and of crystal of LiBr hydrate in the solution storage, the circulating pumps' consumption, heat losses to the ambient, and so on. The details of this model can also be found in N'Tsoukpoe et al. (2012).

3.3.2.1.1   Generator

The internal temperature T_2 in the generator is assumed to be uniform (Figure 3.7) even if it varies along the heat exchanger due to the variation of the concentration of the solution (Patnaik et al., 1993). As in most studies on absorption systems and supported by various experiments, the generator model is based on heat transfer (Banasiak and Koziol, 2009). The power exchanged by the generator internal (solution) and external flows is evaluated by using the logarithmic mean temperature difference approach. The modeling rests on the assumption that the values of the overall heat transfer coefficient of the generator heat exchanger $(UA)_2$ are constant. Actually, the values of UA vary somewhat with the temperature as well as with the mass flows, but this variation

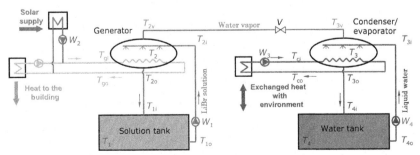

FIGURE 3.7   Schematics of the system and parameters/variables of the dynamic simulation (N'Tsoukpoe et al., 2012).

is relatively small in most cases (Patnaik et al., 1993; Grossman et al., 1995). The exchanged heat in the generator is therefore

$$Q_2 = (UA)_2 \cdot \frac{T_{gi} - T_{go}}{\ln\left(\dfrac{T_{gi} - T_2}{T_{gi} - T_2}\right)} = m_g \cdot Cp_{pw} \cdot (T_{gi} - T_{go}) \qquad (3.3)$$

Equations (3.4) and (3.5) are solved to estimate the generator heat losses to the ambient, assuming that the generator shell is isothermal. Equation (3.4) expresses the generator shell internal energy variation, which is the difference between the internal heat received by convection and the external one.

$$(MCp)_{2\,sh} \cdot \frac{dT_{2\,sh}}{dt} = (UA)_{2\,shint} \cdot (T_2 - T_{2\,sh}) - (UA)_{2\,shext} \cdot (T_{2\,sh} - T_{ref}) \qquad (3.4)$$

$$Q_{2\,loss} = (UA)_{2\,shint} \cdot (T_2 - T_{2\,sh}) \qquad (3.5)$$

In order to take into account the fact that the solution leaving the generator is not in equilibrium conditions, a mass transfer effectiveness α, also called absorption percentage or absorption equilibrium factor (Andberg and Vliet, 1983; Kaushik et al., 1985; George and Murthy, 1989; Patnaik et al., 1993), is introduced in Equation (3.9). The absorption percentage is the ratio of the actual change in concentration of the solution to the maximum possible change that could be obtained with an infinitely long plate. The maximum possible change is the difference between the generator inlet concentration and the equilibrium concentration.

Mass and energy balances in the generator yield Equations (3.6)–(3.15):

$$m_{2i} = m_{1o} = V_{1o} \cdot \rho_{sol}(T_1, x_{1liq}) \quad \text{mass continuity equation} \qquad (3.6)$$

$$m_{2i} = m_{2o} + m_{2v} \quad \text{global mass balance} \qquad (3.7)$$

$$m_{2i} \cdot x_{2i} = m_{2o} \cdot x_{2o} \quad \text{LiBr mass balance} \qquad (3.8)$$

$$x_{2i} - x_{2o} = \alpha \cdot (x_{2i} - x_{2oeq}) \qquad (3.9)$$

$$m_{2i} \cdot h_{2i} + Q_2 = m_{2o} \cdot h_{2o} + m_{2v} \cdot h_{2v} + Q_{2loss} \quad \text{energy balance} \qquad (3.10)$$

$$P_2 = P_{2v} = P_{3v} = P_3 \qquad (3.11)$$

$$P_2 = P_{sol}(T_2, x_{2o}) \quad \text{state balance} \qquad (3.12)$$

$$T_2 = T_{2o} = T_{2v} = T_{3v} \qquad (3.13)$$

$$m_{1o} \cdot h_{1o} + w_1 = m_{2i} \cdot h_{2i} + Q_{1 \to 2loss} \quad \text{energy balance on } 1 \to 2 \qquad (3.14)$$

$$h_{2o} = h_{sol}(T_2, x_{2o}) \quad \text{state equation} \qquad (3.15)$$

3.3.2.1.2 Condenser/Evaporator

The modeling of the condenser is similar to that of the generator. It is added to the previous assumptions that the amount of water vapor desorbed in the generator is completely condensed in the condenser. Conversely, all the water vapor produced in the evaporator is absorbed by the solution in the absorber.

$$m_{3i} = m_{4o} = v_{4o} \cdot \rho_{wliq}(T_4)$$

$$m_{3i} + m_{3v} = m_{3o} \quad \text{global mass balance} \tag{3.16}$$

$$m_{3i} \cdot h_{3i} + m_{3v} \cdot h_{3v} = m_{3o} \cdot h_{3o} + Q_{3loss} + Q_3 \quad \text{energy balance} \tag{3.17}$$

$$P_3 = P_w(T_3) \tag{3.18}$$

$$T_3 = T_{3o} \tag{3.19}$$

$$m_{4o} \cdot h_{4o} + w_4 = m_{3i} \cdot h_{3i} + Q_{4 \to 3loss} \quad \text{energy balance on tube } 4 \to 3 \tag{3.20}$$

$$h_{3o} = h_{wliq}(T_3) \quad \text{state equation (saturation)} \tag{3.21}$$

$$h_{3v} = h_{wv}(T_2, P_2) \quad \text{state equation} \tag{3.22}$$

3.3.2.1.3 Solution Tank

The tank model is based on global mass and energy balances. The solution tank is then assumed to be well mixed (temperature and concentration), and the mass of solution accumulated in the other components of the process is neglected in comparison to the mass of solution in the storage tank. Crystals are considered to be in equilibrium with the saturated solution when they appear. The following equations apply:

$$M_1 = M_1' + (m_{1i} - m_{1o}) \cdot dt \quad \text{global mass balance} \tag{3.23}$$

$$M_1 \cdot x_1 = M_{1liq} \cdot x_{1liq} + M_{cr} \cdot k = M_{LiBr} \quad \text{LiBr mass balance} \tag{3.24}$$

$$\begin{aligned} M_{1liq} \cdot h_{1liq} + M_{cr} \cdot h_{cr} &= M_{1liq}' \cdot h_{1liq}' \\ &+ M_{cr}' \cdot h_{cr}' + (m_{1i} \cdot h_{1i} - Q_{1loss} - m_{1o} \cdot h_{1o}) \cdot dt \quad \text{energy balance} \end{aligned} \tag{3.25}$$

$$Q_{1loss} = UA_1 \cdot (T_1, T_{tank}) \quad \text{heat loss to the ambient} \tag{3.26}$$

$$h_{1o} = h_{1liq} = h_{sol}(T_1, x_{1liq}) \quad \text{state equation} \tag{3.27}$$

$$P_1 = P_{sol}(T_1, x_{1liq}) \quad \text{state equation} \tag{3.28}$$

$$m_{1i} \cdot h_{1i} + Q_{2 \to 1loss} = m_{2o} \cdot h_{2o} \quad \text{energy balance on tube } 2 \to 1 \tag{3.29}$$

$$h_{cr} = h_{1liq} + \Delta h_{cr} \quad \Delta h_{cr} \approx 64.5 \text{kJkg}^{-1} (\text{Apelblat and Tamir, 1986}) \tag{3.30}$$

3.3.2.1.4 Water Tank

Assumptions in the water tank modeling are the same as in the case of the solution tank, except that there is no crystal:

$$M_4 = M_4' + (m_{4i} - m_{4o}) \cdot dt \quad \text{global mass balance} \tag{3.31}$$

$$M_4 \cdot h_4 = M'_4 \cdot h'_4 + (m_{4i} \cdot h_{4i} - Q_{4loss} - m_{4o} \cdot h_{4o}) \cdot dt \quad \text{energy balance} \qquad (3.32)$$

$$Q_{4loss} = UA_4 \cdot (T_4 - T_{tank}) \quad \text{heat loss to the ambient} \qquad (3.33)$$

$$h_{4o} = h_4 = h_{w\,liq}(T_4) \quad \text{state equation (saturation)} \qquad (3.34)$$

$$P_4 = P_w(T_4) \quad \text{state equation} \qquad (3.35)$$

$$m_{4i} \cdot h_{4i} + Q_{3 \to 4loss} = m_{3o} \cdot h_{3o} \quad \text{energy balance on tube } 3 \to 4 \qquad (3.36)$$

3.3.2.1.5 Connection Tubes

The evaluation of the heat losses of tubes that connect the process components is made, granted that the specific heat of the fluid in the tube is constant along the tube. Equations (3.37) and (3.38), for example, give the heat losses to the ambient for the connection tube between the generator 2 and the solution tank 1.

$$T_{1i} = T_{ref} + (T_{2o} - T_{ref}) \cdot \exp\left(\frac{-(UA)_{2 \to 1}}{m_{1i} \cdot Cp_{sol}}\right) \qquad (3.37)$$

$$Q_{2 \to 1loss} = m_{2o} \cdot (h_{2o} - h_{1i}) \qquad (3.38)$$

3.3.2.1.6 Circulating Pumps

The energy transferred to the solution and water by the circulating pumps (W_1 and W_2) is estimated by considering an isentropic efficiency η_{is} for each pump, and the fact that the density of the liquid is constant between the tank outlet and the heat exchanger inlet. Thus, Equation (3.39) gives the work w_1 provided by circulating pump W_1 to the solution.

$$w_1 = v_{1o} \cdot \frac{(P_2 - P_1)}{\eta_{is\,W_1}} \qquad (3.39)$$

A value of 0.8, which is relatively low, is used as the isentropic efficiency of the pumps, as its influence is low on the system performances.

3.3.2.1.7 Environment: Heat Sink/Low-Temperature Heat Source

Simplified models for the condenser heat sink (charging period) and the evaporator low-temperature heat source (discharging period) are introduced. The temperature of the cooling fluid at the entrance of the condenser T_{ci} is set to 3 °C below the outdoor air temperature when the latter is higher than 10 °C; otherwise, T_{ci} is set to 5 °C. The HTF temperature entering the evaporator is constant and equal to 10 °C.

3.3.2.2 Inputs and Assumptions of the Simulations

The main input data of the model are the meteorological data and building heating need. Solar radiation and outdoor air temperatures measured during 2005 in Chambéry, in the alpine region of France, are used. To evaluate the heating need, a simplified method is applied. The heating need given by Equation (3.40) is estimated by considering an overall heat loss coefficient for the building and a base temperature T_{ext} of 10 °C.

$$Q_b = (UA)_b \cdot (T_{int} - T_{ext}) \qquad (3.40)$$

The comfort temperature T_{int} is set to 20 °C at daytime and 16 °C at nighttime.

Simulations have been performed for a single-family house of 120 m^2 that meets passive house standards in order to achieve 100% solar fraction. Its annual heating need is about 1800 kWh, with a peak heat load of 1.2 kW.

The generator inlet temperature is limited to 90 °C in order to limit crystallization risk in the generator. Storage tanks are presumed to be buried underground or put in a basement, where the ambient temperature T_{tank} is assumed constant at 5 °C. As for the condenser and the generator, they are in a nonheated space where the temperature T_{ref} is also supposed to be constant at 15 °C.

3.3.2.3 Simulation Results

The process performances are strongly influenced by the operating conditions and the process design. The numerical model can be used to identify the parameters that have major impacts on the key performance indicators of the system: storage density and thermal efficiency.

3.3.2.3.1 Effects of the Heat Exchanger Sizes

The generator and condenser overall heat transfer coefficients UA_g and UA_c respectively were varied from 0.2 to 0.8 kW K^{-1}, and results are reported in Figure 3.8. This UA range is chosen based on the power (<2 kW) necessary to cover the passive house needs. The storage density can be improved by about 60% when increasing the heat exchanger performance within the considered UA range. This is due to a lower minimum concentration of the cycle (points 1 and 2 in Figure 3.6). Indeed, the smaller temperature difference between the heat exchanger sides for high UA values for a given power has two consequences on the cycle. One is that the evaporator temperature increases (point E in Figure 3.6). The other is that the required absorber temperature is lower (points 6-7 in Figure 3.6).

The required solar thermal collector area also decreases when the UA value increases (Figure 3.8) because desorption is more efficient. This is due to a higher generator temperature and lower condenser temperature on average. For very low UA values (for both $UA_g = UA_c = 0.2$ kW K^{-1} in the present case), the system performance drops sharply.

Further simulations with a solution heat exchanger (SHX) between the solution tank and the generator (Figure 3.9) show that it strongly improves the process thermal efficiency and its storage density. However, this SHX could lead to technical issues (crystal formation and tube clogging), as will be discussed in Section 3.3.3.

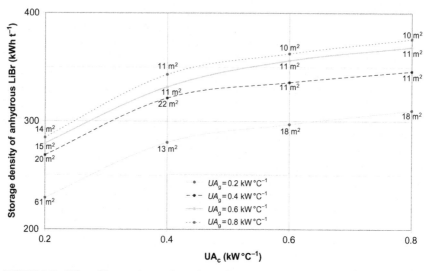

FIGURE 3.8 Effect of heat exchangers' transfer coefficients on the storage density and the required solar collector area (the surface indicated on the line is the necessary solar collector area for the system charging) (N'Tsoukpoe et al., 2012).

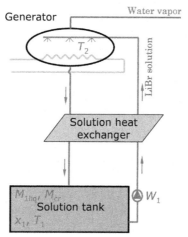

FIGURE 3.9 Solution heat exchanger (SHX) between the solution tank and the generator (N'Tsoukpoe et al., 2012).

3.3.2.3.2 Effects of the Absorption Percentage

The effect of the absorption percentage on the storage density is also studied. The requirements to reach a mean heat supply temperature of 30 °C and a maximum crystallization ratio of 67% are summarized in Table 3.2. The storage density decreases with a low absorption percentage. Indeed, the solution

TABLE 3.2 Main Simulation Parameters and Results for Various Absorption Percentages (N'Tsoukpoe et al., 2012)

Simulation No.	9	10	11	12	13	14
Absorption percentage α	0.75	0.80	0.85	0.90	0.95	1.00
Solution minimum concentration in the storage tank (m%)	52.7	52.1	51.7	51.1	50.7	50.4
Surface area of flat solar collectors (m^2)	16	15	15	13	13	13
Storage density of anhydrous LiBr salt (kWh t^{-1})	269	281	292	300	312	321
Storage density of LiBr aqueous solution (kWh m^{-3})	226	233	239	241	247	252

temperature in the absorber is lower than the equilibrium temperature. Thus, a higher minimum concentration of the cycle is needed. Similarly, during desorption, with low absorption percentage, the generator outlet bulk concentration is lower than the interface concentration, so that the actual amount of generated water is lower than in equilibrium conditions. There are then greater sensible heat losses and greater heat needs for desorption, which result in a larger solar collector area.

Special attention should thus be paid to the generator design, because the absorption percentage depends on the exchanger design and the solution flow rate.

3.3.2.3.3   Effects of the Maximum Crystallization Ratio

Figure 3.10 shows the storage density as a function of the maximum crystallization ratio at average heat supply temperatures of 30 °C and 33 °C. Crystallization can increase the storage density more than three times. However, above a maximum crystallization ratio of 70%, savings become less and less significant, while the technical complexity of the process increases. Thus, a compromise has to be made for the choice of the optimal crystallization ratio.

3.3.3   Process Experimentations

This process has been tested and the feasibility of the crystal formation and dissolution in the solution tank has been proven, as well as the possibility to store and produce heat at the desired temperature levels (N'Tsoukpoe et al., 2013, 2014).

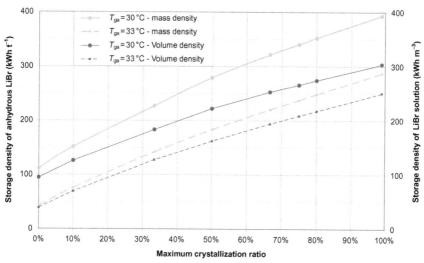

FIGURE 3.10 Effect of the maximum crystallization ratio and the average heat supply temperature to the building on the storage density (N'Tsoukpoe et al., 2012).

3.3.3.1 Prototype Design

A demonstration prototype has been designed, which could theoretically store 8 kWh of heat and produce a heating power of 1 kW. This prototype for demonstration of the feasibility of the concept has been built as shown in Figure 3.11. It mainly consists of two storage tanks and a reactor containing two heat exchangers.

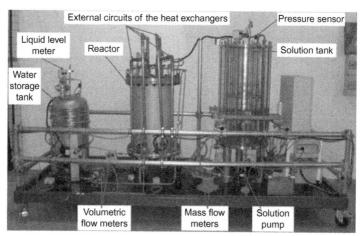

FIGURE 3.11 LiBr/H_2O absorption prototype for long-term energy storage (N'Tsoukpoe et al., 2013).

Two identical vertical falling film heat exchangers (shell-and-tubes) have been chosen (Figure 3.12, Table 3.3). This type of exchanger is compact compared to some related technologies and is most frequently used in absorption machines, as it is seen to be most appropriate because of its small pressure losses (Fujita, 1993; Jani et al., 2004; Bo et al., 2010). Furthermore, it is suitable for processes with low-temperature heat sources, such as solar energy, as they offer high heat transfer coefficients. In the charging phase, the water desorption and condensation take place inside the tubes when the HTF flows on the shell side (Figure 3.12b). Similarly, absorption and evaporation take place on the tube side. These tubes are in brass aluminum (CuZn22Al2), which is a copper alloy particularly renowned for its resistance to corrosion. Indeed, LiBr aqueous solution is particularly corrosive (Behrens et al., 1992; Arnaud et al., 1985; Herold et al., 1996; Florides et al., 2003). A radiation shield (Figure 3.12a) has been installed in the reactor in order to reduce direct heat exchanges between the two heat exchangers. Under each heat exchanger, a collector receives the liquid leaving the heat exchanger before it is pumped into the corresponding storage tank.

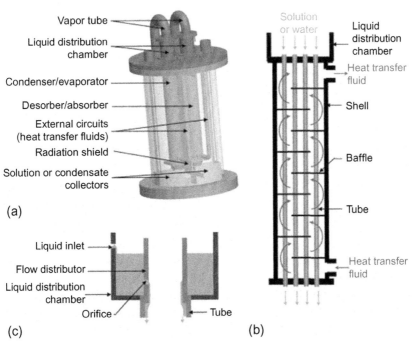

FIGURE 3.12   Reactor design and operating principle of the heat exchangers: (a) reactor, (b) fluid circulation in the heat exchanger (principle), and (c) liquid distribution chamber (principle) (N'Tsoukpoe et al., 2013).

TABLE 3.3 Main Features of Each Heat Exchanger (Total Internal Surface of Tubes: $A = 0.33$ m^2) (N'Tsoukpoe et al., 2013)

	Value
Length of a tube (mm)	620
Inside diameter of tubes (mm)	12
Number of tubes	14
Number of orifices per tube	3
Diameter of each orifice (mm)	0.4

① Temperature sensor   • Liquid level sensor   ⊂⊃ Float
② Pressure sensor   ⊗ Pump   ⊣⊳ Valve clap
⊖ Volumetric flowmeter   ○ Vacuum pump   — Shield
○ Mass flowmeter   ⊶ Valve   ⨝ Safety valve

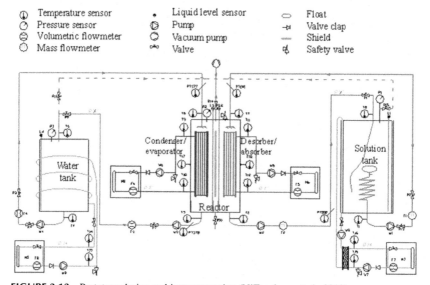

FIGURE 3.13 Prototype design and instrumentation (N'Tsoukpoe et al., 2013).

Four pumps ensure fluid circulation (solution and absorbate) between the components (Figures 3.11 and 3.13). They are of the gear type and magnetically coupled so that seal leaks are normally not possible. To avoid obstruction in the outlet tube of the solution storage tank when crystallization occurs, the solution is pumped near its surface (upper part of the tank). This is achieved by a floating intake with a spiral tube that is connected to the solution pump (Figure 3.13). The desorber/absorber outlet enters the solution storage tank at its bottom in order to favor mixing in the tank. The components are connected with stainless steel tubes.

The water storage tank is made of stainless steel (316 L), whereas the solution storage tank is a glass column (Figure 3.11). This allows observing the crystallization in this tank and possibly other phenomena such as convection or stratification in concentration. The storage tank environment temperature is set through heat exchanges around the tank surfaces (Figure 3.13). Each heat exchanger is connected to a thermal module that can provide the assigned temperature and flow rate, either in static or dynamic operating mode.

Ultra-Torr vacuum fittings from Swagelok are used to ensure vacuum tightness of the prototype. Where necessary, a sealant tape (LSM1310 provided by DIATEX) is added.

3.3.3.2   Measurements and Experimental Procedure

The prototype has been strongly instrumented (Figure 3.13) to be able to measure temperatures (accuracy: ± 0.2 °C), pressures (± 0.25 mbar), mass flow rates ($\pm 0.05\%$ for the solution and $\pm 0.1\%$ for the HTF in external circuits), volume flows (± 1 l h^{-1}), and liquid levels inside the components (± 2 mm). The solution mass flow rates are measured by Coriolis flowmeters that also measure temperature (± 0.2 °C) and density (± 1 kg m^{-3}). The solution concentration at the desorber inlet/outlet is derived from these measurements, using a correlation between the solution concentration, temperature, and density (Lee et al., 1990). The prototype is computer-monitored and controlled using a Labview® program.

Experiments are carried out under static and dynamic conditions. The desorber HTF flow rates, the solution flow rates, the condenser and desorber HTF inlet temperatures, as well as storage tanks' surrounding temperatures are selected as values that can be reached under a French climate for a passive house equipped with flat solar collectors and a geothermal pipe.

The initial solution prepared and introduced in the solution storage tank contains 46 kg of anhydrous LiBr and 42.5 kg of water. This means an initial concentration of 52 m%.

Preliminary tests indicate that the solution distributors at the top of the tubes of the desorber do not operate as expected and seem to be undersized: the solution is distributed mainly by overflow weirs at the top of the tubes instead of the orifices intended for this (Figure 3.12c). This is known to give very bad liquid distribution (dry patch, flooding, etc.) and lead to bad heat and mass transfers (Whalley, 1978a,b).

3.3.3.3   Experimental Results and Discussion in Charging Mode

In this section, a summary of all the results of charging tests is given and discussed.

Table 3.4 presents the main results of the charging tests (experimental design and additional tests) in chronological order.

Three values (measured or derived quantities) are considered most important:

-   The water desorption rate or the mass transfer in the desorber: the amount of desorbed water is an indicator of the amount of stored energy and is therefore considered the most important response

TABLE 3.4 Charging Test Results (N'Tsoukpoe et al., 2013)

Test No.	1	2	17	18	3	4	5	6	7	8	19
Date	02-Sep	02-Sep	02-Sep	13-Sep	13-Sep	15-Sep	15-Sep	16-Sep	19-Sep	19-Sep	21-Sep
Start of the test (h:min)	09:23	13:50	17:50	11:30	14:57	10:43	13:07	09:28	09:50	12:04	10:00
End of the test (h:min)	13:30	17:50	19:00	14:39	16:01	13:02	15:33	11:31	11:50	14:07	14:30
Duration (h:min)	04:07	04:00	01:10	03:09	01:04	02:19	02:26	02:03	02:00	02:03	04:30
Solution initial concentration (m%)	52.2	54.8	56.9	53.3	56.5	53.8	54.9	56.0	54.9	56.3	54.9
Solution final concentration (m%)	54.7	56.9	59.5	55.7	57.7	54.9	56.0	57.1	56.3	58.1	57.4
Desorber inlet HTF temperature (°C)	90.0	90.0	90.0	90.0	89.6	75.0	75.3	88.9	75.0	75.3	-
Condenser inlet HTF temperature (°C)	20.1	10.1	10.1	30.1	10.1	20.1	20.1	20.1	10.2	10.2	20.1
Desorber HTF flow rate (kg h^{-1})	360	720	720	720	360	360	720	720	360	720	360
Condenser HTF flow rate (kg h^{-1})	360	720	720	720	720	720	720	360	360	360	360
Solution flow rate (l h^{-1})	38.2	38.2	18.3	41.1	24.1	44.5	21.7	22.0	22.7	43.5	22.5
Mass of desorbed water (kg)	5.4	4.1	1.3	4.4	2.0	3	2.3	2.2	2.4	3.1	4.0

Mass rate of desorbed water (kg h^{-1})	1.31	1.02	1.13	1.39	1.89	1.30	0.95	1.01	1.18	1.49	0.89
Average desorber heat transfer coefficient (W °C^{-1})	78	74	64	138	72	145	81	72	75	126	75
Average equilibrium factor	0.46	0.25	0.92	1.00	0.73	0.28	1.25	0.80	0.70	0.35	0.57
Average power of the desorber (kW)	2.3	2.2	1.7	3.0	2.2	2.4	1.5	2.0	1.8	2.4	1.6
Average power of the condenser (kW)	−0.9	−1.1	−1.0	−1.1	−1.5	−0.8	−0.7	−0.8	−0.9	−1.0	−0.7
Total exchanged energy in the desorber (kWh)	9.5	8.7	1.7	9.4	2.4	5.5	3.6	3.6	3.6	4.9	8.1
Total exchanged energy in the condenser (kWh)	−3.8	−4.2	−0.9	−3.5	−1.6	−1.9	−1.8	−1.6	−1.7	−2.1	−3.2

- The average heat transfer in the desorber: the heat exchanger performance analysis is important because it strongly affects the process storage density (N'Tsoukpoe et al., 2012)
- The equilibrium factor: as previously stated (Equation (3.9)), it indicates the mass transfer effectiveness in the reactor (N'Tsoukpoe et al., 2012)

3.3.3.3.1 The Water Desorption Rate in the Desorber

An average value of the water desorption rate is calculated for each test by dividing the total mass of the desorbed water during the test by the test duration (Table 3.4). The mean water desorption rate is 1.3 kg h^{-1}. This visibly low rate implies a small concentration change in the desorber (1-2 m%).

The tests suggest that the main variables affecting the water desorption rate are

- The solution flow rate: its increase results in a better heat transfer in the desorber, and thus, a higher water flow release
- The inlet temperature of the HTF in the condenser, which increases the water desorption rate when decreasing

These behaviors can be derived from simple mass and heat transfer analysis and are as foreseen in the process simulation (N'Tsoukpoe et al., 2012).

3.3.3.3.2 Heat Transfer in the Desorber

The observed power is acceptable, according to the process design for a real plant (2-5 kW; Table 3.4).

Average values of two indicators are evaluated and used for the heat transfer analysis: the overall heat transfer coefficient UA and the Reynolds number. For a test, $\overline{UA}$ (Equation (3.42)) is the average value of the calculated UA values at each measurement step during the test:

$$Q_2 = UA \cdot \left(\frac{T_{gi} - T_{go}}{\ln\left(\frac{T_{gi} - T_2}{T_{go} - T_2}\right)} \right) = m_g \cdot Cp_g \cdot (T_{gi} - T_{go}) \tag{3.41}$$

$$\overline{UA} = \text{average}\left(m_g \cdot Cp_g \cdot \ln\left(\frac{T_{gi} - T_2}{T_{go} - T_2}\right) \right) \tag{3.42}$$

Similarly, the average Reynolds number $\overline{Re}$ is defined as follows (Equation (3.43)):

$$\overline{Re} = \text{average}\left(\frac{4\Gamma_{2i}}{\mu}\right) = \text{average}\left(\frac{4m_{2i}}{\pi \cdot d \cdot n \cdot \mu}\right) \tag{3.43}$$

The solution properties are calculated at the desorber inlet.

The solution flow rate appears to be the variable that mostly affects the heat transfer rate in the explored experimental range (Figure 3.14): $\overline{UA}$ roughly doubles when the solution flow rate doubles. This significant change is not in accordance with the process simulation hypotheses, which assume that the solution side thermal resistance does not change significantly with the solution flow rate (N'Tsoukpoe et al., 2012). Anyway, the solution side remains the heat transfer controlling side.

The plot of $\overline{UA}$ with respect to $\overline{Re}$ allows further analysis: there is a significant change in the trend for $\overline{Re}$ above 30-35 (Figure 3.15). The observed $\overline{Re}$ range is from 10 to 40. According to studies on absorption falling film heat exchangers operating with LiBr aqueous solution, this yields two different flow regimes: the smooth laminar flow and the wavy laminar flow. Indeed, the smooth-wavy transition occurs between $Re = 20$ and 30 (Morioka and Kiyota, 1991; Kim et al., 1995; Medrano et al., 2002). The significant $\overline{UA}$ increase can then be explained by the change in the flow pattern. The low values observed during tests no. 1 and 2 (Figure 3.15) can be explained by the fact that they are quite in the transition zone.

The occurrence of the wavy laminar regime, which is the most common in absorption falling film heat exchangers (Morioka and Kiyota, 1991) because it promotes the mixing of the film, has enhanced somewhat the heat transfer, but the design overall heat transfer coefficient ($UA = 0.4 \text{ kW } °\text{C}^{-1}$, that is $U = 1.2 \text{ kW m}^{-2} °\text{C}^{-1}$) was not reached. The low heat transfer yields a large temperature difference between the solution and the HTF in the desorber (25-30 °C), when considering the corresponding power range (2-5 kW).

The main reason for this low performance is probably the maldistribution of the solution in the tubes due to the overflow weir distribution at the top of the tubes (see Section 3.3.3.1). A possible lack of verticality of the heat exchanger in the reactor could be another cause of poor distribution.

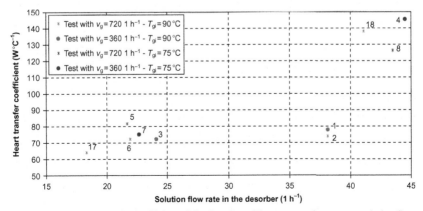

FIGURE 3.14 Heat transfer coefficient of the desorber with respect to the average solution flow rate (The numbers correspond to the test numbers listed in Table 3.4.) (N'Tsoukpoe et al., 2013).

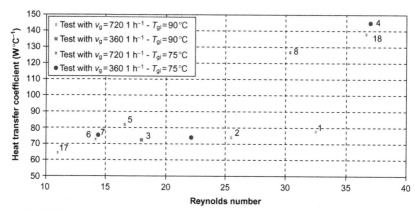

FIGURE 3.15 Heat transfer coefficient of the desorber with respect to the average Reynolds number (The numbers correspond to the test numbers listed in Table 3.4.) (N'Tsoukpoe et al., 2013).

The maximum amount of heat that is charged during the different tests is 13 kWh. Thermal losses from the reactor are huge, as they can reach 25% of the total heat transferred. A better insulation of the reactor would reduce these losses.

3.3.3.3.3 Equilibrium Factor

The observed equilibrium factor values range from 0.2 to 0.9. This equilibrium factor decreases with the solution flow rate increase (Figure 3.16). The film thickness increase, due to the solution flow rate increase, results in a slower water diffusion within the film to the interface, which is theoretically in equilibrium conditions. The waves induced by the Reynolds number increase have not improved the equilibrium factor.

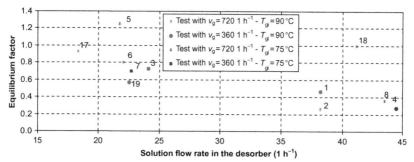

FIGURE 3.16 Evolution of the equilibrium factor with respect to the average solution flow rate (The numbers correspond to the test numbers listed in Table 3.4.) (N'Tsoukpoe et al., 2013).

3.3.3.4   Experimental Results and Discussion in Discharging Mode

3.3.3.4.1   Base Tests

The observed results are summarized in Table 3.5.

The solution entering the absorber at 10-20 °C leaves it with a temperature of about 30-40 °C. Absorption occurs in the absorber and the produced temperature is sufficient for heating purposes (32-40 °C).

Generally, by disregarding the thermal loss to the ambient and assuming that there is no heat exchange between the solution and the HTF (balance of the overall heat exchange is zero), the absorption power is about 0.5 kW only (calculated with the solution flow temperature increase, that is, sensible heat).

3.3.3.4.2   Use of a Heat and Mass Transfer Enhancement Additive

It is an established fact that adding 2-ethyl-1-hexanol (2EH) in a LiBr aqueous solution improves its absorption heat and mass transfer characteristics (Ziegler and Grossman, 1996). Some tests were carried out after adding 2EH in the solution storage tank (100-200 ppm). The results show a slight improvement in the water absorption rate ($\approx$+20%, see in Table 3.5: test no. 11 vs. test no. 23; test no. 12 vs. test no. 24). An increase in the equilibrium factor (Equation (3.9)) from 0.23 (tests without additive) up to 0.85 was also observed. However, no improvement in the overall heat transferred to the HTF was noticed.

3.3.3.4.3   Rise of the Absorber Inlet Solution Temperature

Two tests (no. 22 and no. 27) have been performed in order to know if the increase of the entering solution into the absorber had a major impact on the system performance. The temperature of the water bath around the solution tank was increased to 20 °C in order to feed the absorber with a higher temperature solution (basically, the storage tanks' surrounding temperature was maintained around 10 °C for the basic tests). This would be comparable, for instance, to the use of a SHX, as stated previously (see Figure 3.9), which would recover heat from the absorber outlet solution to increase the temperature of the absorber inlet solution. The corresponding tests have not yielded more satisfactory heat transfer to the HTF (power <0.1 kW).

3.3.3.4.4   Possible Improvement Paths for the Absorber

Improving the performance of the absorber means improving its heat transfer coefficient.

The poor heat transfer coefficient seems mostly due to the hydraulic problems mentioned in Section 3.3.3.4.1. The size or the number of the orifices should be adjusted.

To address this issue, a SHX can also be used between the absorber and the solution tank for heat recovery (Figure 3.9). In absorption chiller and heat pump processes, this heat recovery unit is generally put between the absorber and the

TABLE 3.5 Discharging Test Results Summary (N'Tsoukpoe et al., 2013)

Test No.	9	10	11	12	13	14	20[a]	21[a]	22[a]	15	16	23[b]	24[b]	25[b]	26[a,b]	27[b]
Date	13-Sep	14-Sep	14-Sep	14-Sep	16-Sep	16-Sep	19-Sep	20-Sep	20-Sep	21-Sep	21-Sep	22-Sep	22-Sep	23-Sep	23-Sep	23-Sep
Start of the test (h:min)	17:45	10:20	13:57	17:35	13:38	16:54	16:49	09:26	15:30	17:02	18:42	14:37	17:27	09:24	13:23	16:02
End of the test (h:min)	19:15	13:30	16:40	19:37	16:12	18:56	19:07	12:27	19:18	18:04	19:43	16:45	19:32	11:36	14:58	17:28
Duration (h:min)	1.50	3.17	2.72	2.03	2.57	2.03	0.10	0.13	0.16	1.03	1.02	0.09	0.09	0.09	0.07	0.06
Solution initial concentration (m%)	57.7	57.2	56.1	54.7	57.1	55.5	58.0	57.2	56.7	57.4	56.8	57.6	56.0	55.0	55.0	54.1
Solution final concentration (m%)	57.1	56.1	54.7	53.9	55.5	54.9	57.2	56.7	55.0	56.8	56.0	56.0	55.0	54.1	53.2	52.7
Absorber inlet HTF temperature (°C)	26	26.2	22.1	22.1	26.2	22.1	22.2	26.3	26.4	26.2	22	22.1	22.1	26.1	26.2	26.2
Evaporator inlet HTF temperature (°C)	15	15.1	15.1	15.1	20.1	20.1	15.1	15.1	20.1	20.1	20.1	15.1	15.1	20.1	20.1	20.1
Solution tank surrounding temperature (°C)	8.1	8.1	8.1	8.1	8.1	8.1	8.1	8.0	19.0	8.1	8.1	8.1	8.1	8.1	8.1	19.8
Water tank surrounding temperature (°C)	10.4	10.4	10.4	10.4	10.4	10.4	10.4	10.3	10.3	10.4	10.4	10.5	10.5	10.4	10.3	10.3
Absorber HTF flow rate (kg h^{-1})	720	360	720	360	720	360	360	360	360	360	720	720	360	360	360	360

Evaporator HTF flow rate (kg h^{-1})	360	360	720	720	720	360	720	360	720	720	360	720	720	720	720	720
Solution flow rate (l h^{-1})	14.3	43.5	42.9	22.6	44.3	44.3	20.3	43.2	43.7	22.2	22.9	42.4	22.1	22.4	22.0	23.5
Absorbate (water) flow rate (l h^{-1})	47.0	22.4	21.7	42.4	46.2	46.2	44.7	23.0	21.6	21.2	22.6	22.5	46.4	22.6	21.8	21.4
Mass of absorbed water (kg)	−0.8	−1.5	−2.2	−1.1	−2.3	−0.9	−1.0	−0.7	−2.6	−0.9	−1.2	−2.3	−1.4	−1.4	−2.9	−2.3
Mass rate of absorbed water (kg h^{-1})	−0.53	−0.47	−0.81	−0.54	−0.90	−0.44	−0.43	−0.23	−0.68	−0.87	−1.18	−1.08	−0.67	−0.64	−1.83	−1.60
Average absorber heat transfer coefficient (W °C^{-1})	271	203	209	263	323	287	320	300	164	148	130	234	253	167	147	160
Average equilibrium factor	0.44	0.13	0.23	0.29	0.12	0.06	0.20	0.05	0.13	0.46	0.47	0.43	0.69	0.96	1.11	1.06
Average power of the absorber (kW)	−0.05	0.07	−0.07	0.00	0.00	−0.03	−0.02	−0.25	0.04	−0.11	−0.19	−0.20	−0.04	0.02	−0.06	−0.06
Average power of the evaporator (kW)	0.25	0.36	0.51	0.34	0.59	0.40	0.31	0.14	0.49	0.59	0.55	0.66	0.40	0.57	0.49	0.46
Total exchanged energy in the absorber (kWh)	−0.1	0.2	−0.2	0.0	0.0	0.1	0.0	0.7	0.2	−0.1	−0.2	−0.4	−0.1	0.0	−0.1	−0.1
Total exchanged energy in the evaporator (kWh)	0.41	1.14	1.38	0.70	1.55	0.81	0.71	0.42	1.85	0.62	0.56	1.40	0.83	1.25	0.78	0.66

[a] Absorber concurrent operating mode.
[b] Use of the 2-ethyl-1-hexanol.

desorber in order to recover heat from the desorber outlet solution and preheat its inlet solution. This increases significantly the process thermal efficiency. The SHX would increase the absorber inlet solution temperature. Simulations (N'Tsoukpoe, 2012) show that the storage density is improved by about 30% when a SHX with an effectiveness $\varepsilon = 0.8$ is added to the process. The required solar collector area is also reduced significantly (50%). However, during desorption periods, there is a risk of crystallization in the SHX on the desorber outlet side. Indeed, the solution LiBr mass fraction is high and its temperature decreases in the SHX. To avoid such an eventuality, the SHX could be used only in discharging periods. The effect of this prevention measure on the storage density would be marginal, but the benefit of the required solar collector surface reduction would be lost.

Generally speaking, although the falling film type absorbers are the most commonly encountered absorber type in absorption machines (Kim et al., 1999), they feature several problems: low mass and heat transfer and large volume. An adiabatic absorber appears to be a solution to overcome these points and could be a better solution for this storage process. In an adiabatic absorber, an adiabatic part is introduced prior to the heat transfer part: the mass and heat transfer phenomena are separated (Grossman, 1982). The adiabatic and the heat exchange parts can be designed as two separate chambers in series or as one chamber. A specific constraint is the reversibility of the absorber unit so that it could be used in desorption mode, too.

3.4 CONCLUSION

In recent years, storage of thermal energy has become a very important topic in many engineering applications and has been the subject of a great deal of research activity. This chapter reviews the system and materials for thermal storage of solar energy. The chapter provides insight into recent developments on systems, materials, their classification, their limitations, and possible improvements for their use. Three major thermal energy storage modes (sensible, latent, thermochemical and sorption heat storage) are described, emphasizing the main characteristics of the most suitable heat storage materials for each.

Water remains the most widely used material in sensible heat storage systems used for solar energy storage. It is the material that presents the best compromise between cost, heat storage capacity, density, and environmental impact. Most of the technological projects are today focusing on the development of tools for the optimization of efficient energy consumption more than on the development of new systems with water for energy storage management.

Considering latent heat storage, numerous PCMs have been developed in the 0-80 °C temperature range, which is suitable for low-temperature solar heat needs. The advantages of latent heat storage materials on sensible heat storage materials are their high storage energy density and the various melting temperatures that allow different levels of use. The investigations on these materials are still underway and are attempting to overcome a number of their limitations (low thermal conductivities, supercooling, incongruent melting, etc.) and improve their viability.

Considering sorption/thermochemical heat storage, the technique theoretically offers the greatest heat storage capacity and does not suffer from heat losses during the storage period. But the literature on these processes is marginal compared to that on sensible and latent heat storage materials. The reason is that the research in this field is still in an early stage. There has been an intensification of the research during the last decade to develop more efficient sorption heat storage systems. New thermochemical materials have been characterized and innovative system designs have been developed and tested.

This chapter presents a focus on a long-term solar heat storage system based on water absorption by an LiBr aqueous solution. A detailed dynamic model of the system has been developed for the system simulation in order to evaluate the process performance and optimize its design. It shows the need to allow the crystallization of the solution in the solution tank, as this crystallization greatly enhances the storage capacity of the system. It also brings to light the need for efficient heat and mass exchangers both for the solution (for the absorber and the desorber) and the water (evaporator, mostly), as the exchangers' efficiency is directly linked to the process efficiency and heat storage capacity. A prototype of the system has also been presented in detail, as well as experimental results in the process charging and discharging modes. The prototype has been charged under practical conditions. Although the obtained heat transfer coefficients in the desorber are less than expected, mainly for hydraulic problems, the charging process has been proved with a charging power of 2-5 kW and heat storage up to 13 kWh. Crystallization has been achieved in the storage tank and the crystal dissolution/formation has been observed during several cycles. Discharging tests indicate absorption temperatures of about 30-40 °C, which may enable house heating. However, the absorption heat recovery by the HTF is not effective due to an inadequate design of the absorber. This proves that the design and optimization of the absorber and evaporator is a key research issue for the future of this process. Another research path should lead to innovative sorption couples less expensive than $LiBr/H_2O$, as the amount of sorbent that would be used in a real building to cover its annual needs is too high to allow the use of this costly compound.

The thermal energy storage field is a complicated and multidisciplinary research domain. The scientific and technical barriers are not only concerned with heat and mass transfer, materials, component design and systems, but also with environmental impact, economical constraint, social acceptability, citizen sensibility, and so on. Ongoing research and development studies show that the challenges of the technology focus on the aspects of different types of storage technologies, process and system development, the innovative components, the configurations of storage cycles, and new and advanced materials. Booming progress illustrates that thermal storage is a realistic and sustainable option for storing solar energy, both for high temperature or low temperature, for short- or long-term applications. To bring the storage solutions into the market, more intensive studies in fields of evaluation of advanced materials and development of efficient and compact prototypes are still required.

REFERENCES

Abhat, A., 1980. Short term thermal energy storage. Rev. Phys. Appl. 15, 477–501.

Andberg, J.W., Vliet, G.C., 1983. Design guidelines for water-lithium bromide absorbers. ASHRAE Trans. 89 (Part 1B), 220–232.

Apelblat, A., Tamir, A., 1986. Enthalpy of solution of lithium bromide, lithium bromide monohydrate, and lithium bromide dihydrate, in water at 298.15 K. J. Chem. Thermodyn. 18 (3), 201–212.

Arnaud, D., Barbery, J., Biais, R., Fargette, B., Naudot, P., 1985. Propriétés du cuivre et de ses alliages. Techniques de l'Ingénieur, traité Matériaux métalliques.

Asif, M., Muneer, T., 2007. Solar thermal technologies. In: Capehart, B.L. (Ed.), Encyclopedia of Energy Engineering and Technology. CRC Press, USA, p. 1321.

Banasiak, K., Koziol, J., 2009. Mathematical modelling of a LiBr-H$_2$O absorption chiller including two-dimensional distributions of temperature and concentration fields for heat and mass exchangers. Int. J. Therm. Sci. 48 (9), 1755–1764.

Behrens, B., Kreysa, G., Eckermann, R., 1992. Dechema corrosion handbook: corrosive agents and their interaction with materials. Chlorine Dioxide, Seawater. VCH, Frankfurt.

Bo, S., Ma, X., Lan, Z., Chen, J., Chen, H., 2010. Numerical simulation on the falling film absorption process in a counter-flow absorber. Chem. Eng. J. 156 (3), 607–612.

Dincer, I., Rosen, M.A., 2002. Thermal Energy Storage: Systems and Applications. John Wiley & Sons, UK.

Duffie, J.A., Beckman, W.A., 1989. Solar Energy Thermal Processes. John Wiley & Sons, USA.

Farid, M.M., Khudhair, A.M., Razack, S.A.K., Al-Hallaj, S., 2004. A review on phase change energy storage: materials and applications. Energy Convers. Manag. 45, 1597–1615.

Fath, H.E.S., 1998. Technical assessment of solar thermal energy storage technologies. Renew. Energy 14, 35–40.

Fernandez, A.I., Martínez, M., Segarra, M., Martorell, I., Cabeza, L.F., 2010. Selection of materials with potential in sensible thermal energy storage. Sol. Energy Mater. Sol. Cells 94, 1723–1729.

Florides, G.A., Kalogirou, S.A., Tassou, S.A., Wrobel, L.C., 2003. Design and construction of a LiBr-water absorption machine. Energy Convers. Manag. 44 (15), 2483–2508.

Fujita, T., 1993. Falling liquid films in absorption machines. Int. J. Refrig. 16 (4), 282–294.

Garg, H.P., Mullick, S.C., Bhargava, A.K., 1985. Solar Thermal Energy Storage. Reidel Publishing Company, Dordrecht.

George, J.M., Murthy, S.S., 1989. Influence of absorber effectiveness on performance of vapour absorption heat transformers. Int. J. Energy Res. 13 (6), 629–638.

Gil, A., Medrano, M., Martorell, I., Lázaro, A., Dolado, P., Zalba, B., Cabeza, L.F., 2010. State of the art on high temperature thermal energy storage for power generation. Part 1—concepts, materials and modellization. Renew. Sust. Energ. Rev. 14 (1), 31–55.

Grossman, G., 1982. Adiabatic absorption and desorption for improvement of temperature-boosting absorption heat pumps. ASHRAE Trans. 88 (2), 359–367.

Grossman, G., Zaltash, A., DeVault, R.C., 1995. Simulation and performance analysis of a 4-effect lithium bromide-water absorption chiller. ASHRAE Trans. 101 (1), 1302–1312.

Goswami, D., Kreith, F., Kreider, J.F., 2000. Principles of Solar Engineering, second ed. Taylor & Francis, UK.

Hadorn, J.C., 2008. Advanced storage concepts for active solar energy-IEA SHC Task32 2003-2007. In: Proceedings of First International Conference on Solar Heating, Cooling and Buildings, Lisbon, Portugal.

Hasnain, S.M., 1998. Review on sustainable thermal energy storage technologies. Part I: heat storage materials and techniques. Energy Convers. Manag. 39, 1127–1138.

Hellmann, H.-M., Grossman, G., 1996. Improved property data correlations of absorption fluids for computer simulation of heat pump cycles. ASHRAE Trans. 102 (1), 980–997.

Herold, K.E., Radermacher, R., Klein, S.A., 1996. Absorption Chillers and Heat Pumps. CRC Press, Inc., Boca Raton

Jani, S., Saidi, M.H., Mozaffari, A.A., Heydari, A., 2004. Modeling heat and mass transfer in falling film absorption generators. Sci. Iran. 11, 81–91.

Kaushik, S., Sheridan, N., Lam, K., Kaul, S., 1985. Dynamic simulation of an ammonia-water absorption cycle solar heat pump with integral refrigerant storage. J. Heat Recover. Syst. 5 (2), 101–116.

Kim, K.J., Berman, N.S., Chau, D.S.C., Wood, B.D., 1995. Absorption of water vapour into falling films of aqueous lithium bromide. Int. J. Refrig. 18 (7), 486–494.

Kim, J.-S., Lee, H., Yu, S.I., 1999. Absorption of water vapour into lithium bromide-based solutions with additives using a simple stagnant pool absorber. Int. J. Refrig. 22 (3), 188–193.

Lee, R.J., DiGuilio, R.M., Jeter, S.M., Teja, A.S., 1990. Properties of lithium bromide-water solutions at high temperatures and concentrations. II. Density and viscosity. ASHRAE Trans. 96, 709–714.

Liu, H., N'Tsoukpoe, E., Le Pierrès, N., Luo, L., 2011. Evaluation of a seasonal storage system of solar energy for house heating using different absorption couples. Energy Convers. Manag. 52, 2427–2436.

Medrano, M., Bourouis, M., Coronas, A., 2002. Absorption of water vapour in the falling film of water lithium bromide inside a vertical tube at air-cooling thermal conditions. Int. J. Therm. Sci. 41 (9), 891–898.

Morioka, I., Kiyota, M., 1991. Absorption of water vapor into a wavy film of an aqueous solution of LiBr. JSME Int. J. Ser. B: Fluids Therm. Eng. 34 (2), 183–188.

N'Tsoukpoe, K.E., 2012. Etude du stockage à long terme de l'énergie solaire thermique par procédé d'absorption LiBr-H$_2$O pour le chauffage de l'habitat. PhD thesis, Université de Grenoble (in French).

N'Tsoukpoe, K.E., Liu, H., Le Pierrès, N., Luo, L., 2009. A review on long-term sorption storage. Renew. Sust. Energ. Rev. 13 (9), 2385–2396.

N'Tsoukpoe, K.E., Le Pierrès, N., Luo, L., 2012. Numerical dynamic simulation and analysis of a lithium bromide/water long-term solar heat storage system. Energy 37, 346–358.

N'Tsoukpoe, K.E., Le Pierrès, N., Luo, L., 2013. Experimentation of a LiBr/H$_2$O absorption process for long term solar thermal storage: prototype design and first results. Energy 53, 179–198.

N'Tsoukpoe, K.E., Perier-Muzet, M., Le Pierrès, N., Luo, L., Mangin, D., 2014. Thermodynamic study of a LiBr-H$_2$O absorption process with crystallisation of the solution for solar thermal storage. Sol. Energy 104, 2–15.

Odru, P., 2010. Le stockage de l'énergie. Dunod ed., Paris

Patnaik, V., Perez-Blanco, H., Ryan, W.A., 1993. A simple analytical model for the design of vertical tube absorbers. ASHRAE Trans. 99 (2), 69–80.

Pillai, K.K., Brinkwarth, B.J., 1976. The storage of low grade thermal energy using phase change materials. Appl. Energy 2, 205–216.

Saul, A., Wagner, W., 1987. International equations for the saturation properties of ordinary water substance. J. Phys. Chem. Ref. Data 16 (4), 893–901.

Sharma, A., Tyagi, V.V., Chen, C.R., Buddhi, D., 2009. Review on thermal energy storage with phase change materials and applications. Renew. Sust. Energ. Rev. 13, 318–345.

Tatsidjodoung, P., Le Pierrès, N., Luo, L., 2013. A review of potential materials for thermal energy storage in building applications. Renew. Sust. Energ. Rev. 18, 327–349.

Whalley, P.B., 1978a. Falling film evaporation. Introduction and design principles. HTFS-DR47, Part 1 (AERE-R8990).

Whalley, P.B., 1978b. Falling film evaporation. Liquid distribution in falling film evaporators. HTFS-DR47, Part 6 (AERE-R8995).

Wyman, C., Castle, J., Kreith, F., 1980. A review of collector and energy storage technology for intermediate temperature applications. Sol. Energy 24, 517–540.

Yu, N., Wang, R.Z., Wang, L.W., 2013. Sorption thermal storage for solar energy. Prog. Energy Combust. Sci. 39 (5), 489–514.

Ziegler, F., Grossman, G., 1996. Heat-transfer enhancement by additives. Int. J. Refrig. 19 (5), 301–309.

Chapter 4

Assessment of Electricity Storage Systems

F. Rahman[1,2], M.A. Baseer[3,4] and S. Rehman[5]

[1]*Center for Refining & Petrochemicals, Research Institute, King Fahd University of Petroleum & Minerals, Dhahran, Saudi Arabia*
[2]*Center of Research Excellence in Renewable Energy, King Fahd University of Petroleum & Minerals, Dhahran, Saudi Arabia*
[3]*Mechanical and Aeronautical Engineering Department, University of Pretoria, Pretoria, South Africa*
[4]*Jubail Industrial College, Jubail, Saudi Arabia*
[5]*Center for Engineering Research, Research Institute, King Fahd University of Petroleum and Minerals, Dhahran, Saudi Arabia*

Chapter Contents

Solar Energy Storage. http://dx.doi.org/10.1016/B978-0-12-409540-3.00004-9

4.1 INTRODUCTION

The development of efficient and environmentally safe energy generation is an important and urgent issue to reduce global warming and potentially serious damage due to various pollutants in the atmosphere. Electrical energy that can be generated from renewable sources, such as solar or wind, offers enormous potential for meeting future energy demands, which are expected to be double that of current consumption in the next 50 years. However, the use of electricity generated from these intermittent, renewable sources requires efficient advanced *electricity storage systems* (ESSs).

Electrical energy storage refers to a process of converting electrical energy from a power network into a form that can be stored for converting back to electrical energy when needed. Such a process enables electricity to be produced at times of either low demand or low generation cost, or from intermittent renewable energy sources (RESs), and to be used at times of high demand, high generation cost, or when no other generation means is available.

Due to the exponentially growing population and limited means of fossil fuel, new and renewable sources of energy are being promoted to meet demands both in grid-connected and in off-grid areas. These sources include mainly solar photovoltaic (PV) and wind power.

Therefore, solar and wind resources are being exploited on a large scale, and the available technology is also commercially developed and accessible. Solar PV is a very good source of electricity generation in this part of the world, that is, the Middle East in general and Saudi Arabia in particular. The only barriers are its storage, relatively high cost, low efficiency, and the local environmental problem of high temperature and suspended dust particles, which puts it at a further disadvantage.

The study was initiated by conducting online literature searches for the period 1990-2014, using electronic databases. The literature was reviewed, and relevant materials have been collected for further evaluation and analysis. Numerous types of storage systems are available, or are becoming available, to meet these needs. It is important to identify a suitable match between requirements and the performance of various types of technologies. The overall goal of this study is to address this match by examining both performance characteristics and cost of ESS that are suitable for storing electricity generated from RESs.

4.2 WHY ESS

Electricity is generated by thermal/nuclear power plants and is consumed instantaneously. Unlike other energy generation systems like oil, natural gas, or coal, the power grid or the generation plants essentially have no storage or "surge" capacity to smooth out peaks and valleys in demand or to provide reserve capacity during sudden spikes in demand.

Nuclear power plants, oil-fired steam power plants, gas turbines, as well as hydroelectric plants have turbine generators that operate continuously and hence deliver firm, continuous, and dispatchable power required by consumers. With such a steady base load supply, power plants can meet demand shifts and daily cycles either by adding peaking generators or utilizing the available spinning reserve. Thus, the balancing of power supply and demand, required for the operation of a safe and reliable grid, is now done on line in real time.

Power generated by solar PVs or wind is inherently intermittent. The sun generates power approximately 10-12 h a day, and in the rainy season, clouds can cover the sun intermittently. The wind is also highly fluctuating, both in terms of direction and magnitude over 24 h of the day, and the day of the year. ESS are the critical technology needed by renewable power if it is to become a major source of base load dispatchable power to replace conventional thermal/nuclear power plants. For stable and smooth power supply, large-scale ESS are essential and needed to convert the intermittent and fluctuating renewable power that is generated into dispatchable power. Without sufficient ESS, accessible online, solar/wind power cannot serve as a stable base load supplier (Lee and Gushee, 2008).

According to Dr. Imre Gyuk, energy storage program manager at the U.S. Department of Energy, energy storage can (Gyuk, 2009)

- Provide spinning reserve and energy management to accommodate renewable resources
- Provide power quality and digital reliability
- Provide voltage and frequency regulation and smooth renewable contributions
- Allow better asset utilization of generation and transmission

ESS has several applications in the electrical power system, as shown in Figure 4.1. They are divided into three main categories:

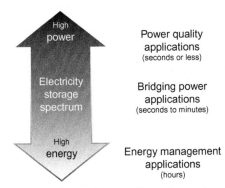

FIGURE 4.1 Electricity storage spectrum that covers from sub-second power quality to multi-hour energy management applications (Nourai, 2002).

- *Power quality*: In these applications, ESS *is only applied for seconds or less*, as needed, to ensure continuity of quality power.
- *Bridging power*: In these applications, ESS is used for *seconds to minutes* to ensure continuity of service when switching from one source of energy generation to another.
- *Energy management*: In these applications, ESS is used to decouple the timing of generation and consumption of electric energy. A typical application is utilization of solar PV/wind power, which involves the charging of ESS when energy is available and utilization as needed. This would also enable consumers to be grid-independent for *many hours*.

These applications require energy discharges from a fraction of a second in high-power applications to hours in high-energy applications (see Figure 4.1). Large-scale electricity storage technologies cover a wide spectrum of applications, ranging from fast power quality applications to improve reliability all the way to slow energy management applications to improve profitability.

There are a number of applications in the aforementioned three main categories. Moreover, the deregulation of the electrical power industry in Saudi Arabia will require unbundling of the generation, transmission, and distribution sectors; some of the sectors will be privatized, and the private industry would like to see an efficient power industry. The main focus of this study is to review and assess the ESS that are suitable for energy management applications.

Lior (2010) presented a comprehensive review of the present situation and possible paths to the future of sustainable energy development. He remarked that wind and solar PV are experiencing exponential growth with decrease in cost. Improvements and technological advances in the distribution and storage of electric power will continue and should be advanced much faster. He indicated that the U.S. government budget in 2009 emphasized clean, renewable energy generation and storage, among other technologies.

DOE/EPRI (2013) published the Electricity Storage Handbook in collaboration with the NRECA, funded through Dr. Imre Gyuk, U.S. Department of Energy (DOE) and Haresh Kamath, Electric Power Research Institute (EPRI). The handbook presents information on stationary energy storage systems. These include batteries, flywheels, compressed air energy storage (CAES), and pumped hydropower, and exclude thermal, hydrogen, and other forms of energy storage that could also support the grid.

4.3 THE POTENTIAL FOR ESSs

The level of storage on electricity networks varies considerable around the world. In the United States, around 2.5% of all the power delivered on the network has passed through a storage plant. In Europe, by contrast, that figure is 10%, and in Japan it is 15% (EPRI-DOE, 2003).

Generally, the public view is that solar PV/wind power can replace thermal/nuclear power plants as sources of power if enough wind farms and solar PV systems are built. Therefore, research and development is focused on improving the performance and cost of solar PV/wind electricity generation. As one of the critical components of renewable power supply systems, ESS is not getting due attention in its development. To obtain an estimate of the order of magnitude of the size of the ESS needed and the associated capital investment, three scenarios of renewable power penetration into the US grid are discussed:

a. 20%—ESS required for grid stability
b. 50%—Renewable becomes the principal power source
c. 75%—Ultimate renewable penetration

Table 4.1 summarizes the current "upper bound" conservative estimates for the needed ESS capacity for the United States under the three scenarios of grid penetration. The conventional electricity value chain has been considered to consist of five links, namely fuel/energy source, generation, transmission, distribution, and customer side energy service, as shown in Figure 4.2. A typical configuration for integration of ESS with renewables is illustrated in Figure 4.3.

4.4 REQUIREMENTS OF ESS FOR SAUDI ARABIA

Demand for new energy storage systems is increasing for applications such as remote area power supply systems (e.g., offshore platforms, telecommunication installations), stressed electricity supply systems (e.g., Hajj operations), emergency backup applications, and mobile applications. The supply of electric power to remote areas is becoming more attractive due to advancements in

TABLE 4.1 "Upper Bound" Estimate of U.S. ESS Size and Cost

Estimate of U.S. ESS Size and Cost			
	Option 1	Option 2	Option 3
Grid penetration by renewable power (%)	20	50	75
Firm renewable demand, GW	200	500	750
Nameplate renewable installed capacity, GW	570	1430	2150
Capital invest. for installed capacity, $ billion	860	2150	3220
ESS power capacity, GW	114	285	428
ESS energy capacity, GWh	912	2280	3424
ESS capital investment, $ billion	342	855	1284
Investment ratio of ESS to renewable power	0.40	0.40	0.40

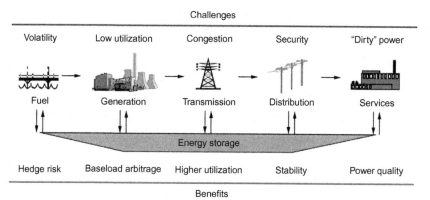

FIGURE 4.2 Energy storage applications (Chen et al., 2009).

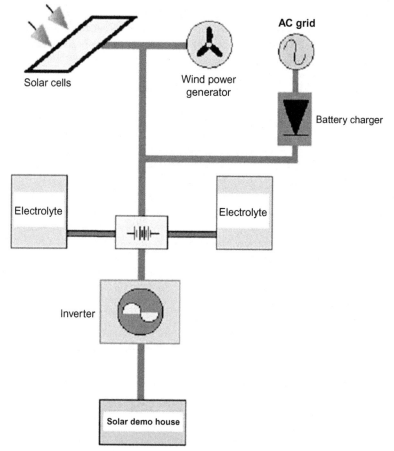

FIGURE 4.3 Typical configuration for storing electricity from renewables.

PV technologies and wind power generation systems, along with the development of advanced storage batteries. The principal limitation of renewable energy production is its fluctuating and unpredictable nature, requiring it to be "balanced" with reliable fossil fuel derived energy so that electricity can be supplied constantly to industry and society. Energy storage systems can be used to balance fluctuations in the supply and demand of electricity.

Due to the exponentially growing population and limited means of fossil fuel energy, new and renewable sources of energy are being promoted to meet energy demands both in grid-connected areas and in off-grid areas. These sources include solar PV, solar thermal, wind power, geothermal, tidal, wave, biofuels, and so on. Of these, solar and wind resources are being exploited on a large scale, and the available technology is also commercially developed and accessible. Solar PV is a very good source of electricity generation in this part of the world, that is, the Middle East in general and Saudi Arabia in particular. The only barriers are its storage, relatively high cost, low efficiency, and the local environmental problem of high temperature and suspended dust particles, which puts it at a further disadvantage. The status of the local environment in which the ESS has to operate needs to be thoroughly investigated to understand the limitations or advantages of local conditions for ESS operations.

4.4.1 Climate of Saudi Arabia

Usually, batteries are used to store the energy produced by the sun or the wind to ensure continuous, 24/7 supply. The batteries are very sensitive to weather conditions (temperature, relative humidity, barometric pressure, wind speed, etc.) and need to be evaluated both for efficiency and for working life degradation in the harsh environment of Saudi Arabia. Among all the weather parameters, temperature and relative humidity are very critical for battery efficiency and working life, and hence should be considered when selecting a battery for energy storage purposes.

Saudi Arabia has a desert climate characterized by extreme heat during the day, an abrupt drop in temperature at night, and slight, erratic rainfall. Because of the influence of a subtropical high-pressure system and the many fluctuations in elevation, there is considerable variation in temperature and humidity. The two main extremes in climate are felt between the coastal lands and the interior.

Temperatures are different in each part of the country. Particularly in the central area and the north, temperatures can be very high. From June through August, midday temperatures in the desert can soar to 50 °C (122 °F) or more. The south has moderate temperatures, which can go as low as 10 °C (50 °F) during the summer in the mountains of Sarawat in Asir. Along the coastal regions of the Red Sea and the Persian Gulf, the desert temperature is moderated by the proximity of these large bodies of water. Temperatures seldom rise above 38 °C, but the relative humidity is usually more than 85%, and frequently 100% for extended periods. This combination produces a hot mist during the day and a warm fog at night. A uniform climate prevails in Najd, Al-Qasim Province, and the great deserts. The average summer temperature is 45 °C, but readings of up to 51 °C are common. The heat becomes intense shortly after sunrise and lasts until sunset, followed by comparatively cool nights.

During the winter, the temperatures are moderate in general but turn cold at night, sometimes descending below freezing, especially in mountainous areas of the west and along the northern border. Torrential rains fall along the Red Sea coast during March and April. In Najd, Al-Qasim Province, in the winter, the temperature seldom drops below 0 °C, but the almost total absence of humidity and the high wind-chill factor make a bitterly cold atmosphere. In the spring and autumn, temperatures average 29 °C.

The entire year's rainfall may consist of one or two torrential outbursts that flood the wadis and then rapidly disappear into the soil to be trapped above the layers of impervious rock. This is sufficient, however, to sustain forage growth. Although the average rainfall is 100 mm per year, the whole region may not experience rainfall for several years. When such droughts occur, as they did in the north in 1957 and 1958, affected areas may become incapable of sustaining either livestock or agriculture. The region of Asir is subject to Indian Ocean monsoons, usually occurring between October and March. An average of 300 mm of rainfall occurs during this period, which is about 60% of the annual total. Additionally, in Asir and the southern Hijaz, condensation caused by the higher mountain slopes contributes to the total rainfall.

Prevailing winds are from the north, and when they blow, coastal areas become bearable in the summer and even pleasant in winter. A southerly wind is accompanied invariably by an increase in temperature and humidity and by a particular kind of storm known in the gulf area as a kauf. In late spring and early summer, a strong northwesterly wind, the shamal, blows; it is particularly severe in eastern Arabia and continues for almost three months. The shamal produces sandstorms and dust storms that can decrease visibility to a few meters.

The mean maximum and minimum values of temperature over a period of 35 years (i.e., from 1970 to 2006) are summarized in Table 4.2. This is a very important parameter that affects the performance of some ESS. The long-term mean temperature was found to vary between a minimum of 18.6 °C at Abha and a maximum of 30.2 °C at Gizan. On the other hand, the maximum and minimum temperatures recorded were 51 and −9.4 °C, corresponding to Qaisumah and Hail stations, respectively. So any ESS being considered for Saudi Arabia should have a tolerance of withstanding a temperature of about −9 to 51 °C.

4.4.2 Supply-Demand Situation of Power in Saudi Arabia

Electric energy in the Kingdom of Saudi Arabia is provided mainly by the Saudi Electricity Company (SEC). SEC is divided into four operating areas: the eastern, central, western, and southern operating areas. Residential and

TABLE 4.2 Long-Term Statistics of Weather Parameters for 20 Meteorological Stations

S. No.	Station	Pressure (mb)	Rain (mm)	Temperature (°C)			Relative Humidity (%)	Wind Speed (m/s)	
		Mean	Max	Mean	Min	Max	Mean	Mean	Max
1.	Dhahran	1006.7	125.0	26.4	2.5	49.0	52.5	4.38	11.8
2.	Gizan	1007.7	90.0	30.2	11.8	45.3	68.4	3.24	7.7
3.	Guriat	954.8	36.5	19.5	−8.0	47.6	43.5	4.22	16.5
4.	Jeddah	1007.3	55.0	28.2	9.8	49.0	61.4	3.71	11.3
5.	Turaif	916.9	25.7	19.0	−8.0	45.5	40.3	4.33	14.4
6.	Riyadh	942.4	70.0	26.7	0.0	47.8	26.2	3.09	8.8
7.	Yanbu	1007.8	73.2	27.7	4.7	49.0	53.8	3.76	10.3
8.	Abha	794.0	119.9	18.6	0.0	34.1	54.6	2.94	14.9
9.	Hail	901.3	47.5	22.4	−9.4	44.5	33.2	3.24	10.8
10.	Al-Jouf	936.1	34.0	22.0	−6.0	46.7	32.1	4.02	15.9
11.	Al-Wejh	1007.9	116.0	25.1	5.1	46.0	64.6	4.43	11.8
12.	Arar	949.6	38.0	22.2	−5.6	48.2	36.2	3.61	12.9
13.	Bisha	884.0	40.0	25.9	0.3	44.8	29.3	2.47	10.3
14.	Gassim	937.6	86.0	25.1	−4.0	49.0	29.0	2.78	9.3
15.	Khamis	797.9	99.0	19.8	1.5	35.0	51.4	3.14	12.9
16.	Nejran	879.4	157.0	25.7	1.0	44.0	30.1	2.10	8.8
17.	Qaisumah	969.5	64.0	25.5	−4.0	51.0	31.4	3.55	11.8
18.	Rafina	960.3	121.0	23.4	−5.8	49.0	38.4	3.86	12.4
19.	Tabouk	926.0	36.0	22.0	−3.5	46.4	34.0	2.73	15.5
20.	Taif	855.4	169.0	23.1	−1.5	40.2	43.3	3.66	10.3

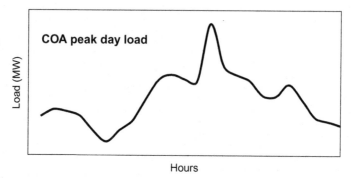

FIGURE 4.4 Load variations during the peak load day in the central operating area.

commercial loads represent more than 60% of the SEC total load. A large portion of the loads is mainly from air conditioners; therefore, reducing the use of energy at peak hours and building it at off-peak hours looks like a viable option.

To study the viability, the hourly load data for SEC's four operating areas for 2006 were obtained. In this report, the preliminary assessment is conducted for the central operating area (COA). COA was chosen for the assessment because the load in COA is mostly residential and commercial, and the difference between the daily peak and minimum load is quite large. The peak load recorded in 2006 was 9725 MW and occurred in the month of July; the total annual energy was 52,794 GWh. The minimum load recorded was 2133 MW and occurred in the month of January. The value of the peak-to-minimum ratio for COA was 4.56 and the annual load factor was 0.62. Figure 4.4 shows the peak day load for COA: the maximum load during the day was 9725 MW, and it occurred at 1400 h on July 1, 2006. The minimum load for the day was 7290 MW, and the average load for the day was 8169 MW.

It can be seen from Figure 4.4 that the load remains below 8500 MW for around 16 h and above the 8500 MW mark for around 8 h. If the load during peak hours is supplied through any ESS, the ESS could be charged during off-peak hours. It should be noted that this is a preliminary analysis, and a detailed techno-economic analysis is required before recommending ESS for load shifting in COA.

4.4.3 Local Raw Materials for ESS

Sulfur is abundantly available and cheap in Saudi Arabia due to large refining capacity. ESS that utilize sulfur or any form of sulfur-based product may be a step toward utilization of sulfur, which is a disposal problem for Saudi Aramco.

4.4.4   Global and Local PV/Wind Power Installed Capacities

According to the latest update, global investment on the development and utilization of renewable sources of power increased to $244 billion in 2012 compared to $279 billion in 2011 (Weblink1, 2014). Global wind power installed capacity increased by 12.4% to more than 318 GW in 2013 due to greater participation by China and Canada. However, installations slowed in 2013 to about 35.5 GW, almost 10 GW less than in 2012 (Weblink2, 2014). China's installed wind capacity was 75.3 GW at the end of 2012, and it reached 91.4 GW in 2013. Canada, the other major contributor to wind power capacity, added 1.6 GW of new capacity. Europe's installed wind capacity rose to almost 121.5 GW in 2013, compared to 110 GW at the end of 2012. The world's cumulative PV installed capacity surpassed the impressive 100 GW electrical power mark, achieving just over 102 GW at the end of 2012 (Weblink3, 2014). The global cumulative PV installed capacity in 2011 was 71 GW. This capacity is capable of producing as much annual electrical energy as 16 coal power plants or nuclear reactors of 1 GW each. Each year, these PV installations save more than 53 million tons of CO_2 equivalent greenhouse gases from entering into the atmosphere. The total hydro installed capacity reached 990 GW in 2012 compared to 960 GW in year 2011, an increase of 3.12% (Weblink1, 2014).

In this part of the world (i.e., in Saudi Arabia), higher intensities of global solar radiation are available, with longer hours of sunshine. Global solar radiation varies between a minimum of 1.63 MWh/m^2/year at Tabuk and a maximum of 2.56 MWh/m^2/year at Bisha, and an overall mean of 2.06 MWh/m^2/year (Rehman et al., 2007). The duration of sunshine varied between 7.4 and 9.4 h, with an overall mean of 8.89 h. The seasonal variation of global solar radiation and sunshine duration obtained using respective data from 41 locations in the kingdom is shown in Figures 4.5 and 4.6, respectively. It is evident from these figures that higher intensities of global solar radiation and longer durations of sunshine are observed during summer, when the water requirement is comparatively higher.

The utilization of solar PVs in Saudi Arabia goes back to early 1980s, when King Abdulaziz City for Science and Technology installed the world's largest grid-connected 5 kW PV plant at Solar Village in Riyadh. PV panels are being widely used for remote applications such as cathodic protection and communication towers. Per available information, cumulative PV installed capacity in Saudi Arabia is depicted in Figure 4.7. A 500 kW PV plant was realized in 2011 in the Farasan Island power plant in Gizan, while another 10,200 kW grid-connected PV power plant was installed in 2012 at Aramco Campus in Dhahran, as can be seen in Figure 4.7. In 2013, KASPAREC in Riyadh brought a 3500 kW PV plant on line, which boosted the total PV installed capacity of Saudi Arabia to more than 20 MW. The present trends show that utilization of PV will increase, and hence for smooth and quality power supply to consumers, appropriate energy storage systems will be required.

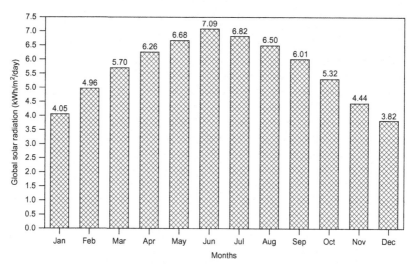

FIGURE 4.5 Seasonal variation of global solar radiation intensity over Saudi Arabia, [Rehman et al. 2007].

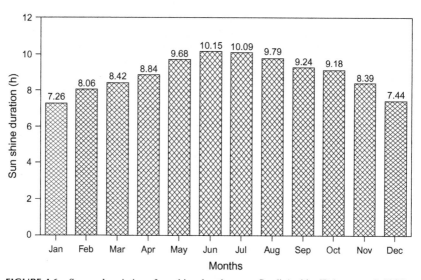

FIGURE 4.6 Seasonal variation of sunshine duration over Saudi Arabia, [Rehman et al. 2007].

The typical overall diurnal variation of wind speed with time at five different locations in Saudi Arabia is illustrated in Figure 4.8. The coastal sites (i.e., Yanbu and Dhahran) experienced similar characteristics, where wind speed is high during the hours 13:00-18:00. That makes wind energy a potential tool for electrical peak shaving during the hours of high air-conditioning demand load. Figure 4.8 shows similar profiles for inland sites, Gassim and Dhulum. These sites are located in the mid and mid-south of the country. At these sites,

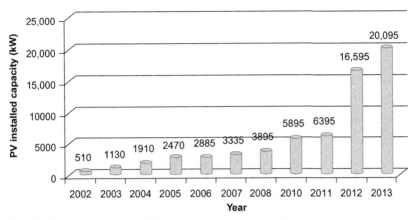

FIGURE 4.7   Cumulative solar PV installed capacity, Saudi Arabia.

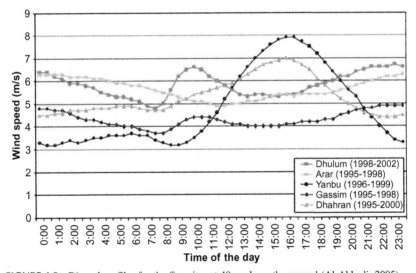

FIGURE 4.8   Diurnal profiles for the five sites at 40 m above the ground (Al-Abbadi, 2005).

the wind speed peaks twice: the highest peak occurs around midnight, and the second occurs during the late morning hours (8:00-12:00). The wind speed profile of the site Arar showed peculiar behavior compared to the others.

An estimated annual energy that can be generated at the five sites by a typical wind turbine, Nordex N43/600, was obtained using the wind power curve of the turbine (Nordex, 2003) and 1-year wind duration data recorded at the sites. Table 4.3 summarizes the technical data of the wind machine used. Because the hub height of the wind machine is 50 m, 1/7th power law was used to calculate the half-hourly mean wind speed values. Figure 4.9 illustrates a comparison of

TABLE 4.3 Technical Data of Wind Machine from Nordex N43/600

Cut-in speed (m/s)	3
Cut-out speed (m/s)	25
Rates speed (m/s)	13.5
Survival speed (m/s)	70
Rated output (kW)	600
Hub height (m)	50
Rotor diameter (m)	43

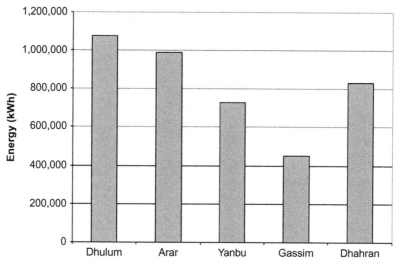

FIGURE 4.9   The annual wind energy produced by the Nordex N43 at the sites (Al-Abbadi, 2005).

the annual energy that can be generated using the Nordex N43. It shows that at the Dhulum site, the calculated annual energy is about 1080 and 990 MWh at Arar. For Dhahran and Yanbu, the annual energy values are 833, 730 MWh, respectively, while at the Gassim site, the Nordex N43 only produces about 454 MWh.

As of today, Saudi Arabia does not have any wind power installed capacity, but is expected to achieve a cumulative capacity of 11 GW by 2032.

4.5 DESCRIPTION OF MAJOR ESS

The ESS can be divided into two categories based on function and form. In terms of function, ESS technologies can be categorized into those that are intended first for high power ratings with a relatively small energy content, making them suitable for power quality or UPS, and those designed for energy management, as shown in Figure 4.1.

Because this report concentrates on energy storage devices for stationary applications needed for energy management based on the renewable energy program, ESS for power quality management that last for few seconds to minutes as well as portable and transport applications are excluded, with exceptions made for technologies that may have dual uses. Figure 4.10 compares the various energy storage technologies for the largest power and duration periods for which they are expected to be applied. Note that superconducting magnetic energy storage (SMES) and a few other devices that are only used for high power applications are also included to indicate the differences between the power quality management and energy management applications.

Large-scale energy storage systems provide significant advantages to electric power systems, including load following, peak power, and standby reserve. The role of ESS becomes more important in renewable energy systems, as PV and wind are not available during extended periods of time. It is generally agreed that no more than about 20% of a region's demand can be provided by intermittent renewables without energy storage (Cavallo, 2001).

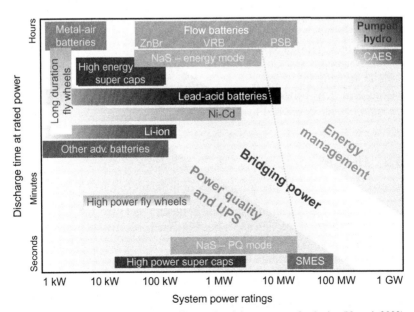

FIGURE 4.10 Application ranges for different electricity storage technologies (Nourai, 2002).

Denholm et al. (2004) presented the results of a life cycle assessment that evaluates the energy requirements and greenhouse gas emissions resulting from the construction and operation of energy storage systems. Kaldellis and Zafirakis (2007) remarked that to bypass the electrical grid stability constraints arising from extensive RES utilization, the adaptation of an appropriate energy storage system is essential.

The electricity storage technologies suitable for "energy management applications" that are available in the market and needs assessment for storing electrical energy from renewable sources are described in the following paragraphs.

4.5.1 Chemical/Electrochemical Systems

The oldest and most established way of storing electrical energy is in the form of chemical energy storage systems. One of the main features of this form of energy storage is that it has very high efficiency due to no moving parts in the system. However, the biggest challenges that have been experienced by electrochemical storage technologies are their inability to cycle repeatedly in a deep discharge fashion, with sufficient reliability, efficiency, and better economics.

1. Flow batteries:
   - Zinc bromine (ZnBr) batteries
   - Polysulfide bromide (PSB) batteries
   - Vanadium redox battery (VRB)
2. Sodium-sulfur (NaS) batteries
3. Lithium-ion (Li-ion) batteries
4. Nickel-cadmium (Ni-Cd) batteries
5. Metal air batteries
6. Lead-acid (LA) batteries
7. Liquid battery developed by MIT

4.5.2 Electrical Systems

This involves storing electrostatic energy in different forms.

- SMES is a niche technology for power quality and especially high-power distribution or transmission networks. Projected costs for bulk storage, however, show it to be expensive.
- Electrochemical capacitors (supercapacitors)

Because SMES and supercapacitors are generally meant for power quality purposes, as can be observed in Figure 4.10, these ESS will not be further discussed.

4.5.3   Mechanical Systems

Electricity is stored in the form of kinetic energy or potential energy.

- Flywheel energy storage is a good match for a range of short-term applications up to a size of several MW.
- CAES requires some type of geologic feature for storage.
- Pumped hydroelectricity storage (PHS) is best for load management when geology is available and response time in the order of minutes is acceptable. However, PHS of electricity facilities requires a lot of space and a significant water resource.

4.5.4   Thermal Systems

The principle of this type of energy storage is that the electricity available from renewables or from the grid at off-peak hours is used to heat a material (e.g., molten salts) that has thermal properties that allow the heat to be stored. This heat is utilized when the demand is high, usually by generating steam, which is in turn used to power a steam turbine or it could be used in heating/cooling systems.

4.5.5   Salient Features of Selected ESS for Renewables

Schoenung (2001) of Sandia National Laboratory presented characteristics and technologies for long- versus short-term energy storage. Applications of energy storage have a wide range of performance requirements. One important feature is storage time or discharge duration. In this study, applications and technologies have been evaluated to determine how storage time requirements match technology characteristics.

The advantages and disadvantages of various ESS are presented in Table 4.4 along with the application type, whether power quality application or energy management application. To store electrical energy generated from renewable sources such as PV cells or wind farms, ESS required are for the duration of hours to days. Therefore, suitable ESS for energy management applications that can be integrated with renewables based on Figure 4.3 and Table 4.4 are as follows:

1. PHS
2. CAES
3. Flow batteries
4. NaS battery
5. NiCd battery
6. LA battery

The main features of the selected ESS are described in the following paragraphs.

TABLE 4.4 Main Features of ESS for Power Quality Applications and Energy Management Applications (ESA, 2009)

Storage Technologies	Main Advantages (relative)	Disadvantages (Relative)	Power Application	Energy Application
Pumped storage	High capacity, low cost	Special site requirement		●
CAES	High capacity, low cost	Special site requirement, need gas fuel		●
Flow batteries: PSB, VRB, ZnBr	High capacity, independent power and energy ratings	Low energy density	◐	●
Metal-air	Very high energy density	Electric charging is different		●
NaS	High power and energy densities, high efficiency	Production cost, safety concerns (addressed in design)	●	●
Li-ion	High power and energy densities, high efficiency	High production cost, requires special charging circuit	●	○
Ni-Cd	High power and energy densities, efficiency		●	◐
Other advanced batteries	High power and energy densities, high efficiency	High production cost	●	○
Lead-acid	Low capital cost	Limited cycle life when deeply discharged	●	○
Flywheels	High power	Low energy density	●	○
SMES, DSMES	High power	Low energy density, high production cost	●	
E.C. capacitors	Long cycle life, high efficiency	Low energy density	●	◐

4.5.5.1 Pumped Hydroelectricity Storage

Generally, the pumped hydroelectric storage system is used in power plants for load balancing or peak load shaving. This method stores energy in the form of water, pumped from a lower elevation reservoir to a higher elevation. In pumped hydroelectric energy storage systems, water is pumped to a higher elevation and then released and gravity-fed through a turbine that generates electricity. Conventional hydroelectric storage systems rely on natural elevation differentials between water bodies on the Earth's surface to store energy. Most large hydroelectric installations rely on hydraulic heads of at least 150 ft, with average head of about 400 ft. Because head height is proportional to energy, power, and efficiency, a larger head is desirable.

4.5.5.1.1 Process Description

Conventional PHS uses two water reservoirs, separated vertically. During off-peak hours, water is pumped from the lower reservoir to the upper reservoir. When required, the water flow is reversed to generate electricity. Some high-dam hydro plants have a storage capability and can be dispatched as a pumped hydro. Underground pumped storage, using flooded mine shafts or other cavities, is also technically possible. Open sea can also be used as the lower reservoir.

A schematic view of a pumped hydroelectric storage system is shown in Figure 4.11 (Kaldellis et al., 2009). It consists of (i) two reservoirs located at different elevations, (ii) a unit to pump water to the reservoir at high elevation (to store electricity in the form of hydraulic potential energy during availability of renewable power and/or during off-peak hours), and (iii) a turbine to generate electricity with the water flowing from the higher reservoir to the reservoir at low elevation (converting the potential energy to electricity). Clearly, the

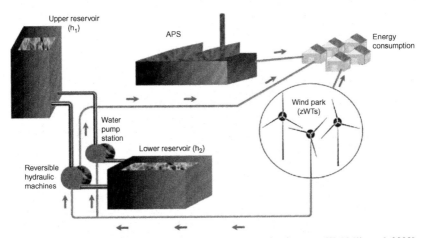

FIGURE 4.11 Schematic of a conventional pumped-storage development, [Kaldellis et al, 2009].

amount of stored energy is proportional to the difference in height between the two reservoirs and the volume of water stored. Some high-dam hydroelectric power plants have a storage capability and can be dispatched as a PHS. Underground pumped storage, using flooded mine shafts or other cavities, is also technically possible. If the PHS is located near seaside, open sea can be treated as the lower reservoir (Ahearne, 2004).

4.5.5.1.2 Performance Characteristics

- The self-discharge (energy dissipation) per day for PHS has a very small self-discharge ratio, so it is suitable for a long storage period.
- PHS has a cycle efficiency of 60-90%.
- The energy density of PHS is among the lowest, below ~30 Wh/kg.
- PHS has a long cycle life. This technology is based on conventional mechanical engineering, and the lifetime is mainly determined by the lifetime of the mechanical components.

4.5.5.1.3 Pumped Hydro Storage Applications

The typical rating of PHS is about 1000 MW (100-3000 MW), and facilities continue to be installed worldwide at a rate of up to 5 GW per year. The rating of PHS is the highest all over the available ESS, hence it is generally applied for energy management, frequency control, and provision of reserve. Since the first use in Italy and Switzerland in the 1890s and the first large-scale commercial application in the United States in 1929 (Rocky River PHS plant, Hartford), there are now over 200 units and 100 GW of PHS in operation worldwide (Ahearne, 2004; Linden, 2003) (32 GW installed in Europe, 21 GW in Japan, 19.5 GW in the United States, and others in Asia and Latin America), which is about 3% of global generation capacity (ESA, 2009). Pumped storage is the most widespread energy storage system in use on power networks. Its main applications are for energy management, frequency control, and provision of reserve.

4.5.5.1.4 Advantages/Disadvantages

PHS is a mature technology with large volume, long storage period, high efficiency, and relatively low capital cost per unit of energy. Owing to the small evaporation and penetration, the storage period of PHS can be varied, from typically hours to days and even years. Taking into account the evaporation and conversion losses, 71-85% of the electrical energy used to pump the water into the elevated reservoir can be regained.

The major drawback of PHS lies in the scarcity of available sites for two large reservoirs and one or two dams. Long lead time (typically 10 years) and high cost (typically hundreds to thousands of million US dollars) for construction and environmental issues (e.g., removing trees and vegetation from the large amounts of land prior to the reservoir being flooded) (Denholm et al., 2004;

Denholm and Holloway, 2005) are the other three major constraints in the deployment of PHS. Many pumped hydroelectric systems can have negative impacts on land and wildlife. Disruption of fish spawning routes or creation of large reservoirs that fill canyons or gorges are common concerns. In general, in geographically flat places, PHS may be difficult to use or may not be used at all.

The construction of PHS systems inevitably involves destruction of trees and green land for building reservoirs. The construction of reservoirs could also change the local ecological system, which may have environmental consequences.

4.5.5.1.5 Commercial Maturity

PHS is a matured technology and has been in use for more than 100 years. Pump hydroelectricity installations that are 1000 MW and larger are shown in Table 4.5.

4.5.5.1.6 Cost

Capital cost is one of the most important factors for the industrial take-up of the ESS. It is expressed in terms of cost per kW, per kWh, and per kWh per cycle. All the costs per unit energy shown here have been divided by the storage efficiency to obtain the cost per unit of output energy. The per-cycle cost is defined as the cost per unit energy divided by the cycle life, which is one of the best ways to evaluate the cost of energy storage in a frequent charge/ discharge application such as load leveling. The costs of operation and maintenance, disposal, replacement, and other ownership expenses are not considered, because they are not available for some emerging technologies (Haisheng et al., 2009).

$600-2000/kW, $5-100/kWh,   0.1-1.4 cents/kWh/cycle

4.5.5.1.7 Developers/Suppliers

MWH
Corporate Headquarters
380 Interlocken Crescent
Suite 200, Broomfield, Colorado USA 80021

4.5.5.2 *Compressed Air Energy Storage*

Off-peak (low-cost) electrical power is used to compress air into an underground air-storage "vessel" (the Norton mine), and later the air is used to feed a gas-fired turbine generator complex to generate electricity during on-peak (high-price) times.

TABLE 4.5 Pumped HydroElectricity Installations Worldwide (1000 MW and Larger)

Location	Plant Name	On-Line Data	Hydraulic Head (m)	Max Total Rating (MW)	Hours of Discharge	Plant Cost
Australia	Tumut 3	1973		1690		
China	Tianhuangping	2001	590	1800		$1080 M
	Guangzhu	2000	554	2400		
France	Grand Maison	1987	955	1800		
Germany	Marker sbach	1981		1050		
	Goldisthal	2002		1060		$700 M
Iran	Siah Bisheh	1996		1140		
Italy	Piastra Edolo	1982	1260	1020		
	Chiotas	1981	1070	1184		
	Presenzano	1992		1000		
	Lago Delio	1971		1040		
Japan	Imaichi	1991	524	1050	7.2	
	Okuyoshino	1978	505	1240		
	Kazunogowa	2001	714	1600	8.2	$3200 M
	Matanogawa	1999	489	1200		
	Ohkawachi	1995	411	1280	6	
	Okukiyotsu	1982	470	1040		
	Okumino	1995	485	1036		
	Okutataragi	1998	387	1240		

Country	Name	Year				
	Shimogo	1991	387	1040		
	Shin Takesagawa	1981	229	1280	7	
	Shin Toyne	1973	203	1150		
	Tamahare	1986	518	1200	13	
Luxembourg	Vianden	1964	287	1096		
Russia	Zagorsk	1994	539	1200		
	Kaishador	1993		1600		
	Dneister	1996		2268		
South Africa	Drakensbergs	1983	473	1200		$866 M
Taiwan	Minghu	1985	310	1008		$1338 M
	Mingtan	1994	380	1620		$310 M
UK/Wales	Dinorwig	1984	545	1890	5	
USA/CA	Castaic	1978	350	1566	10	
USA/CA	Helms	1984	520	1212	153	$416 M
USA/MA	Northfield MT	1973	240	1080	10	$685 M
USA/MI	Ludington	1973	110	1980	9	$327 M
USA/NY	Blenhein-Gilboa	1973	340	1200	12	$212 M
USA/NY	Lewiston (Niagara)	1961	33	2880	20	
USA/SC	Bad Creek	1991	370	1065	24	$652 M
USA/TN	Racoon Mt	1979	310	1900	21	$288 M
USA/VA	Bath County	1985	380	2700	11	$1650 M

Source: ESA (2009)

4.5.5.2.1 Process Description

Off-peak electricity is used to power a motor/generator that drives compressors to force air into an underground storage reservoir. Figure 4.12 shows a schematic diagram of a CAES system (Jewitt, 2005; McDowall, 2004). It consists of five major components: (1) a motor/generator that employs clutches to provide alternate engagement to the compressor or turbine trains; (2) an air compressor of two or more stages with intercoolers and aftercoolers, to achieve economy of compression and reduce the moisture content of the compressed air; (3) a turbine train, containing both high- and low-pressure turbines; (4) a cavity/container for storing compressed air, which can be underground rock caverns created by excavating comparatively hard and impervious rock formations, salt caverns created by solution- or dry-mining of salt formations, and porous media reservoirs made by water-bearing aquifers or depleted gas or oil fields (e.g., sandstone and fissured lime); and (5) equipment controls and auxiliaries such as fuel storage and heat exchanger units.

CAES works on the basis of conventional gas turbine generation. It decouples the compression and expansion cycles of a conventional gas turbine into two separated processes and stores the energy in the form of elastic potential energy of compressed air (Haisheng et al., 2009). During low demand, electrical energy is stored by compressing air into an airtight space, typically 4.0-8.0 MPa. To extract the stored electrical energy, compressed air is withdrawn

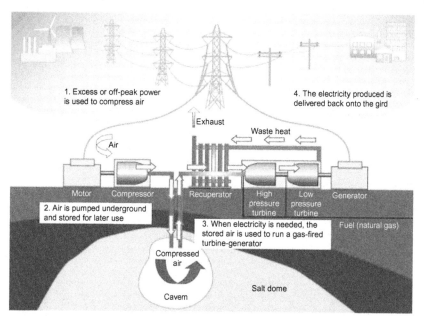

FIGURE 4.12 Schematic diagram of a compressed-air energy storage setup [Jewitt, 2005; McDowall, 2004].

from the storage vessel, heated, and then expanded through a high-pressure turbine. Thus, it captures some of the energy in the compressed air. The air is then mixed with fuel and combusted with the exhaust-expanded air through a low-pressure turbine. Both the high- and low-pressure turbines are connected to a generator to produce electricity.

4.5.5.2.2 Performance Characteristics

- The self-discharge (energy dissipation) per day for CAES has a very small self-discharge ratio, so it is suitable for a long storage period.
- CAES has a cycle efficiency of 60-90%.
- CAES has medium energy density.
- CAES has a long cycle life. This technology is based on conventional mechanical engineering, and the lifetime is mainly determined by the lifetime of the mechanical components.

4.5.5.2.3 Applications

The concept of CAES to help generate electricity is more than 30 years old. Two plants currently exist: an 11-year-old plant (110 MW) in McIntosh, Alabama, and a 23-year-old plant (290 MW) in Germany, both in caverns created by salt deposits. The construction took 30 months and cost $65 M (about $591/kW). It takes about 1.5-2 years to create such a cavern by dissolving salt. This unit comes on line within 14 min. The third commercial CAES, the largest ever, is a 2700 MW plant that is planned for construction in Norton, Ohio. This 9-unit plant will compress air to 1500 psi in an existing limestone mine some 2200 ft underground.

4.5.5.2.4 Advantages/Disadvantages

In a conventional power plant, nearly two-thirds of the natural gas is consumed by a typical natural gas turbine because the gas is used to drive the machine's compressor. In contrast, a compressed air storage plant uses low-cost heated compressed air to power the turbines and create off-peak electricity, conserving some natural gas.

However, compressed air has safety concerns, mainly the catastrophic rupture of the tank. Highly conservative safety codes make this a rare occurrence at the trade-off of higher weight. Codes may limit the legal working pressure to less than 40% of the rupture pressure for steel bottles (safety factor of 2.5), and less than 20% for fiber-wound bottles (safety factor of 5).

CAES is based on conventional gas turbine technology and involves combustion of fossil fuel, hence emissions can be an environmental concern.

4.5.5.2.5 Commercial Maturity

CAES is a developed technology and is commercially available. However, the actual applications, especially for large-scale utility, are still not widespread.

Their competitiveness and reliability still need more trials by the electricity industry and the market.

4.5.5.2.6 Cost

Capital cost is one of the most important factors for the industrial take-up of the ESS. It is expressed in terms of cost per kW, per kWh, and per kWh per cycle. The costs of operation and maintenance, disposal, replacement, and other ownership expenses are not considered, because they are not available for some emerging technologies (Haisheng et al., 2009).

$400-800/kW, $2-50/kWh,   2-4 cents/kWh/cycle

4.5.5.2.7 Developers/Suppliers

CAES Development Company
Dresser-Rand Company
Energy Storage and Power LLC
Ridge Energy Storage

4.5.5.3 Flow Batteries

Redox flow batteries have received considerable interest over the last 20 years as potentially low-cost and highly efficient energy storage systems for large-scale applications. Preliminary work on redox cells was conducted initially by Kangro and Pieper (1962) and Boeke (1970) to provide an electrically rechargeable bulk energy storage system that is economically feasible and has a high overall efficiency, extended cycle life, and high reliability. Thaller (1976) proposed a practical redox flow battery based on the redox couples Fe^{2+}/Fe^{3+} and Cr^{2+}/Cr^{3+}. Unfortunately, the Fe/Cr flow cell had problems like cross-contamination of the electrolyte and poor reversibility of the chromium half-cell. To overcome these problems, Skyllas-Kazacos and Grossmith (1987), Skyllas-Kazacos et al. (1986a,b, 1988a,b), and Skyllas-Kazacos and Kazacos (1994)) suggested an all-vanadium redox flow battery employing V(II)/V(III) and V(IV)/V(V) redox couples in the negative and positive half-cell electrolytes, respectively. The vanadium redox cell developed at the University of New South Wales (UNSW) by Skyllas-Kazacos and coworkers (Rahman et al., 2004) is showing great promise as an efficient new energy storage system for a wide range of applications. They identified sulfuric acid as a suitable supporting electrolyte to prepare concentrated vanadium solutions for the VRB.

4.5.5.3.1 Main Features of Vanadium Flow Batteries

The main features of the VRB are shown in Figure 4.13. The cell consists of two compartments separated by a proton exchange membrane that allows

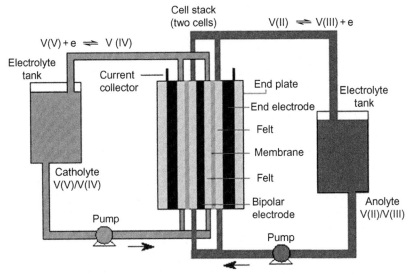

FIGURE 4.13 Schematic of All Vanadium Redox Flow Battery (Rahman, 1998)

protons (H^+ ions) to pass but prevents vanadium ions from passing through the membrane. The electrolyte on each side consists of vanadium dissolved in a sulfuric acid solution of concentration 2-3 M. At one electrode, V^{4+} ion is converted into V^{5+} ion during discharge of the cell, releasing an electron that is transferred to the electrode in the second compartment, where V^{3+} ion is converted into V^{2+} ion with the absorption of an electron. The membrane prevents cross mixing of the electrolytes. Each side of the cell contains an inert electrode made of highly porous carbon felt. The electrolytes, both anolyte and catholyte, are stored in two large external reservoirs. The rechargeable electrolyte is pumped through the inert electrode, where the electrochemical reactions occur.

During charging, electrochemical reactions within the battery stack change the valence of the vanadium, the negative reaction change V(III) to V(II), and the positive reaction change V(IV) to V(V). This process is reversed during the discharge cycle.

In the vanadium redox cell, the following half-cell reactions are involved:
At the negative electrode:

$$V^{3+} + e^- \Rightarrow V^{2+} \quad E^\circ = -0.26\,V$$

At the positive electrode:

$$VO^{2+} + 2H^+ + e^- \Rightarrow VO_2{}^+ + H_2O \quad E^\circ = 1.00\,V$$

This gives a standard cell potential $E°$ (cell) $= 1.26$ V at concentrations of 1 mol per liter and at 25 °C. Under actual cell conditions, an open-circuit cell voltage of 1.4 V is observed at 50% state of charge (SOC), while a fully charged cell produces more than 1.6 V at open circuit. The electrolyte for the vanadium battery is 2 M vanadium sulfate in 2.5 M H_2SO_4, the vanadium sulfate (initially 1 M V (III) + 1 M V (IV)) being prepared by chemical reduction or electrolytic dissolution of V_2O_5 powder.

The stability of the electrolyte is affected by the operating temperature, SOC, vanadium concentrations, sulfuric acid concentrations, and so on.

A stack of these energy-producing cells can be connected in series in a bipolar manner. The physical size of the battery stack determines the power kW available from the battery, and the volume of the electrolyte in the reservoirs determines the kWh energy storage of the battery (Rahman and Skyllas-Kazacos, 2000).

4.5.5.3.2   Performance Characteristics

Cell lifetimes are expected to be up to 10 years, though cells have run for up to 14,000 cycles. A VRB can be refurbished by replacing the cell stack alone. The reagents and storage tanks should be reusable. During operation, round-trip efficiency is about 75%. Units can run up from zero to full power in several milliseconds. However, other equipment associated with the cell may limit its response time to around 20 mS. A unit can provide instant pulses of energy without the reservoir pumps running, provided it is charged. Charged units can supply up to five times the rated output for limited periods.

4.5.5.3.3   Advantages

The VRB system offers many advantages over conventional LA batteries.

- It has very high efficiency.
- It has reasonable energy density.
- It has high charge/discharge rates.
- It has a long lifespan, independent of state-of-charge and load profiles.
- Instant recharge is possible by replacing the discharged electrolyte with fresh-charged solution, or the solution can be electrically recharged at high rates about 8-10 times faster than for LA batteries.
- The capacity of the battery can be simply increased by adding extra electrolyte to the reservoirs.
- The system can be fully discharged with no adverse effects to the battery.
- The solution life is indefinite and can be recycled continuously, so replacement costs are low and there are no waste disposal problems.
- The vanadium battery system is considered environmentally friendly.
- It requires low maintenance.
- The shelf life is theoretically unlimited.

The battery is insensitive to atmospheric oxygen, has a high 1.4 V cell voltage, and the electrolytes are not mutually destructive. The battery stack electrochemical reactions are all highly efficient with the energy, with efficiencies ranging from 75 to 80%. An accurate state-of-charge determination is possible by measuring the potential of a small open-circuit cell attached to the battery, with some portion of the electrolytes being pumped through it. No complex solid phase changes are involved during charging and discharging, which lead to shedding or shorting in conventional batteries.

Redox flow batteries, and to a lesser extent hybrid flow batteries, have the advantages of flexible layout (due to separation of the power and energy components), long cycle life (because there are no solid-solid *phase changes*), quick response times (similar to nearly all batteries), no need for "equalization" charging, and no harmful emissions (similar to nearly all batteries). Some types also offer easy state-of-charge determination (through voltage dependence on charge), low maintenance, and tolerance to overcharge/overdischarge.

Additionally, this battery has a feature that allows for some new options not available with LA technology. It is possible to simultaneously charge the battery at one voltage while discharging it at another voltage. This feature can be utilized to make a minimum cost, high efficiency, and maximum power point tracker, or allow the battery to operate as a DC transformer, electrochemically transforming a current and a voltage into a different current and voltage.

Significant interest has been observed recently in VRB system research and applications in different countries. In 2002, a research and development laboratory for VRB was established in Central South University in China with the financial support of Panzhihua Steel Corporation (Huang et al., 2008). Vetter et al. (2010) reported research activities on various redox flow batteries, including VRB, at Fraunhofer Institute for Solar Energy Systems (ISE), Heidenhofstraße, Freiburg, Germany.

Vanadium redox flow battery energy storage research and development work has made new progress. Dalian Institute of Chemical Physics, Chinese Academy of Sciences recently initiated vanadium redox flow battery energy storage research and development to achieve new progress in the successful development of a 1 kW-class vanadium redox flow battery from the storage battery module composed of a 10 kW-class battery system, based on the successfully developed 5 kW all-vanadium flow storage battery module. The module is running stable, with energy efficiency of 78% (Miller, 2009).

4.5.5.3.4 Disadvantages

On the negative side, flow batteries are rather complicated in comparison with standard batteries, as they may require pumps, sensors, control units, and secondary containment vessels. The energy densities vary considerably but are, in general, rather low compared to portable batteries, such as the Li-ion.

4.5.5.3.5 Use of Local Raw Materials

One of the main components of VRB is vanadium electrolyte prepared in 2-3 M sulfuric acid solution. This will utilize sulfur or in turn sulfuric acid as a local raw material. Another component is the conducting polymer that is used in the electrode material.

4.5.5.3.6 Commercialization Status of VRB

Several demonstration VRBs of various sizes have been evaluated in Australia, Japan, and Thailand, where atmospheric temperature does not exceed 40 °C. Vanadium electrolyte optimization studies are being performed to improve the stability of the electrolyte at higher temperatures (Rahman et al., 1996, 2004; Rahman, 1998; Rahman and Skyllas, 1998; Rahman and Skyllas-Kazacos, 2009).

A 5 kW/13kWh vanadium battery system has been installed in a solar demonstration house in collaboration with the Centre for Photovoltaic Devices and Systems, UNSW, and Thai Gypsum Products Co. Ltd., Thailand, and its suitability for application in energy self-sufficient remote housing has been evaluated. Another vanadium battery of 4.1 kW peak power and 3.9 kWh capacity has recently been evaluated as a submarine backup battery for the Department of Defense in Australia, while a golf cart with the 5 kW battery stack mounted on the back and electrolyte tanks under the seat has been undergoing testing at UNSW since 1994. Extensive off-road trials of this battery showed excellent performance over a range of terrains. With 40 l of 1.85 M vanadium electrolyte per half-cell tank, a driving range of 17 km was obtained. The total driving range expected for 3 M vanadium solutions and full tanks containing 60 l each side, therefore, is around 40 km (Menictas et al., 1994).

In 1997, a 200 kW/800 kWh grid-connected vanadium battery was commissioned at the Kashima-Kita Electric Power station in Japan, where it is currently undergoing long-term testing as a load-leveling system. By the beginning of 1998, it had already undergone 150 charge-discharge cycles and was continuing to show high energy efficiencies of close to 80% at current densities of 80-100 mA/cm^2.

Skyllas-Kazacos and Kazacos (2009) reported that VRB technology has been technically proven in systems up to 6 MWh; its more widespread commercialization in large grid-connected applications necessitated further cost reductions in membrane materials, cell stack design, and control systems. Brief information about several VRB units currently under evaluation by Sumitomo is given below.

Substation Demonstration System (Load leveling DC 450 kW × 2 h)

This demonstration system installed in Tatsumi Substation of the Kansai Electric Power Co., Inc. is linked to the commercial systems. The large amount of demonstration data obtained through this system has become the basis of the present systems. (Operation start: Dec. 1996)

System for Office Building (Load Leveling AC 100 kW × 8 h)

The system is installed in the head office building of Sumitomo Densetsu Co., Ltd. The electrolyte tank is installed in an underground cistern (dead space). (Operation start: Feb. 2000)

Wind Power Generation Output Stabilization: (Output Stabilization AC 170 kW × 6 h)

The facilities were delivered to NEDO and IAE for the demonstration testing of wind power generation output stabilization. The facilities were constructed next to the wind power generator of Hokkaido Electric Power Co., Ltd. (Operation start: Mar. 2001)

Voltage Sag Protection (Voltage Sag Protection AC 3000kW × 1.5 s; Load Leveling AC 1500 kW × 1 h)

This ESS is constructed for the liquid crystal plant of Tottori SANYO Electric Co., Ltd. When instantaneous voltage sag is caused by lightning or other causes, the system protects low voltage and prevents the manufacturing line from stopping. Under normal conditions, load-leveling operation is conducted. (Operation start: Apr. 2001)

PowerPedia: Vanadium Redox Batteries (2009)

The extremely large capacities possible from vanadium redox batteries make them well suited to use in large power storage applications, helping to average out the production of highly variable generation sources such as wind or solar power, or to help generators cope with large surges in demand. Their extremely rapid response times also make them superbly well suited to UPS-type applications, where they can be used to replace LA batteries and even diesel generators.

Currently installed vanadium batteries include

- A 1.5 MW UPS system in a semiconductor fabrication plant in Japan
- A 275 kW output balancer in use on a wind power project in the Tomari Wind Hills of Hokkaido
- A 200 kW, 800kWh output leveler in use at the Huxley Hill Wind Farm on King Island, Tasmania
- A 250 kW, 2 MWh load leveler in use at Castle Valley, Utah

4.5.5.3.7 Cost

Cost estimates of VRBs by the UNSW group and independent consulting groups place mass production costs between $300 and $500 per kW for the cell stack and $30-50 per kWh for the electrolyte (Skyllas-Kazacos, 2006; Hennessy, 2008). It should be noted that apart from cost, the VRB has some additional advantages that were previously discussed. Variation in capital cost of VRBs with storage time is illustrated in Figure 4.14, and a comparison of cost of energy ($/MWh) for VRBs and LA batteries is shown in Figure 4.15. The important feature of this graph is that the cost of each kWh generated by the

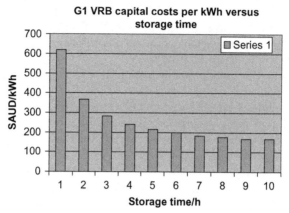

FIGURE 4.14 Vanadium Redox Flow Battery Capital costs vs storage time (Skyllas-Kazacos, 2006) (Total battery costs per kWh versus storage time for stack cost of $AUD 500/kW and V2O5 price of $US5/lb)

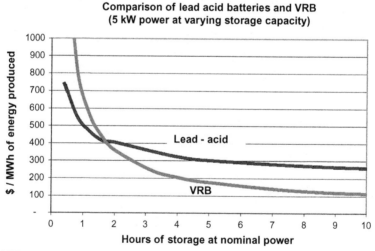

FIGURE 4.15 Comparison of Lead acid and Vanadium Redox Flow Battery energy costs vs storage time (V-Fuel, 2008)

VRB over its life is considerably lower than that provided by LA batteries for storage times in excess of 1.5 h. This confirms the significant cost benefit of the VRB in applications that require long storage times.

(Total battery costs per kWh versus storage time for stack cost of $AUD 500/kW and V_2O_5 price of $US5/lb)

4.5.5.4 NaS Battery

4.5.5.4.1 Process Description

An NaS battery consists of liquid (molten) sulfur at the positive electrode and liquid (molten) sodium at the negative electrode as active materials. They are separated by a solid beta alumina ceramic electrolyte (see Figure 4.16). The electrolyte allows only the positive sodium ions to go through it and combines with sulfur to form sodium polysulfides according to the following reaction (ESA, 2009):

$$2Na + 4S \rightarrow Na_2S_4$$

The NGK design for the NaS battery is cylindrical. In the center is a pool of molten sodium, which is maintained in a molten state by keeping the cell temperature around 300-350 °C. This pool is contained and surrounded by a cylindrical tube of solid electrolyte, a material called beta alumina, which is surrounded by molten sulfur and the cell container. During discharge, sodium atoms give up electrons to the sodium electrode, becoming sodium ions, which then migrate through the solid beta alumina to the sulfur, where they react to form NaS compounds with the take-up of electrons. Cell reactions are reversed during charging, regenerating the liquid sodium and the liquid sulfur. Open circuit voltage for this cell remains constant for up to 75% of the discharge. The cell is capable of providing short-term currents of up to five times its rated output.

The typical energy and power density of NaS batteries are in the range of 150-240 W/kg. They have a typical cycle life of 2500 cycles. NaS battery cells are efficient (75-90%) and have a pulse power capability over six times their continuous rating (for 30 s). This attribute enables NaS batteries to be economically used in combined power quality and peak shaving applications (Chen et al., 2009).

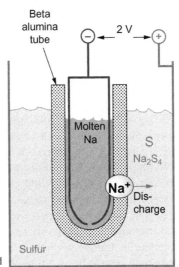

FIGURE 4.16 Schematic of a sodium-sulfur battery [ESA, 2009, DOE/EPRI, 2013].

4.5.5.4.2 Advantages/Disadvantages

The NaS battery is unique among secondary cells in using a solid electrolyte. The cell operates at a high temperature, which enhances transport of ions through the solid. It has up to three times the energy density of a LA battery, its lifetime is longer, and it requires little maintenance. It can utilize sulfur, which is a disposal problem in Saudi Arabia.

The major drawback is that a heat source is required, which uses the battery's own stored energy. This partially reducing battery performance, because the NaS battery needs to operate at a high temperature (300-350 °C), as previously mentioned. High-temperature operation is more costly than operating at ambient temperature, but this cell has several advantages that may outweigh the additional operating costs.

4.5.5.4.3 Cost

Initial capital cost is another issue ($2000/kW and $350/kWh), but it is expected to fall as the manufacturing capacity is expanding.

4.5.5.4.4 Developers/Suppliers and Deployment Status

NaS battery is an advanced secondary battery that was developed by Tokyo Electric Power Company (TEPCO) and NGK Insulators, Ltd. in 1983 (Kamibayashi and Tanaka, 2001). Feasibility studies of various demonstration projects show that the NAS battery technology is attractive for use in relatively large-scale battery energy storage system applications due to its outstanding energy density, efficiency, and zero maintenance.

Typical NaS battery modules developed by NGK have a rated power of 50 kW and rated capacity of 360-430 kWh. A 50 kW unit can supply 100 kW for 2 h and 250 kW for 30 s. The typical rated lifetime is 15 years or 4500 cycles to 90% capacity. DC-DC efficiency is 85%. A NaS module can reach full power output in 1mS. When in standby, a 50 kW unit uses 2-3 kW to maintain its temperature. Table 4.6 shows NaS batteries deployed at various locations in Japan.

NaS battery technology has been demonstrated at more than 190 sites in Japan, totaling more than 270 MW with stored energy suitable for 6 h daily peak shaving. The largest NaS installation is a 34 MW, 245 MWh unit for wind stabilization in Northern Japan. US utilities have deployed 9MW battery for peak shaving, backup power, firming wind capacity, and other applications, and project development is in progress for an amount equal to Japan's.

The demand for NaS batteries as an effective means of stabilizing renewable energy output and providing ancillary services is expanding. Several projects are under development in Europe, as well as in Japan and the United States. The annual production capacity is 90 MW, 150 MW, planned in 2010 (ESA, 2009).

TABLE 4.6 Utility Scale Sodium Sulfur Batteries Deployed in Japan

Name	Startup	Power (kW)	Storage Capacity (kWh)
TEPCO, Tsunashima substation	1997	6000	48,000
TEPCO, Ohito substation	1999	6000	48,000
TEPCO, Saitama substation	1999	2000	16,000
TEPCO, Tsunashima	2000	2000	14,400
TEPCO, Pacifico	2002	2000	14,400
TEPCO, Fujitsu	2002	3000	7200

Efforts are being made to reduce the operating temperature of the NaS battery. Inside Ceramatec's wonder battery is a chunk of solid sodium metal mated to a sulfur compound by an extraordinary, paper-thin ceramic membrane. The membrane conducts ions—electrically charged particles—back and forth to generate a current. The company calculates that the battery will cram 20-40 kWh of energy into a package about the size of a refrigerator, and operate below 90 °C.

4.5.5.5 Ni-Cd Battery

4.5.5.5.1 Process Description

NiCd batteries consists of a positive electrode plate made up of nickel hydroxide, a cadmium hydroxide negative electrode plate, a separator, and an alkaline electrolyte. NiCd batteries generally have a metal case with a sealing plate equipped with a self-sealing safety valve. The positive and negative electrode plates, isolated from each other by the separator, are rolled in a spiral shape inside the case. Ni-Cd batteries rank alongside LA batteries in terms of their commercial maturity ($\sim$100 years) and popularity.

The chemical reaction is

$$2NiO(OH) + Cd + 2H_2O \leftrightarrow 2Ni(OH)_2 + Cd(OH)_2$$

4.5.5.5.2 Performance Characteristics

The efficiency of a Ni-Cd battery is around 70% for DC-to-DC, although others in the family can provide efficiencies as high as 85%. The cells will self-discharge more rapidly than LA batteries will, and can lose up to 5% of their charge per month. Their lifetime varies, depending on the type of duty for which they are used. With low levels of discharge, Ni-Cd cells can operate for up to 50,000 cycles. Typical cells are rated for lifetimes of 10-15 years but can operate for longer. This drops, however, if the operating temperature becomes elevated. However, they are good for operations at low temperatures.

4.5.5.5.3 Cost

Ni-Cd battery system costs are expected to be higher than the cost of LA batteries because the cells themselves are more expensive. The Fairbanks battery energy storage plant costs around $30 m for a plant with a rated power of 27 MW, providing a unit cost of $1110/kW. EPRI-DOE estimates suggest that in 2003, the cost per kWh for batteries was around $1200/kWh for a 5 MW plant, while total plant cost was $1780/kW. This is significantly higher than the estimated cost of LA cells previously quoted.

4.5.5.5.4 Advantages/Disadvantages

NiCd batteries are robust and reliable, and they have reasonable energy density (50-75 Wh/kg) and very low maintenance requirements, but relatively low cycle life (2000-2500). These advantages over LA batteries make them favored for power tools, portable devices, emergency lighting, UPS, telecoms, and generator starting. However, over the past decade, portable devices such as mobile telephones, tablets, and laptops have been effectively displaced from these markets by other electrochemistries.

The main drawback of NiCd batteries is their relatively high cost ($\sim$$1000/ kWh), due to the expensive manufacturing process. Besides, cadmium is a toxic heavy metal and has issues associated with the disposal of Cd from waste NiCd batteries. NiCd batteries also suffer from the "memory effect," where the batteries will only take full charge after a series of full discharges. Proper battery management procedures can help mitigate this effect.

4.5.5.5.5 Developers/Suppliers and Deployment Status

The NiCd system is the "world's largest (most powerful) battery," installed at Golden Valley, Fairbanks, Alaska, United States (Breeze, 2009). The system is rated at 27 MW for 15 min, 40 MW for 7 min, and with an ultimate 46 MVA limitation imposed by the power converter. The batteries are expected to perform 100 complete and 500 partial discharges in the system's 20-year design life. The system provides critical spinning reserve functionality in what is effectively an "electrical island."

4.5.5.6 LA Battery
4.5.5.6.1 Process Description

LA batteries, invented in 1859, are the oldest and most widely used rechargeable electrochemical devices. A LA battery consists of (in the charged state) electrodes of lead metal and lead oxide in an electrolyte of about 37% (5.99 M) sulfuric acid.

The chemical reactions are (charged to discharged)

Anode (oxidation):

$$Pb(s) + HSO_4^-(aq) + H_2O(l) \leftrightarrow PbSO_4(s) + H_3O^+(aq) + 2e^- \quad E° = -0.356\,V$$

Cathode (reduction):

$$PbO_2(s) + 3H_3O^+(aq) + HSO_4^-(aq) + 2e^- \leftrightarrow PbSO_4(s) + 5H_2O(l) \quad E° = 1.685\,V$$

Because of the open cells with liquid electrolyte in most LA batteries, over-charging with excessive charging voltages will generate oxygen and hydrogen gas by electrolysis of water, forming an explosive mix. The acid electrolyte is also corrosive.

The cell reaction in a LA battery involves lead (Pb), lead oxide (PbO_2), and sulfuric acid (H_2SO_4). One electrode of the cell is composed of lead oxide and the other of lead (usually in an alloy form to make it more tractable). During dis-charge of the cell, lead dissolves into the electrolyte at one electrode, releasing electrons that are transferred to the second electrode via an external circuit, where they are taken up by lead oxide, which also thereby dissolves into the sulfuric acid electrolyte. This depletes the active material on both the electrodes and the con-centration of the sulfuric acid, and eventually the cell can produce no more elec-tricity. Charging the cell reverses both these processes, depositing lead on one electrode and lead oxide on the other, while regenerating the sulfuric acid.

4.5.5.6.2 Performance Characteristics

Efficiency of LA batteries depends on a number of factors, including temperature variation and depth of charge. Typical efficiencies range between 75% and 85% for DC-DC cycling. Cells will lose charge over time due to a small but significant electrical conductivity of the electrolyte. This will increase with temperature, increasing the self-discharge rate. At low temperatures, performance will drop off too, with capacity falling as the temperature falls. At very low temperatures the electrolyte can freeze, so the optimum temperature of operation is 25 °C. The typical lifetime for a LA battery is around 5 years or up to 1000 charge-discharge cycles, though this depends on the way the cells are used. Batteries designed for utility applications can have lifetimes of 15-30 years.

There are several types of LA batteries, including the flooded battery requir-ing regular topping up with distilled water, the sealed maintenance-free battery having a gelled/absorbed electrolyte, and the valve-regulated battery. It is a popular storage choice for power quality, UPS, and some spinning reserve applications. Its application for energy management, however, has been very limited due to its short cycle life (500-1000 cycles) and low energy density (30-50 Wh/kg) due to the inherently high density of lead.

4.5.5.6.3 Cost

The current cost of utility-scale LA battery facilities is a matter for conjecture. Recent estimates suggest that such plants can be built for $200-580/kW.

TABLE 4.7 Utility Scale Lead-Acid Batteries Deployed World Wide

Location	Year of Installation	Rated Power Output (MW)	Storage Capacity (MWh)	Cost ($/kW)	Cost ($/kW)
BEWAG, Berlin, Germany	1986	8.5	8.5	-	-
Chino, California	1988	10	40	456	1823
PREPA, Puerto Rico	1994	20	14	1574	1102
Vernon, California	1995	3	4.5	944	1416

However, the costs for the plants in Table 4.7 were between $1100 and $1800/kW, significantly higher. All the plants in the table were first-of-kind designs, and costs might be expected to be high. More significant for a storage plant is the cost per kWh. For the plants listed in Table 4.7, this was between $456 and $1574/kWh, an extremely wide range that probably again reflects their first-of-kind status. More recent cost estimates obtained by EPRI-DOE, based on 2003 prices, suggest that costs were between $320 and $1260/kW for the batteries and basic interconnections and racks. The power converter system added a further $150-220/kW and balance of plant another $100/kW, putting the total cost between $570 and $1580/kW. The estimated cost/kWh for systems with capacities of 10 MW, 10 MWh and 10 MW, 40 MWh in 2003 was $590/kWh and $393/kWh.

4.5.5.6.4   Advantages/Disadvantages

LA batteries have high reliability. One of the major problems with LA batteries is that they produce hydrogen and oxygen during charging (by electrolysis of water), once the charging voltage exceeds a certain value. Because a rise in voltage is inevitable as the cell charges, the generation of gas cannot be avoided. Some cell designs simply allow the gas to escape. Such cells are normally called flooded cells because the electrodes are "flooded" with sulfuric acid electrolyte, and the water level is topped up regularly to take account of the loss through electrolysis. The alternative is the valve-regulated LA cell, sometimes also called the sealed LA cell. These use special design features to encourage the hydrogen and oxygen generated by the cell to recombine into water, while releasing any excess gas that does not recombine through a control valve. LA batteries also have a poor low-temperature performance, and therefore require a thermal management system.

4.5.5.6.5   Developers/Suppliers and Deployment Status

GNB Industrial Power/Exide
Delco
East Penn
Teledyne
Optima Batteries
Winston Salem
JCI Battery Group
Trojan
Crown Battery

A number of large-scale LA battery-based utility storage plants have been built, though none recently. Details of these are collected in Table 4.7. The earliest modern LA storage plant was built by the Berlin utility BEWAG to provide peak power and grid support on what was then an island grid in West Berlin. This unit has a rated power of 8.5 MW, which it could supply for up to 1 h. Output could be doubled to 15 MW, but this could only be maintained for 20 min. The plant operated successfully, with few problems. The largest one is a 40 MWh system in Chino, California (USA), which works with a rated power of 10 MW for 4 h (Moore and Douglas, 2006).

4.6   ASSESSMENT OF ESS TECHNOLOGIES

Ibrahim and Ilinca (2013) presented a techno-economic analysis of different energy storage technologies. They indicated that the study of complete systems (energy storage, associated transformation of electricity, power electronics, control systems, etc.) will lead to the optimization of techniques in terms of cost, efficiency, reliability, maintenance, social and environmental impacts, and so on.

Schoenung (2001) of Sandia National Laboratory presented characteristics and technologies for long- versus short-term energy storage. Applications of energy storage have a wide range of performance requirements. One important feature is storage time or discharge duration. *This study focuses on technologies that are more relevant to renewable electrical energy storage for Saudi Arabian conditions.*

Figure 4.10, presented in Section 4.5 compares the various energy storage technologies for the largest power and duration periods for which they are expected to be applied. It can be observed that for energy management, PHS, CAES, NaS, flow batteries, NiCd, and LA are suitable for applications in scales between 1 and 100 MW, with hourly to daily output durations. Note that SMES and a few other devices that are only used for high-power applications are also included to indicate the differences between the high-power and high-energy storage devices. Figure 4.17 illustrates the commercial maturity of some of the ESS discussed above, and Figure 4.18 shows the installed power capacities as of 2008.

Table 4.8 presents the main features of different ESS. For Saudi Arabia, the PHS and CAES that have special site requirements besides water and gas fuel

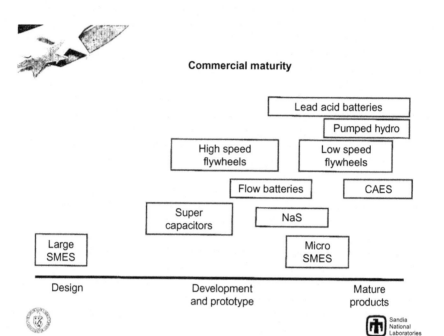

FIGURE 4.17 Status of commercial maturity of different ESS [ESA, 2009].

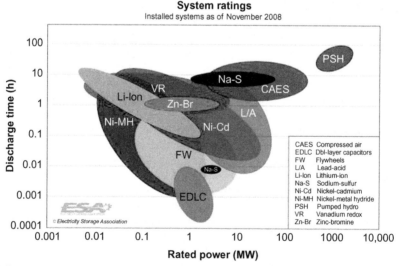

FIGURE 4.18 Rate installed capacities of different electricity storage technologies (ESA, 2009).

TABLE 4.8 Main Features of Different ESS [ESA, 2009]

Technology		Advantages	Disadvantages	Commercial Maturity	Application	Cost	Remarks
Mechanical energy storage	Pumped storage hydro (PSH)	High capacity, low efficiency	Special site requirement	High	Energy management	Low cost	Suitable for load leveling
	Compressed air energy storage (CAES)	High capacity	Special site requirement need gas fuel	High	Energy management	Low cost	Suitable for load leveling
	Fly wheel	High power	Low energy density	High	Power quality management		Suitable for short duration in sec or min (voltage dip)
Electrical energy storage	Capacitors/supercapacitors	Long cycle life/high efficiency	Low energy density	Medium	Power quality management		Suitable for short duration in sec or min (voltage dip)
	Superconducting magnetic energy storage (SMES)	High power	Low energy density and high production cost	Medium	Power quality management	High production cost	Suitable for short duration in sec or min (voltage dip)

Continued

TABLE 4.8 Main Features of Different ESS [ESA, 2009]—Cont'd

Technology		Advantages	Disadvantages	Commercial Maturity	Application	Cost	Remarks
Chemical energy storage	Lead-acid		Limited cycle life when deeply discharged	High		Low capital cost	Renewables & power plants
	Flow batteries: VRB, ZnBr, PSB	High capacity, independent power and energy ratings	Low energy density	Medium	Power quality & energy management		Renewables & power plants
	NiCd	High power/high efficiency	High energy density, high cost	High	Power quality & energy management		Renewables & power plants
	Li-ion	High power/high efficiency	High cost and needs special charging circuit	Medium	Power quality management		Power quality management
	Metal-air		Electrically not re-chargeable				
	NaS			Medium	Power quality & energy management		Renewables & power plants

are not a viable option, and are therefore eliminated from further discussion. SMES and supercapacitors are more suited for power quality applications for short duration, not for the long duration needed in renewables. Li-ion batteries are expensive and need a special circuit for charging, whereas renewables require large-scale, low-cost ESS.

The competing ESS technologies for stationary applications to utilize renewables are LA, flow batteries, NaS, and NiCd. These ESS will be further assessed in terms of their properties. Table 4.9 presents major characteristics of the six ESS. Among the flow batteries, VRBs appear to be better than PSB and ZnBr, as they are of low capacity and low cycle life. Efficiency-wise, VRBs are also slightly better. VRBs are distinct from hybrid flow batteries (e.g., ZnBr, NaS) that have one reactive electrode and therefore suffer from the degradation drawbacks of conventional batteries. Using only vanadium in the electrolyte, as opposed to a blend of electrochemical elements, gives VRB systems the most competitive advantage in terms of operating cost, system life, maintenance, and safety. Size and weight of storage devices are important factors for certain applications, but for large-scale stationary applications such as renewables, they are not considered.

The efficiency and lifetime of selected ESS are illustrated in Figure 4.19. Efficiency and cycle life are two important parameters to consider along with other parameters before selecting a storage technology. Both of these parameters affect the overall storage cost. Low efficiency increases the effective energy cost, as only a fraction of the stored energy could be utilized. Low cycle life also increases the total cost, as the storage device needs to be replaced more often. The present values of these expenses need to be considered along with the capital cost and operating expenses to obtain a better picture of the total ownership cost for a storage technology.

4.7 ECONOMIC EVALUATION OF SELECTED ESS

The competing ESS technologies for stationary applications to utilize renewables based on technology evaluation appear to be LA, VRB, NaS, and NiCd. The three batteries are basically advanced batteries, and the LA battery is included in the analysis to compare with the currently used ESS. A preliminary economic evaluation was performed based on available cost data.

The capital costs for different ESS are shown in Figure 4.20. The capital costs have been adjusted to exclude the cost of power conversion electronics. The cost per unit energy has also been divided by the storage efficiency to obtain the cost per output (useful) energy. Installation cost also varies with the type and size of the storage and with location. The information here should only be used as a guide, not as detailed data. The costs of storage technologies are changing as they evolve. The cost data in Figure 4.20 are based on approximate values in 2002 and the expected mature values in a few years.

TABLE 4.9 Main Features of Competing ESS Technologies [Chen, 2009]

Technology	VRB	PSB	ZnBr	NaS	NiCD	Lead Acid
Efficiency (%)	85	75	75	89	60-65	65
Cycle life charge/discharge	13,000	NA	25,000	3000	2500	1000
Size range (MW)	0.5-100	1-15	0.05-1	0.15-10	1-10	0.001-40
Operation temp (°C)	0-40	50	50	350	Ambient	−5 to 40
Energy density (Wh/kg)	30	NA	50	150-240	50-75	50
Self Discharge	Small	Small	Small	20% per day	0.6% per day	0.3% per day
Green	Yes		Yes	No	No	No
Disadvantages	Low energy density	Low cycle life	low cycle life	High temp. system, complex safety design	Heavy metal Cd recycling issues	Issues with mining and processing of lead. Limited cycle life when deeply discharges
Remarks	Most promising			2nd Best		

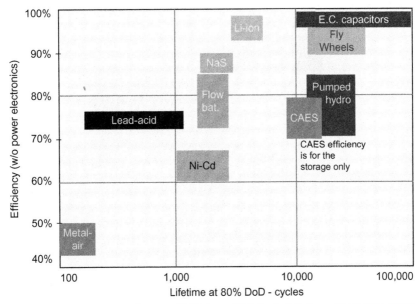

FIGURE 4.19 Variation in efficiency with lifetime at 80 % depth of discharge (DoD) (ESA, 2009).

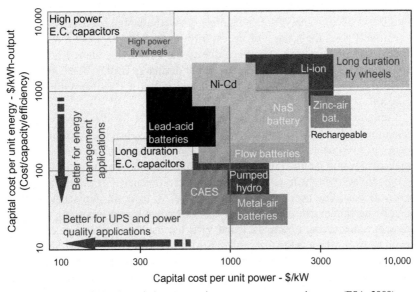

FIGURE 4.20 Variation in capital cost per unit energy vs cost per unit power (ESA, 2009).

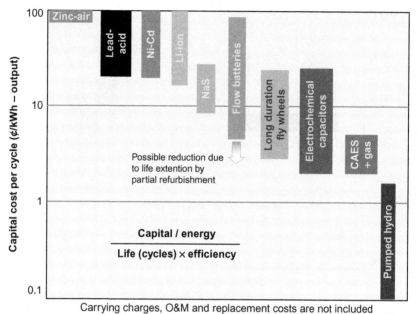

Carrying charges, O&M and replacement costs are not included

FIGURE 4.21   Capital cost per cycle for selected ESS (ESA, 2009).

While capital cost is an important economic parameter, it should be realized that the total ownership cost (including the impact of equipment life and operation and maintenance (O&M) costs) is a much more meaningful index for a complete economic analysis. For example, while the capital cost of LA batteries is relatively low, they may not necessarily be the least expensive option for energy management (load leveling) due to their relatively short life for this type of application and their limited depth of discharge (DoD). The capital cost of NiCd is higher than NaS and VRB, and it has a Cd disposal/recycling problem.

Capital cost per cycle is illustrated in Figure 4.21 for the competing ESS technologies. Per-cycle cost can be the best way to evaluate the cost of storing energy in a frequent charge/discharge application, such as renewables and load leveling in power plants. Figure 4.21 shows the capital component of this cost, taking into account the impact of cycle life and efficiency. For a more complete per-cycle cost, one needs to also consider O&M, disposal, replacement, and other ownership expenses, which may not be known for the emerging technologies. It is clear from Figure 4.21 that VRB has the lowest per-cycle cost, followed by NaS, whereas the per-cycle costs for NiCd and LA are relatively high. Because the electrolyte of VRB has indefinite life, and partial refurbishment can increase the battery life, the overall economics of VRB are further improved

TABLE 4.10 Cost Comparison of Competing ESS

Technology	VRB	NaS	NiCd	Lead Acid	Reference
Power cost ($/kW)	600-1500	1000-3000	500-1500	300-600	Chen et al., 2009
Energy cost ($/kWh)	150-1000	300-500	800-1500	200-400	Chen et al., 2009
Cents/kWh-per cycle	5-80	8-20	20-100	20-100	Chen et al., 2009
O&M costs ($/kWh)	0.001	0.02		0.02	Huang, 2008
Overall assessment	Most promising	2nd best			

compared to NaS. Table 4.11 presents the cost data for the selected ESS technologies in terms of $/kW, $/kWh, and cents/kWh per cycle and O&M cost. Based on data presented in Table 4.10 and Figures 4.20 and 4.21, it can be concluded that VRB is the low-cost battery among the four ESS technologies selected.

4.8   CONCLUSIONS AND RECOMMENDATIONS

The use of electricity generated from intermittent renewable sources (solar or wind) requires efficient and cost- effective advanced ESS. ESS will guarantee high power quality, better utilization of existing power generation systems, and efficient dispatchability of renewable sources of energy. The overall techno-economic evaluation of selected ESS technologies is presented in Table 4.11.

Although there are various commercially available ESS systems, no single storage system meets all the requirements for an ideal ESS: maturity, long lifetime, low cost, high density, high efficiency, and environmentally benignancy. Each ESS has a suitable application range.

This study has identified vanadium redox batteries (VRBs) and NaS batteries as appropriate ESS that match the requirements of Saudi Arabia. Specific findings are highlighted:

- Sulfur or sulfur-based compounds are cheap in Saudi Arabia; if utilized, the economics of ESS will improve.
- The maximum atmospheric temperature was found to be 51 °C at Qaisumah and the minimum temperature recorded was −9 °C at Hail.

TABLE 4.11 Overall Techno-Economic Evaluation of Competing ESS Technologies

Overall Comparison of Electricity Storage Technologies

Technology	VRB	NaS	NiCd	Lead Acid
Efficiency (%)	85	89	60-65	65
Cycle life charge/discharge	13,000	3000	2500	1000
Size range (MW)	0.5-100	0.15-10	1-10	0.001-40
Operation temp (°C)	0-40	350	Ambient	−5 to 40
Energy density (Wh/kg)	30	150-240	50-75	50
Self discharge	Small	20% per day	0.6% per day	0.3% per day
Disadvantages	low energy density	High temp. system, complex safety design	Heavy mental Cadmium recycling issues	Issues with lead and limited cycle life
Power cost ($/kW)	600-1500	1000-3000	500-1500	300-600
Cents/kWh-per cycle	5-80	8-20	20-100	20-100
O&M costs ($/kW h)	0.001	0.02		0.02
Overall assessment	Most promising	2nd Best		

- In 2008, the SEC capacities were as follows: peak load, 38,028 MW; installed capacity, 38,808 MW; firm capacity, 34,470 MW; deficit, 3558 MW. This deficiency can be partially supported by ESS.
- The VRB was found to be promising for Saudi conditions. However, the operating temperature range of the electrolyte needs to be increased above 40 °C through further research, by developing additives to improve electrolyte stability.
- NaS is also a suitable battery for Saudi conditions, as its operating temperature is high, around 350 °C.

Based on comprehensive technology assessment, economic evaluation, and Saudi Arabian conditions, it is recommended to initiate research and development work to further improve technology as well as economics. Two ESS that can be considered for further research are VRB and NaS. These advanced energy storage systems can be integrated for storing electrical energy generated from renewable sources to improve dispatchability and reliability.

The future use of electrical energy depends on the development of the next generation of batteries. Batteries are needed to realize the full potential of RESs as part of the electric distribution grid. Existing battery technologies have cost and performance limitations that hinder the rapid transition to efficient use of RESs. Other technical bottlenecks include the limited energy storage capacity of individual battery cells and the lack of fast recharge cycles with long cell lifetimes. To increase the energy storage capacity of each cell significantly, increases in cell voltage and/or in the amount of charge stored reversibly per unit weight and volume are needed. Increasing the cell voltage requires the development of new electrolytes for thermodynamic stability or of surface passivation layers that adjust rapidly to changes in electrode morphology during a fast charge and discharge. The need to mitigate the volume and structural changes of the active electrode sites in a charge-discharge cycle calls for the exploration of new materials that have nanoscale features that could enhance reversible charge storage. Especially exciting is the potential for designing novel multifunctional materials that, for example, would increase the level of energy storage per unit volume and decrease dead weight.

Specifically, recommended research areas for future work are

- Develop novel additives to stabilize electrolyte for VRB to increase the operating temperature range. Optimize the electrolyte composition for Saudi conditions.
- Improve electrode and membrane efficiency for VRB.
- Reduce operating temperature, minimize self-discharge, and improve cycle life in NaS battery.

ACKNOWLEDGMENTS

The authors are grateful to the Center of Research Excellence in Renewable Energy (CoRE-RE) at King Fahd University of Petroleum & Minerals, Dhahran, established by the Ministry of Higher Education, Saudi Arabia, for providing the financial grant under Project No. CoRE-RE03.

REFERENCES

Ahearne, J., 2004. Storage of electric energy. Report on research and development of energy technologies. International Union of Pure and Applied Physics, pp. 76–86. Available online http://www.iupap.org/wg/energy/report-a.pdf (20.03.07).

Al-Abbadi, N.M., 2005. Wind energy resource assessment for five locations in Saudi Arabia. Renew. Energy 30, 1489–1499.

Boeke, J., 1970. Redox flow cell. United States Patent No. 3,540,933.

Breeze, P., 2009. Business insights, management report. The future of electrical storage, the economics and potential of new technologies.

Cavallo, A., 2001. Energy storage technologies for utility scale intermittent renewable energy system. J. Solar Energy Eng. 123, 387–389.

Chen, H., Cong, T.N., Yang, W., Tan, C., Li, Y., Ding, Y., 2009. Progress in electrical energy storage system: a critical review. Prog. Nat. Sci. 19, 291–312.

Denholm, P., Holloway, T., 2005. Improved accounting of emissions from utility energy storage system operation. Environ. Sci. Technol. 39, 9016–9022.

Denholm, P., et al., 2004. Life cycle energy requirement and greenhouse gas emission from large scale energy storage system. Energy Convers. Manag. 45, 2153–2172.

DOE/EPRI, 2013. Electricity storage handbook in collaboration with NRECA.

EPRI-DOE, 2003. Handbook of energy storage for transmission and distribution application.

ESA, 2009. http://electricitystorage.org/tech/technologies_technologies.

Gyuk, I., 2009. Survey of energy storage options for stationary utility scale storage. In: Energy Storage: Materials, Systems, and Applications Symposium, Materials Science & Technology 2009 Meeting, USA.

Haisheng, C., Thang, N.C., Wei, Y., Chunqing, T., Yongliang, Li, Yulong, D., 2009. Progress in electrical energy storage system: a critical review. Prog. Nat. Sci. 19, 291–312.

Hennessy, T., 2008. Storage options in planning, prepared for, green energy coalition (David Suzuki Foundation, Eneract, Greenpeace Canada, Sierra Club of Canada, World Wildlife Fund Canada). Pembina Institute, Ontario Sustainable Energy Association Energy Association.

Huang, K.L., Li, X., Liu, S., Tan, N., Chen, L., 2008. Research progress of vanadium redox flow battery for energy storage in China. Renew. Energy 33, 186–192.

Ibrahim, H., Ilinca, A., 2013. Techno-Economic Analysis of Different Energy Storage Technologies. (Chapter 1), http://dx.doi.org/10.5772/52220.

Jewitt J. "Impact of CAES on Wind in Tx, OK and NM", In: Annual peer review meeting of DOE energy storage systems research. San Francisco, USA, Oct. 20, 2005; p. 1–16.

Kaldellis, J.K., Zafirakis, D., 2007. Optimum energy storage techniques for the improvement of renewable energy sources-based electricity generation economic efficiency. Energy 32, 2295–2305.

Kaldellis, J.K., Zafirakis, D., Kavadias, K., 2009. Techno-economic comparison of energy storage systems for island autonomous electrical networks. Renew. Sust. Energ. Rev. 13, 378–392.

Kamibayashi, M., Tanaka, K., 2001. Recent sodium sulfur battery applications. In: Proc. IEEE PES Transmission and Distribution Conference and Exposition, USA, vol. 2, 28 Oct-2 Nov, 2001, pp. 1169–1173.

Kangro, W., Pieper, H., 1962. Zur frage de speicherung von elektrischer energie. Flussigk. Acta 7, 435.

Lee, B., Gushee, D., 2008. Massive electricity storage, an AIChE white paper. AIChE, 3 Park Avenue, New York, NY 10016, June 2008.

Linden, S., 2003. The commercial world of energy storage: a review of operating facilities (under construction or planned). In: Proceeding of 1st Annual Conference of the Energy Storage Council, Houston, Texas, March 3.

Lior, Noam, 2010. Sustainable energy development: The present (2009) situation and possible paths to the future. Energy 35 (2010), 3976–3994.

McDowall, J.A., 2004. High power batteries for utilities—the world's most powerful battery and other developments. In: Power Engineering Society General Meeting. IEEE, Denver, USA, 6–10 June, 2004, pp. 2034–2037.

Menictas, C., Hong, D.R., Wilson, Y.J., Kazacos, M., Skyllas-Kazacos, M., 1994. Status of the vanadium redox battery development program. In: Electrical Engineering Congress, 24-30 November, Sydney, Australia.

Miller, R., 2009. Vanadium redox flow battery energy storage research and development work has made new progress. http://milkbatteries.over-blog.com/article-35532833.html.

Moore, T., Douglas, J., 2006. Energy Storage: Big Opportunities on a Smaller Scale. EPRI Journal (Spring 2006), 16–23. http://mydocs.epri.com/docs/CorporateDocuments/EPRI_Journal/2006-Spring/1013289_storage.pdf.

Nordex Publication, 2003. N43/600 kW technical description document, Rev. http://www.nordex-online.com/_e/produkte_und_service/index.html.

Nourai, A., 2002. Large scale energy storage technologies for energy management. In: Power Engineering Society Summer Meeting, 2002 IEEE Publication, vol. 1, 25-25 July, 2002, pp. 310–315.http://ieeexplore.ieee.org/iel5/8076/22348/01043240.pdf?arnumber=1043240.

PowerPedia: Vanadium Redox Batteries, 2009. http://peswiki.com/index.php/Vanadium_redox_batteries.

Rahman, F., 1998. Stability and properties of supersaturated vanadium electrolytes for high energy density vanadium redox battery. Ph.D. dissertation, The University of New South Wales, Sydney, Australia.

Rahman, F., Skyllas-Kazacos, M., 1998. Solubility of vanadyl sulfate in concentrated sulfuric acid solutions. J. Power Sources 72, 105–110.

Rahman, F., Skyllas-Kazacos, M., 2000. Vanadium redox battery for large scale energy storage applications. In: 7th IEEE Technical Exchange Meeting, April 18-19, 2000, KFUPM, Dhahran, Saudi Arabia.

Rahman, F., Skyllas-Kazacos, M., 2009. Vanadium redox battery: positive half-cell electrolyte studies. J. Power Sources 189 (2009), 1212–1219.

Rahman, F., Peng, C.Z.X., Skyllas-Kazacos, M., 1996. Stability of supersaturated vanadium electrolytes for high energy density redox cell. In: Chemeca '96, 24th Australian and New Zealand Chemical Engineering Conference, Sydney, Australia, September 30-October 2, 1996.

Rahman, F., Habiballah, I.O., Skyllas-Kazacos, M., 2004. Electrochemical behavior of vanadium electrolyte for vanadium redox battery—a new technology for large scale energy storage systems. In: Proceedings of CIGRE Session 2004, 29 August–3 September 2004, France.

Rehman, S., Bader, M.A., Al-Moallem, S.A., 2007. Cost of solar energy generated using PV panels. Renew. Sust. Energ. Rev. 11, 1843–1857.

Schoenung, S.M., 2001. Characteristics and technologies for long vs short term energy storage. A study by the DOE energy storage systems program, prepared by Sandia National Laboratory, SAND2001-0765, Unlimited Release, Printed March 2001.

Skyllas-Kazacos, M., 2006. G1 and G2 vanadium redox batteries for renewable energy storage, private communication.

Skyllas-Kazacos, M., Grossmith, F., 1987. Efficient vanadium redox flow cell. J. Electrochem. Soc. 134 (12), 2950–2953.

Skyllas-Kazacos, M., Kazacos, M., 1994. Stabilized electrolyte solutions, methods of preparation thereof and redox cells and batteries containing stabilised electrolyte solutions, PCT/AU94/0077.

Skyllas-Kazacos, M., Kazacos, G.C., 2009. Cost reductions and materials optimization for the vanadium redox flow battery. In: Energy Storage: Materials, Systems, and Applications Symposium, Materials Science & Technology 2009 Meeting, USA, 2009.

Skyllas-Kazacos, M., Rychcik, M., Robbins, R.G., 1986a. All vanadium redox battery. U.S. patent no. 849,094.

Skyllas-Kazacos, M., M. Rychcik, and R.G. Robbins, A.G. Fane and M. A. Green (1986b). New All-Vanadium Redox Flow Cell, J. Electrochem. Soc. 133, (5), 1057–1058, http://dx.doi.org/10.1149/1.2108706.

Skyllas-Kazacos, M., Kazacos, M., McDermott, R., 1988. Patent application PCT appl/AKU 88/000471.

Skyllas-Kazacos, M., Rychcik, M., Robbins, R.G., 1988. All vanadium redox battery. U.S. patent no. 4 786 567.

Thaller, L.H., 1976. Electrically rechargeable redox flow cell. United States patent no. 39,996,064.

Vetter, M., Dennenmoser, M., Schwunk, S., Smolinka, T., Dötsch, C., Berthold, S., Tübke, J., Noack, J., 2010. Redox flow batteries—already an alternative storage solution for hybrid PV mini-grids? In: Proceedings 5th European Conference PV-Hybrid and Mini-Grid, Tarragona, Spain, 29–30 April.

V-Fuel Company and Technology Information Package 2008, http://www.vfuel.com.au/infosheet. pdf.

Weblink1: REN21-renewable energy policy network for the 21st century. http://www.ren21.net/ portals/0/documents/resources/gsr/2013/gsr2013_lowres.pdf (accessed 06.02.14).

Weblink2: Global Wind Energy Council (GWEC): Global wind capacity. http://www. businessspectator.com.au/news/2014/2/6/wind-power/global-wind-capacity-124-2013 (accessed 06.02.14).

Weblink3: global PV installed capacity. http://www.epia.org/fileadmin/user_upload/Publications/ GMO_2013_-_Final_PDF.pdf (accessed 06.02.14).

Chapter 5

Storage of Solar Thermal Energy in Dependency of Geographical and Climatic Boundary Conditions

Roman Marx

Institute of Thermodynamics and Thermal Engineering (ITW), Research and Testing Centre for Thermal Solar Systems (TZS), University of Stuttgart, Stuttgart, Germany

Chapter Outline

5.1 INTRODUCTION

In solar thermal applications, the fluctuating solar irradiation (and thus solar yield) and the actual heating demand of connected consumers are rarely equal by time and magnitude. Hence, storage of thermal energy is important to balance this mismatch. Storing thermal energy to cover the heating demand for domestic hot water (DHW) preparation and space heating for residential buildings can be distinguished by storage duration. There are short-term, long-term, and seasonal

Solar Energy Storage. http://dx.doi.org/10.1016/B978-0-12-409540-3.00005-0

115

thermal energy storage (TES) concepts. The short-term TES concepts are mainly to balance day to night periods up to a few days. The long-term TES concepts can store thermal energy for several weeks. Their intention is to shift some heat from periods of high solar yield to times of high heating demand, and they can bridge some time of low solar irradiation (e.g., during cloudy periods). Seasonal TES is often used within solar district heating (SDH) networks consisting of large volumes in the range of a few 1000 m³ to many 10,000 m³. If these are designed well, they can store the thermal energy from summer at times of high solar thermal yield to winter at times of high heating demand. An additional differentiating factor is the desired scope of application of the TES. It can either be installed in systems providing heat for DHW preparation or for the combination of DHW preparation and space heating. Usually, TES for combined purposes is larger in volume because it needs to cover a higher heating demand.

The incipient statements are generalized for a fixed location of application. The dimensioning of solar thermal installations including TES relies on more than the scope of application. There are several boundary conditions that affect the size of both the solar thermal collector area and the volume of the TES. The main factors are solar irradiation and heating demand. These factors must be taken into account on an annual basis, but even more important, by the course of transient changes during shorter periods of time. One must consider if there is a good match between solar yield and heat demand at an overlapping period of time, or if there is a long period of time in between. This interrelation depends strongly on geographical and climatic boundary conditions. The dependency of all boundary conditions will be discussed in this chapter, focusing on the situation in Europe.

5.2 INFLUENCING BOUNDARY CONDITIONS

The focus of the assessment in this chapter is on the region of Europe. The limitation to Europe is justified because of its large north-south extension, many different countries with individual building stock, and profound databases for all significant boundary conditions for the assessment. Furthermore, ambitious goals in European energy politics toward a predominant renewable energy supply create a huge market in Europe for solar thermal applications. According to the ESTIF (2013), the installed solar thermal capacity was 28 GW$_{th}$ in the 27 EU countries and Switzerland in 2012 (glazed collectors). This corresponds to an annual solar thermal energy supply of 20 TWh$_{th}$. Moreover, the IEA (2012) estimates a solar thermal energy supply of about 280 TW$_{th}$ for Europe in 2050 on building level for DHW preparation and space heating, not taking SDH into account.

5.2.1 European Climate

Europe has a north-south extension of about 3800 km and east-west extension of about 6000 km. It reaches from the 35th to 70th latitude north. Thus, significantly different climates can occur among the European countries. According to the Köppen-Geiger climate classification by Peel et al. (2007), the European climate has a range from arid to polar, with a large share of cold climate. The main factors for the classification are the ambient air temperatures in summer and winter, as well as precipitation during those periods. Ambient air temperature especially influences heating demand. In Europe, the mean ambient air temperature varies between −2 and +19 °C.

A method to correlate ambient air temperature with heating demand can be done by heating degree days (HDDs). An HDD is defined as the sum of daily temperature differences between an effective room temperature and the mean ambient air temperature, when it is lower than a defined limiting temperature. Within the work of Ecoheatcool and Euroheat & Power (2006), a map of Europe has been developed in which HDD isolines have been implemented, interpolated by 80 measurement stations (see Figure 5.1). Internationally, there is no harmonized definition for the limiting temperatures of HDDs. In the case of Ecoheatcool and Euroheat & Power (2006), an effective room temperature of 17 °C has been defined. It is selected relatively low to consider additional

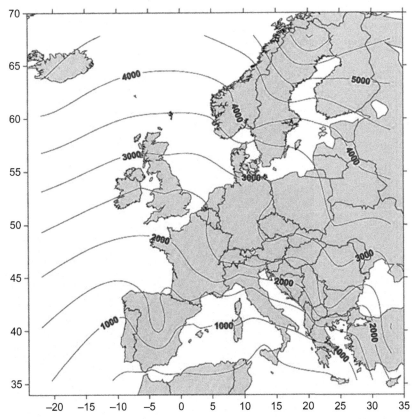

FIGURE 5.1   HDD isolines within Europe (Ecoheatcool and Euroheat & Power, 2006).

internal heat gains (e.g., by persons, electric energy consumption, etc.). The limiting ambient air temperature, when heating starts, is set to 13 °C, taking passive solar gains of the building into account. In Figure 5.1, one can see that HDDs differ significantly within Europe. In south Italy, with about 500 Kd, the HDD is 11 times smaller than in the north of Scandinavia. Hence, it could be assumed that the heating demand is also much higher in Scandinavia than in the Mediterranean region. This is not the case, because there are large differences in the building construction design between those regions.

A further factor related to climate and latitude is solar irradiation. It not only determines the potential solar yield by solar thermal collectors, but also influences significantly the ambient air temperature and thus the heating demand. Hence, the solar thermal collector area needs to be dimensioned according to solar irradiation and the kind of application (DHW preparation or a combination of DHW preparation and space heating). Because the thermal energy store levels out the fluctuations between solar yield and heating demand, the storage capacity must be dimensioned accordingly. Figure 5.2 illustrates the distribution of solar irradiation in Europe. The annual sum of solar irradiation is shown on an optimally incident surface and azimuth directed to the south. The range of potential solar irradiation differs between 800 and 2200 kWh/(m² a). A north-south gradient can clearly be identified, in addition to a west-east gradient.

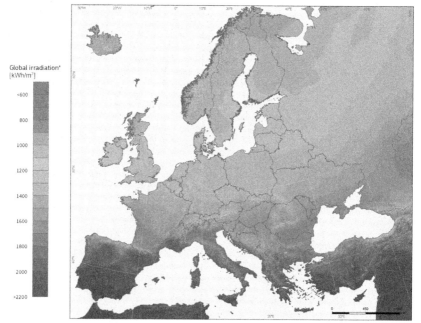

FIGURE 5.2 Solar irradiation in Europe as annual sum (*on optimally incident surface and azimuth directed to the south; Šúri et al. (2007) and Huld et al. (2012)).

Particularly at the coastal regions in central and northern Europe, the solar irradiation is lower than in more continental regions at the same latitude.

5.2.2 European Building Stock

Considering the climatic boundary conditions, a clear tendency for favorable solar thermal applications in southern European countries can be concluded. However, these are not the only factors influencing the dimension and design of the technology.

The construction of buildings, and thus their heating demand, is also a crucial factor. Nowadays, thermally well-insulated buildings are standard in newly built houses. In many European countries, certain standards are even regulated by law in order to reduce the primary consumption. Hence, the heating demand decreases in those buildings. The lowered heating demand can be covered more easily by solar thermal applications. Nevertheless, newly built houses are only a minority in the entire building stock within Europe. Many buildings in Europe are older than 50 years. According to the BPIE (2011), about 40% of all buildings were erected before 1960, which was before any regulations regarding energy savings or efficiency came into effect. Further, the retrofitting rate of buildings is about 1% per year. That means that the change within the building stock is proceeding slowly.

Focusing on European residential buildings is meaningful, having a share of 75% of all buildings (BPIE, 2011). Furthermore, the share of single-family houses is about 64%, compared to a share of 34% of multifamily houses. Single-family houses are characterized by a relatively high surface-to-volume ratio compared to larger buildings. The higher the heat exchanging surface of buildings, the higher the heat losses; thus, the heating demand increases. To reduce the primary energy consumption of such buildings, the heating demand can be reduced by thermally insulating the building's facade, or by including renewable energy sources such as solar thermal application. The large advantage of solar thermal applications for single-family houses is that there is usually enough space for solar thermal collectors on the roof. Thus, including solar thermal in those buildings can be an effective and also more economic measure to reduce the primary energy consumption instead of insulating the facade.

Taking retrofitting of solar thermal applications into account, the heating demand of European buildings must be assessed in order to evaluate their potential. The heating demand of buildings depends on several factors. The main factors are the standard of thermal insulation, air exchange rate, gross area of the building, and individual DHW consumption. These factors can differ significantly throughout Europe and counterbalance advantages of the climate.

The standard of insulation varies among European counties. For example, in Figure 5.3, the U-values ($W/(m^2 K)$) of external walls are shown for Sweden and the Netherlands at different periods of construction. The trend toward lower U-values and thus lower heat losses of the buildings can clearly be seen.

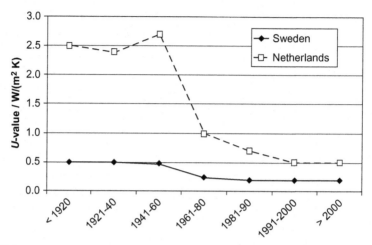

FIGURE 5.3 *U*-values of external walls for different construction periods in Sweden and the Netherlands (BPIE, 2011).

Nevertheless, this example highlights the obvious differences in thermal insulation standards in different countries.

The air exchange rate differs in a similar magnitude as the *U*-values. Often having ventilation systems that do not include heat recovery, a too-high air exchange rate induces unnecessarily high heat losses. According to BPIE (2011), the air exchange rate is for houses constructed since 2000 (e.g., 0.2 1/h in Germany, 0.6 1/h in Denmark, 4 1/h in the Czech Republic).

The *U*-value and air exchange rate influence the specific heating demand (kWh/(m^2 a)) of buildings. In Figure 5.4, the specific heating demands of single-family houses are shown for three different countries from three regions of Europe, taking different periods of construction into account. According to the development of the *U*-values, the specific heating demand also decreases for all depicted countries by time. Although in Sweden and Germany the HDDs (compare Figure 5.1) are several times higher than in Italy, the specific heating demand is in a similar range. This demonstrates that the building stock has already been adapted to the specific climate for a long time in order to reduce heating demand. Thus, the climate has a much smaller influence on the heating demand, as indicated by the HDD.

FIGURE 5.4 Average specific heating demand (kWh/(m^2 a)) of single-family houses for different construction periods in Sweden, Germany, and Italy (BPIE, 2011).

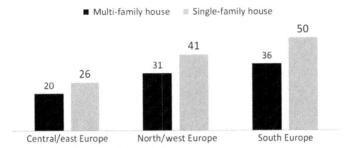

FIGURE 5.5   Average gross floor area (m^2) of the building per capita (BPIE, 2011).

The previously described values are specific values. To quantify these values, they must be combined with the actual gross floor area of the buildings. In Figure 5.5, the average gross floor areas for single and multifamily houses are depicted for different regions of Europe per capita. Again, the distribution is heterogeneous throughout Europe. The fact that space requirements per capita are the largest in southern Europe is important. That means that even though low values occur for the specific heating demand in southern Europe, in combination with the largest space to be heated, the total heating demand increases. The total heating demand of the building can even be higher in the milder south than it is in significantly colder regions of Europe.

In addition to the heating demand for space heating, the heating demand for DHW preparation must be considered. Particularly for new buildings with very sophisticated thermal insulation, the demand for DHW preparation can be a large share of the total heating demand. Also, regarding solar thermal systems for DHW preparation only, this heating demand is a very important boundary condition for the dimensioning of the system. Based on the latest available Eurostat data (1999), the distribution of hot water consumption per capita is shown in Figure 5.6. The average hot water consumption in Europe is about 50 l/d per capita, but large differences between countries can be seen. In the data, a temperature lift from cold to hot water of 50 K is considered, as well as circulation heat losses. A generally lower hot water consumption can be estimated compared to the Eurostat data (1999) using current water saving devices, but the qualitative distribution may persist. Again, a geographical deviation of the hot water consumption can be assessed. Predominantly in the southern regions of Europe, the hot water consumption is lower than the average, and in the northern regions of Europe, it is above the average consumption.

5.3   CLASSIFICATION OF SOLAR THERMAL SYSTEMS WITH TES

Depending on the kind of solar thermal system, different boundary conditions affect the design and dimension of the TES. In this chapter and in Chapters 4, 6, and 10, different TES concepts and designs are described. In order to determine

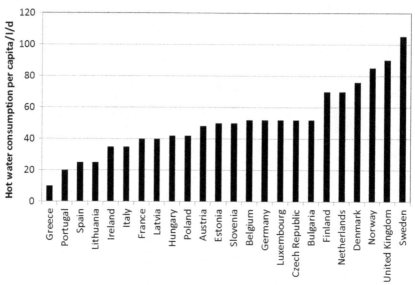

FIGURE 5.6 Hot water consumption per capita (l/d) for different countries in Europe (Eurostat, 1999).

the influence of the different geographical and climatic boundary conditions, three different TES applications have been selected.

The first system is a system for DHW preparation only. It consists of solar thermal collectors and a buffer store separated from the heating loop for space heating. Predominantly in summer, the solar yield can be used to achieve high solar fractions for DHW preparation. The typical solar fraction of this system is within the range of 50-60% of the annual DHW preparation in central/western Europe. The solar fraction is defined in Equation (5.1). In the case of systems for DHW preparation only, the heating demand for DHW preparation is considered; for the other two examples of systems, the sum of both heating demand for DHW preparation and space heating is considered.

$$\text{Solar fraction} = \frac{\text{Solar thermal energy used for heating [kWh/a]}}{\text{Total heating demand [kWh/a]}} \quad (5.1)$$

The second system is a solar combi-system: it combines the use of solar thermal energy for DHW preparation and space heating. Hence, the solar collector area and the TES capacity (and thus volume) are dimensioned larger than for the first system. The solar combi-system assists the conventional heating system for space heating. The system requires a more sophisticated control strategy than the first system to manage storage and distribution of solar heat for DHW preparation and space heating. The buffer tank stores thermal energy for DHW preparation and space heating. The typical solar fraction of solar combi-systems is within the range of 25-35% of the entire annual heating demand in central/western Europe.

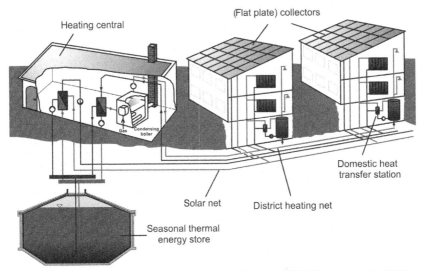

Heating central

(Flat plate) collectors

Gas Condensing boiler

Domestic heat transfer station

Solar net

District heating net

Seasonal thermal energy store

FIGURE 5.7   Typical solar district heating system with seasonal TES (Bodmann et al., 2005).

The third system considered for this assessment is a SDH system with seasonal TES. In comparison to the first two systems, a centralized seasonal TES and heating station are integrated into the system. The buildings are connected to a district heating net. Usually, the minimum amount of dwellings connected to those types of system is 100 (Bauer et al., 2010). The solar thermal collectors can be mounted on the buildings' roofs or on the ground, depending on availability. The collector fields are connected to the heating station by a solar net. The seasonal TES is designed in such a way that it can store enough thermal energy at times of high solar yield in summer for times of high heating demand in fall and winter. Backup heating usually consists of boilers using, for example, oil or gas as conventional energy sources, or biofuels such as wood chips as renewable energy sources. To increase the performance of the solar thermal components, heat pumps are often integrated into the systems (Marx et al., 2013). Those systems usually achieve solar fractions of around 50% of the total annual heating demand. Figure 5.7 shows a simplified scheme of a typical SDH system.

5.4   CASE STUDY TO EVALUATE THE INFLUENCE OF THE DIVERSITY OF BOUNDARY CONDITIONS

As previously described, solar thermal applications can significantly reduce the primary consumption of conventional energy sources in buildings. A large variety of boundary conditions influence the dimension of solar thermal systems. Some boundary conditions are transient (e.g., weather or heating demand). Thus, stationary calculations do not represent the actual behavior of the system correctly. Therefore, transient system simulations need to be carried out. Stepper (2014) has investigated climatic and geographical boundary conditions within his work by such transient simulations carried out with the program TRNSYS (Klein et al., 2010).

In order to reduce the amount of configuration, some simplifications have been made. The investigated building stock is reduced to single-family houses from the 1980 standard with a radiator heating system (max. 70/50 °C supply/return temperature) and the new building standard with under-floor heating (max. 35/28 °C supply/return temperature). For both heat distribution systems, a heating curve depending on the outside temperature has been applied. The gross floor area of the buildings is set to a European mean value of 152 m^2, taking four inhabitants per dwelling into account. Additionally, a district heating net (65/40 °C supply/return temperature) with an annual heat demand of 1500 MWh, consisting of the previously defined new single-family houses, has been chosen.

Five representative locations have been selected within Europe, each representing a different region with its own specific climate, building standard, and user behavior. The locations selected are Stockholm, representing the Scandinavian region; Rome, representing the Mediterranean region; Stuttgart, representing central Europe; London, representing the Gulf Stream influenced region; and Sofia, representing central eastern Europe. In Table 5.1, the individual parameters of each location are listed. The heating demand for DHW preparation is reduced by 20% compared to Eurostat data (1999) to take energy saving measures into account. For the 1980 building standard, Stockholm has the lowest total heating demand of all locations, and London has the highest. For the new building standard, Stuttgart has the lowest heating demand, and Stockholm has the highest.

At each location, the three solar thermal systems defined in Section 5.3 have been applied. The values in Table 5.1 highlight that the heating demand for DHW preparation and space heating differs significantly within Europe. Hence, reaching a certain solar fraction at each location also requires different amounts of solar thermal energy delivered into the system. This can lead to very large solar systems (collector area or TES) at some locations. As it is economically unreasonable to achieve a certain solar fraction at each location, energy saving potentials have been defined for the decentralized small-scale systems. Thus, for systems with DHW preparation only, 1, 1.5, and 2 MWh/a of effective energy savings by the solar thermal system have been defined. For the solar combi-systems, effective energy savings of 2, 3, and 4 MWh/a have been defined. For the simulation only, the collector area and TES volume have been varied. Many combinations of solar collector area and TES volume can meet the defined savings for each location, but not every combination is economically reasonable. To achieve a specific energy savings or solar fraction, for instance, one can either use a system with a very large collector area and a very small capacity of TES, or a system with a smaller collector area and a suitable

TABLE 5.1 Parameters of the Different Locations of the Case Study (According to Meteonorm, Ecoheatcool and Euroheat & Power (2006), and BPIE (2011))

	Stockholm	Rome	Stuttgart	London	Sofia
Global horizontal solar irradiation (kWh/(m^2 a))	1204	1755	1220	1017	1305
HDD (Kd)	4190	1127	2956	2210	2610
Specific hot water consumption per capita (l/d)	84	28	42	72	42
Specific heating demand for 1980 standard (kWh/(m^2 a))	147	180	156	268	255
Specific heating demand for new building standard (kWh/(m^2 a))	124	95	53	103	101
Total heating demand for DHW and space heating for 1980 standard (kWh/a)	22,344	27,360	23,712	40,736	38,760
Heating demand for DHW and space heating for new building standard (kWh/a)	18,848	14,440	8056	15,656	15,352

capacity of TES. There are technical issues to be considered, such as stagnation of the solar thermal collectors and economic factors. Combining the components' dimensions (collector area and TES volume) with their costs enables the possibility of finding the most economical solution. Depending on many factors, the prices for solar collectors and TES can vary (e.g., volume-specific costs for smaller TES are higher than for larger ones). For this assessment, both for the solar collectors and the TES, a price range has been applied. For the solar collectors (flat-plate collectors), prices between 220 and 540€/m^2 have been estimated; for the TES (hot-water buffer tanks), prices between 350 and 1200€/m^3 have been estimated. Both price ranges are without VAT and installation. A minimum case analysis has been carried out to determine the most economical configuration. The most economical configuration using the least expensive collector price and the most expensive TES price and vice versa has been determined. The system configuration fulfilling the minimum and thus least expensive combination for both analyses is the most economical combination within the regarded price range. By this method, the most suitable configuration between solar collector area and storage capacity can be determined for each location under economic considerations. Evaluating the most suitable economic configuration by technical requirements (e.g., minimization of stagnation) shows good agreement between those two criteria, and an overall good configuration has been determined.

For the SDH system, the approach differs from that of the single-family house. Being able to configure the district heating net of a variable number of dwellings, for each location, a district heating net with an annual heating demand of 1500 MWh has been defined. Having the same heating demand at each location leads to equal solar fractions, for the same energy savings by solar thermal energy at every system. Hence, defining a uniform solar fraction makes it possible to compare the systems to each other very well. For this case study, a solar fraction of 50% has been defined. The system consists of a buried hot-water TES as seasonal TES and a hot-water buffer tank. The solar thermal collectors are connected to the buffer tank, either supplying the solar heat directly to the district heating net or charging the surplus into the seasonal TES. In between the buffer tank and the seasonal store, a heat pump is installed. If the temperature level of the seasonal TES decreases below the temperature level for a direct use of the stored thermal energy, the heat pump can further discharge the seasonal TES. If the solar part of the system, including the heat pump, cannot ensure a high enough temperature level in the buffer tank to deliver heat to the district heating net at 65 °C, a gas boiler starts to operate as a backup heating system.

Just like for the system, on a single-family house level, the volume of the TES and the collector area have been varied. Regarding the TES volume, the main focus was on the volume of the seasonal TES, but some variations of the volume of the buffer tank have also been undertaken. Furthermore, the thermal power of the heat pump has been varied. Similar to the small-scale systems, different configurations can lead to a solar fraction of 50%. Thus, the cost minimum has been determined using the following cost estimations:

- Heat pump: $500 €/kW_{th}$
- Flat-plate collector: $220 €/m^2$
- Buffer tank: $350 €/m^3$
- Seasonal TES (logarithmic interpolation based on the realized hot-water tank TES in Germany (Mangold et al., 2011) and taking a nonlinear cost reduction in dependency of the volume into account): $(-129.843 \ln (\text{Volume}_{TES}[m^3]) + 1289.567) €/m^3$

5.4.1 Results of the Case Study for the System for Only DHW Preparation

The system assisting the DHW preparation is the smallest considering the heating demand, but it is also the most common one in Europe (IEA, 2013). In

TABLE 5.2 System Configurations for DHW Preparation with Energy Savings of 1, 1.5, and 2 MWh/a at Representative Locations

Location	Energy Savings (MWh/a)	Collector Area (m²)	Storage Volume (m³)	Mean VA-Ratio (l/m²)	Solar Fraction at DHW (%)
Stockholm	1	3	0.1	43	14
	1.5	4	0.2		21
	2	5.5	0.25		28
Rome	1	1.5	0.15	83	40
	1.5	2.5	0.15		64
	2	4	0.35		84
Stuttgart	1	3	0.15	59	29
	1.5	4.5	0.25		42
	2	7	0.5		57
London	1	3.5	0.15	45	17
	1.5	5	0.25		25
	2	7	0.3		33
Sofia	1	2.5	0.15	61	29
	1.5	4	0.2		44
	2	5.5	0.4		41

Table 5.2, the simulation results are listed for effective energy savings for 1, 1.5, and 2 MWh/a at the representative locations. The building standard is not considered because the DHW consumption is independent from the space heating demand. The results of the case study are that TES volume varies for each location within a similar range. The solar thermal collector area differs more significantly; for example, the required solar thermal collector in London is twice as high as the one in Rome, considering the same amount of saved energy. For higher energy savings, not only larger collector areas, but also larger volumes are required. A ratio between TES volume and solar thermal collector area indicates the dependency of both values at each location. In this case, the ratio (VA-ratio) is defined as TES volume in liters divided by collector area in square meters (see Equation (5.2)):

$$VA = \frac{TES \; volume \, [l]}{Collector \; area \, [m^2]} \quad (5.2)$$

In Table 5.2, the mean values of the VA-ratio for all three energy savings are shown for each location. This approach is valid because the absolute values for each energy saving are within a close range. As indicated by the distribution of the collector area, the VA-ratio differs in a similar manner. In Stockholm, the lowest value of 43 l/m could be determined; in Rome, the highest value of 83 l/m^2 could be determined.

The solar fraction at DHW preparation is also included in Table 5.2. Based on the very different thermal energy demand for DHW preparation and the fixed energy savings, the solar fraction varies significantly. Taking the two extreme examples of Stockholm and Rome, it can be seen that, for example, saving 1 MWh/a of energy leads to a solar fraction of 14% in Stockholm, but to 40% in Rome.

5.4.2 Results of the Case Study for Solar Combi-Systems

Similar to the results in the previous section, the results for the solar combi-systems providing solar thermal energy for DHW preparation and space heating are shown in Table 5.3. The effective energy savings have been defined as 2, 3, and 4 MWh/a of the total heating demand (DHW preparation and space heating) for both the building standard from 1980 and the new building standard. The results demonstrate that buildings from 1980 with a relatively high

TABLE 5.3 System Configurations for DHW Preparation and Space Heating with Energy Savings of 2, 3, and 4 MWh/a for the 1980 Building Standard and the New Building Standard at Representative Locations

Location	Energy Savings (MWh/a)	Collector Area (m^2)	Storage Volume (m^3)	Mean VA-Ratio (l/m^2)	Solar Fraction (%)
Stockholm 1980	2	10	1.1	92	9
	3	14	1.25		14
	4	18	1.4		18
Stockholm new	2	6	0.5	93	12
	3	7	0.8		16
	4	10	0.8		22
Rome 1980	2	7	0.5	77	8
	3	10	0.8		11
	4	14	1.1		15

Continued

TABLE 5.3 System Configurations for DHW Preparation and Space Heating with Energy Savings of 2, 3, and 4 MWh/a for the 1980 Building Standard and the New Building Standard at Representative Locations—Cont'd

Location	Energy Savings (MWh/a)	Collector Area (m²)	Storage Volume (m³)	Mean VA-Ratio (l/m²)	Solar Fraction (%)
Rome new	2	4	0.5	116	15
	3	6	0.8		21
	4	9	0.8		27
Stuttgart 1980	2	12	1.1	81	9
	3	18	1.1		13
	4	22	2.0		17
Stuttgart new	2	6	0.8	119	27
	3	10	0.8		38
	4	14	2.0		49
London 1980	2	12	1.1	75	5
	3	18	1.1		7
	4	24	1.7		10
London new	2	6	0.8	121	13
	3	9	0.8		19
	4	12	1.7		25
Sofia 1980	2	11	0.5	73	5
	3	14	1.1		7
	4	18	1.7		10
Sofia new	2	6	0.5	83	14
	3	8	0.8		21
	4	12	0.8		26

heat-distribution temperature (radiators) are less favorable than the new building standard with an under-floor heating system. The buildings from 1980 have a higher heating demand, and thus longer heating periods. In transit periods, there is a more timely overlapping of heating demand and solar yield; consequently, a better direct use of the solar yield is possible. However, systems with larger collector areas and TES volume are required to achieve the same amount of energy savings as those at the new building standard. Furthermore, higher VA-ratios are required for the solar combi-systems than for systems with only DHW preparation. This is logical because it is necessary to store a higher amount of thermal energy for DHW preparation and space heating than only for DHW preparation. Also, for the new building standard, the VA-ratio is in general higher than for the building standard from 1980, because the same energy savings result in higher solar fractions due to lower total heating demands. Therefore, the solar thermal energy needs to be stored for a longer period of time to make it usable at required periods of demand. Hence, the VA-ratio of combi-systems is not depending as clearly on the latitude of application as DHW systems. A clear geographical tendency cannot be determined by those values. Larger variations can be noticed for the different building standards (e.g., London or Stuttgart).

The solar fractions differ from location to location. In Stuttgart, for new buildings, energy savings of 4 MWh/a equal a solar fraction of 49%, while in Stockholm, the same amount of saved energy equals a solar fraction of 22%.

5.4.3 Results of the Case Study for the SDH System with Seasonal TES

In comparison to the previously discussed systems, a constant heat demand could be defined for the SDH system, consisting of a different number of single-family houses, depending on the location. The number of houses differs between 109 in Stockholm and 132 in London. One exception is Stuttgart, consisting of 270 houses due to the very low heating demand of each house compared to the other locations (see Table 5.1). The simulations have resulted in an optimum buffer TES of a volume of 100 m^3 at each location. All other values are listed in Table 5.4. Additionally, the thermal power of the heat pump integrated into the system has been evaluated. The results of the collector area show

TABLE 5.4 SDH System with Seasonal TES and a Solar Fraction of 50% at Representative Locations

Location	Collector Area (m^2)	Storage Volume Seasonal TES (m^3)	Thermal Power Heat Pump (kW$_{th}$)	VA-Ratio (l/m^2)
Stockholm	4000	5000	265	1250
Rome	3000	2000	135	667
Stuttgart	4000	3000	265	750
London	5000	5500	310	1100
Sofia	3500	5500	310	1571

a dependency to the sum of the annual solar irradiation. In Rome, for the highest solar irradiation the smallest collector area is required, and in London, the opposite is true. Also, a certain correlation between the specific heating demand and the volume of the seasonal TES can be seen. In Stockholm, London, and Sofia, the specific heating demand is higher than $100 \, kWh/(m^2 \, a)$. At these locations, the seasonal TES is also the largest. The reason for this is that having a higher specific space heating demand reduces the heating demand for DHW preparation at constant total heating demand. Hence, the solar thermal energy needs to be stored for a longer period of time, instead of using it time-near for DHW preparation. This requires larger storage capacities, which are realized by larger TES volumes. The heat pump dimension is dependent on the storage capacity because the TES serves as the heat source for the heat pump. The larger the heat source, the higher the thermal power of the heat pump must be in order to be able to use the stored thermal energy for heating purposes. The VA-ratio is significantly higher than for the small-scale systems. Like the size of the volume of the seasonal TES, the VA-ratio depends on the amount and period of time the thermal energy needs to be stored. Furthermore, being able to achieve a 50% solar fraction by a larger share of DHW preparation increases the cycle number of the seasonal TES (numbers of cycles of charging and discharging the TES within a period). Consequently, the usable capacity of the TES is increased, and thus can be designed smaller to store in total the same amount of thermal energy as a TES with a lower cycle number.

5.5   CONCLUSIONS

Assessing relevant boundary conditions, which can influence the design and dimension of solar thermal application, including TES, demonstrates that there is no universal dependency between geographic location and design and dimension of such systems. The climatic favorable conditions in southern regions are often neutralized by less efficient building standards or individual user behavior. The different building standards that are responsible for the buildings' heating demand rely heavily on laws and regulations by the individual authorities in each country or region.

The case study for solar thermal systems for DHW preparation demonstrates a certain geographical dependency on the dimension of such systems. The VA-ratio of those systems depends on the latitude of location. In the south, the value is the highest; in the north, the value is the lowest. That means that, in general, larger collector areas at similar TES sizes are required in the northern regions to achieve comparable effective energy savings as in the southern regions.

Examining solar combi-systems, a geographical dependency cannot be observed. The total heating demand, which depends on the building standard, has a much higher influence on design and dimension. Furthermore, the temperature range of the heat distribution system within the building influences the dimensions significantly. The VA-ratio for the new building standard with a generally lower heating demand for space heating is higher than for the 1980 building standard. Having a shorter heating period for the buildings with the new standard requires a larger TES capacity (and thus volume to store the solar thermal energy for a longer period) to make it then usable for heating purposes.

For the SDH systems with seasonal TES, some dependency can be observed. The size of the solar thermal collector area depends on the amount of solar irradiation at each location. Higher solar irradiations lead to smaller collector areas, and vice versa. The storage volume respectively the storage capacity does not depend on geographical location; the ratio between the share of thermal energy required for the demand of DHW preparation and space heating is much more important. If the share of the space heating compared to DHW preparation is high, the seasonal TES needs to be dimensioned with a larger volume, because the thermal energy needs to be stored for a longer period of time. Contrariwise, if the share for DHW preparation is high and, which is nearly constant of time, the solar yield can be used more directly and the required storage capacity is smaller. The suitable heat pump of those systems depends on the capacity of the seasonal TES. The larger the capacity of the seasonal TES, the larger the thermal power of the heat pump needs to be in order to discharge the seasonal TES in times of heating demand.

Comparing all three different systems to each other shows that the VA-ratio increases also by the size of the system. The smallest VA-ratios are obtained by the systems only for DHW preparation, and the highest ratios by the SDH systems with seasonal TES. The assessment shows further that already with small systems, a significant share of energy can be saved. Nevertheless, considering only the costs for the collectors and the TES, the economics are in favor of the large systems. The specific cost for large TES is smaller than for large collector areas. The VA-ratio is higher for the larger systems than for smaller systems. Thus, the specific costs decrease for the larger systems compared to smaller systems.

This assessment includes some generalizations. Planning solar thermal systems at specific locations makes it necessary to take the specific boundary conditions at each site into consideration. Possible different heating loads and their profiles, as well as specific local investment costs, can make a difference on the results of the assessment carried out in this chapter. Thus, the dependencies and values of this chapter can serve only as a basis for the general evaluation of the potential at different locations. This does not substitute for the detailed design and dimensioning of the solar thermal system that needs to be carried out for each concrete application.

REFERENCES

Bauer, D., Marx, R., Nußbicker-Lux, J., Ochs, F., Heidemann, W., Müller-Steinhagen, H., 2010. German central solar heating plants with seasonal heat storage. Sol. Energy 84, 612–623.

Bodmann, M., Mangold, D., Nußbicker, J., Raab, S., Schenke, A., Schmidt, T., 2005. Solar unterstützte Nahwärme und Langzeit-Wärmespeicher (Februar 2003 bis Mai 2005). Forschungsbericht zum BMWA/BMU-Vorhaben 0329607F, Stuttgart, Germany.

Buildings Performance Institute Europe (BPIE), 2011. Europe's Buildings Under the Microscope: A Country-by-Country Review of the Energy Performance of Buildings. Buildings Performance Institute Europe (BPIE), Brussels, Belgium. ISBN: 9789491143014.

Ecoheatcool and Euroheat & Power, 2006. Ecoheatcool work package 1: the European heat market. Final Report, Brussels, Belgium.

European Solar Thermal Industry Federation (ESTIF), 2013. Solar Thermal Markets in Europe: Trends and Market Statistics 2012. European Solar Thermal Industry Federation (ESTIF), Brussels, Belgium.

Eurostat, 1999. Energy consumption in households. EU and Norway—1995 survey. CEE countries—1996 survey, Luxembourg.

Huld, T., Müller, R., Gambardella, A., 2012. A new solar radiation database for estimating PV performance in Europe and Africa. Sol. Energy 86, 1803–1815. http://dx.doi.org/10.1016/j.solener.2012.03.003.

IEA Solar Heating Cooling Program, 2013. Solar Heat Worldwide: Markets and Contribution to the Energy Supply 2011. IEA, Austria.

International Energy Agency (IEA), 2012. Technology Roadmap: Solar Heating and Cooling. International Energy Agency (IEA), Paris, France.

Klein, S.A., et al., 2010. TRNSYS 17: A Transient System Simulation Program. Solar Energy Laboratory University of Wisconsin, Madison, USA.

Mangold, D., Pauschinger, T., Schmidt, T., 2011. Solare Nahwärme mit saisonaler Wärmespeicherung— Stand der Technik 2010 und Perspektiven bis 2020. In: 21. Symposium Thermische Solarenergie, 11.-13.05.2011, Kloster Banz, Bad Staffestein, Germany. ISBN: 978-3-941785-57-1.

Marx, R., Bauer, D., Drück, H., 2013. Energy efficient integration of heat pumps into solar district heating systems with seasonal thermal energy storage. In: Energy Procedia, ISES Solar World Congress 2013, Cancun, Mexico, 3.-7.11.2013.

Meteonorm weather data for Stockholm, Rome, Stuttgart, London and Sofia. Created with Meteonorm 7.0.20.22267.

Peel, M.C., Finlayson, B.L., McMahon, T.A., 2007. Updated world map of the Köppen-Geiger climate classification. Hydrol. Earth Syst. Sci. 11, 1633–1644. http://dx.doi.org/10.5194/hess-11-1633-2007.

Stepper, R., 2014. Speicherung solarthermischer Energie in Abhängigkeit geographischer Randbedingungen. Master thesis, Institute of Thermodynamics and Thermal Engineering, University of Stuttgart, Stuttgart, Germany (non-published student thesis).

Šúri, M., Huld, T.A., Dunlop, E.D., Ossensbrink, H.A., 2007. Potential of solar electricity generation in the European Union member states and candidate countries. Sol. Energy 81, 1295–1305. http://dx.doi.org/10.1016/j.solener.2006.12.007.http://re.jrc.ec.europa.eu/pvgis/.

Chapter 6

Sorption Heat Storage

H.A. Zondag

Department of Mechanical Engineering, Eindhoven University of Technology, Eindhoven,
The Netherlands

Chapter Outline

6.1 CHARACTERISTICS OF DIFFERENT TYPES OF HEAT STORAGE

6.1.1 Introduction

Heat storage provides a method to overcome the mismatch in thermal demand and thermal supply, both in time and in power. Heat can be stored as sensible heat (increasing the temperature), as latent heat (melting), or as thermochemical heat (endothermic decomposition reaction). If these processes are reversible, it is then possible to discharge the storage by respectively cooling down,

Solar Energy Storage. http://dx.doi.org/10.1016/B978-0-12-409540-3.00006-2

135

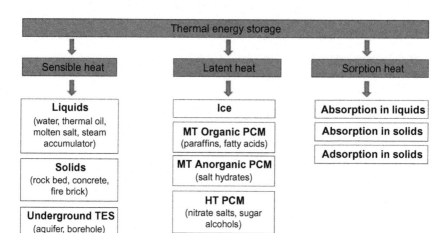

FIGURE 6.1 Overview of thermal energy storage types.

solidification, or the reverse exothermic reaction. For example, if a solid block of $CaCl_2 \cdot 6H_2O(s)$ is raised in temperature, the result is sensible heat storage. If this block of $CaCl_2 \cdot 6H_2O(s)$ is molten according to the reaction $CaCl_2 \cdot 6H_2O(s) \leftrightarrow CaCl_2 \cdot 6H_2O(l)$, the result is latent heat storage. If this block is dehydrated according to the reaction $CaCl_2 \cdot 6H_2O(s) \leftrightarrow CaCl_2 \cdot 2H_2O(s) + 4H_2O(g)$, the result is thermochemical heat storage.

An overview of the different methods is shown in Figure 6.1. This chapter will focus on thermochemical heat storage, which is also known as sorption heat storage. First, the general principles of sorption will be explained. Next, the different sorption materials will be explained, also indicating the difference between adsorption and absorption materials. Finally, an overview of sorption heat storage systems will be presented.

6.1.2   Introduction of Sorption Heat Storage

In sorption heat storage, heat is stored in a reversible (de)composition reaction, by means of the following reaction:

$$A(s/l) + B(g) \Leftrightarrow AB(s/l) + heat$$

In this equation, A is the charged sorbent (solid or liquid), B is the sorbate (vapor), and AB is the discharged sorbent, while the subscripts s, l, and g stand for the solid, liquid, and gaseous state, respectively. By heating the discharged sorbent to a temperature above its equilibrium temperature, the composite material AB is separated into its components (charging the sorbent). If the material is cooled to below its equilibrium temperature, the reverse reaction will take place, and the components A and B will react again under the release of heat (discharging the sorbent). However, the reverse reaction can be postponed by separating the components after charging. The heat is thus stored in chemical form, minimizing heat loss. The charged sorbent can be stored at ambient temperature, and the heat is only generated when the sorbate is taken up again.

For the sorbent material, solid adsorption materials (e.g., zeolites or silicagel), solid absorption materials (e.g., salt hydrates, salt hydroxides, or salt ammoniates), and liquid absorption materials (e.g., concentrated solutions of LiBr, LiCl, or NaOH) can be used. For liquid sorption, on absorption, the liquid is going from a concentrated hygroscopic solution to a weak solution, and the reverse on desorption.

For the sorbate material, water is most often applied. In closed systems, the sorbate is stored after desorption. This storage is normally in condensed form, which strongly reduces the required volume. The condensation heat is released at ambient temperature. On sorption, the sorbate is normally evaporated again with low temperature heat (e.g., ambient temperature heat from a borehole) before it is taken up by the sorbent. In principle, the sorbate can also be applied to the sorbent in liquid form, but this strongly reduces the energy released on the sorption reaction, because in that case a large fraction of the reaction energy is taken up by breaking the bonds between the sorbate molecules. For hydrates, this is typically 70-75% of the total energy released. Therefore, it is important that the sorbate is supplied as vapor; if it were supplied as liquid, the energy released on sorption would be much lower, and typical energy densities for these materials would be strongly reduced.

Therefore, in sorption heat storage, a large part of the energy density results from the evaporation energy of the vapor. In this sense, sorption heat storage can be compared to latent heat storage, but now using the evaporation energy instead of the melting energy. Because evaporation energy is typically much larger than melting energy, sorption heat storage can reach much higher energy densities than are possible for latent heat storage, depending on how compact the vapor can be stored in adsorbed or absorbed form.

6.1.3   Comparison of Different Types of Heat Storage

An overview of typical storage densities of the different materials is given in Figure 6.2. Note that sorption energies and evaporation energies can be of the order 10^3 MJ/m^3, melting enthalpies are of the order 10^2 MJ/m^3, and specific heats of the order 10^0 MJ/m$^3 \cdot$K. In other words, for the case of MgCl$_2 \cdot$6H$_2$O, the sorption energy is equivalent to about 13 times the melting enthalpy, and to an amount of sensible heat equivalent to heating the material over a temperature range in the order of $\sim$1000 °C.

Advantages and disadvantages of sorption heat storage compared to other types of heat storage are shown in Figure 6.3. Sorption materials have the highest energy storage density, and in addition, once the material has been charged, it can be used for long-term loss-free storage of heat. However, sorption systems are relatively complex due to the fact that the storage material consists of two components that have to be stored separately after charging. Also, sorption

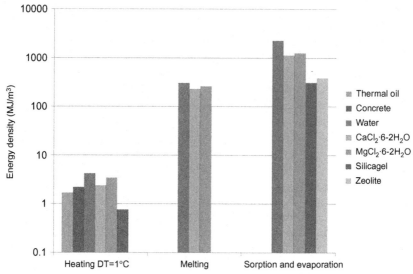

FIGURE 6.2 Overview of typical storage densities comparing sensible heat storage, latent heat storage, and sorption heat storage materials.

	Sensible	Latent	Sorption
Energy storage density	0	+	++
Heat loss	−	−	+
Degradation	0	−	−
System complexity	+	+	−

FIGURE 6.3 Advantages and disadvantages of different heat storage principles.

materials may degrade over time due to changes in chemical or mechanical composition over many sorption/desorption cycles.

6.2 PRINCIPLES OF SORPTION HEAT STORAGE

In sorption heat storage, heat is stored in a reversible (de)composition reaction between vapor and a solid or liquid. An important aspect for sorption materials is the equilibrium temperature, which is a function of vapor pressure. Above this temperature, the sorbent desorbs the sorbate, while below this temperature the sorbent absorbs (or adsorbs) the sorbate. The equilibrium temperature can be calculated by the Clausius-Clapeyron equation, involving the enthalpy ΔH and entropy ΔS of the reaction per mole of sorbate, which for chemical equillibrium can be written in the following form:

$$RT_{eq} \ln \frac{p}{p_0} = -\Delta H + T_{eq}\Delta S \Rightarrow T_{eq} = \frac{\Delta H}{\Delta S - R \ln \frac{p}{p_0}}$$

In the equation, T_{eq} is the equilibrium temperature, p is the vapor pressure, p_0 is the reference pressure of 1 bar, and R is the universal gas constant 8.31 J/mol K.

For a sorbate taken up by a sorbent, the entropy change ΔS is a measure for the change in disorder in the system, and is mostly determined by the transition of the sorbate from the disordered gas phase to the more ordered solid phase (for solid sorbents) or liquid phase (for liquid sorbents). Therefore, on selecting a working pair, ΔS is largely determined by the choice of the sorbate, but almost independent of the choice of sorbent. On the other hand, ΔH strongly depends on the choice of the sorbent, related to how strongly the sorbate is bound to the sorbent. As an illustrative example, for a hydration reaction, ΔS is typically 150 J/mol/K, while ΔH is typically 55-80 kJ/mol. For a water vapor pressure of 1 bar, this gives equilibrium temperatures in the range 100-250 °C. For a water vapor pressure of 10 mbar (corresponding to a source temperature of ~8 °C), this gives equilibrium temperatures in the range of 20-150 °C, as can be calculated from the Clausius-Clapeyron equation.

In Figure 6.4, the vapor pressure curves of different common sorbates (water, ammonia, and methanol) are compared. Because water is a strongly polar liquid with hydrogen bonding between the water molecules, water molecules are relatively strongly bound to each other. The binding between CH_3OH molecules—and even more between NH_3 molecules—is weaker. Consequently, water has a lower vapor pressure than the other two liquids, and a relatively high

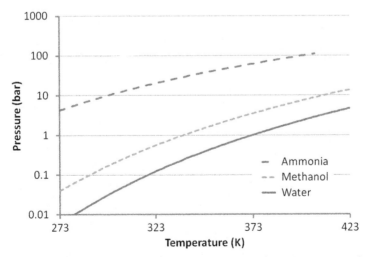

FIGURE 6.4 Comparison of vapor pressures of common sorbates (water, ammonia, and methanol).

heat of evaporation of 44.0 kJ/mol as compared to 37.4 kJ/mol for methanol and 19.9 kJ/mol for ammonia (at 25 °C reference temperature).

In a similar way as for sorbates, vapor pressure curves can be drawn for sorbents such as salt hydrates, hygroscopic liquids, and adsorption materials, as a function of their fractional loading. In this case, the vapor pressure curve gives the equilibrium between sorbate vapor and the sorbate absorbed/adsorbed in the sorbent. The equilibrium pressure depends on the strength of the bonding between the sorbate and the sorbent; a stronger binding results in a lower vapor pressure.

The equilibrium composition in the reversible reaction $A(s) + B(g) \leftrightarrow AB(s) +$ heat is a function of temperature and vapor pressure. This can be plotted in the isostere graph (concentration as a function of temperature and pressure). As an example, the equilibrium curves for $CaCl_2$ are shown in Figure 6.5 for

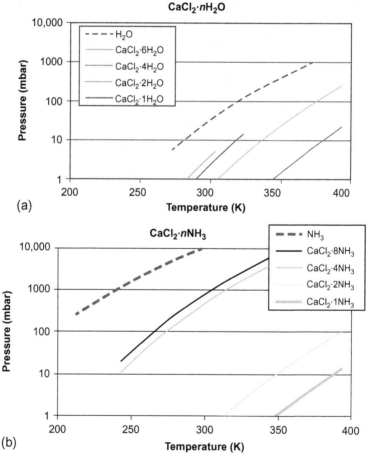

FIGURE 6.5 $p\text{-}T$ curves for absorption in solids: (a) $CaCl_2$ hydrates and (b) $CaCl_2$ ammoniates (calculated from estimated thermodynamic data; melting points are ignored).

absorption of water and for absorption of ammonia. The figures show different equilibrium lines, each line resembling one equilibrium loading of the material. The different lines are due to the different binding energies with which the water is bound to the salt molecule. The first water molecules are bound stronger than subsequent water molecules. As can be seen from Figure 6.5b, if pure ammonia is used in the evaporator, even at ambient temperatures already high vapor pressures of over 8 bar result, leading to high power densities. This makes ammonia very suitable for chilling applications, and as such it is widely applied in industrial chillers. On the other hand, the use of ammonia involves safety issues, restricting the use of this material to industrial environments. For use at higher temperatures, ammonia is usually not applied in its pure form, but is absorbed in other materials, which leads to lower vapor pressure.

For absorption in solids, the sorption process shows discrete steps in the amount of sorbate that can be taken up, related to how many molecules of sorbate can be taken up by one molecule of sorbent. This is in contrast to absorption in liquids and adsorption in solids, which have a more continuous character, with a large amount of equilibrium lines showing the fractional loading of the material. In Figure 6.6, examples are shown for water sorption on silicagel and water sorption on zeolite 4A.

6.3 SORPTION HEAT STORAGE MATERIALS

6.3.1 Introduction

Different types of sorption reactions exist, as shown in Figure 6.7. Absorption in liquids relates to the absorption of water vapor in hygroscopic salt solutions, but also to the absorption of gasses like ammonia vapor in water. For solid sorption, a distinction can be made between absorption in solids, in which the sorbate is integrated into the crystal lattice of the sorbent, and adsorption in solids, in which the sorbate is attached to the internal surface area of the sorbent.

6.3.2 Physisorption Materials—Zeolites, Silicagel

In adsorption, the molecule adheres to the surface of the sorbent by van der Waals interaction. Typical adsorption materials therefore have a large internal surface area, due to their extensive pore structure. Because the sorption takes place at the surface, the structure of the sorbent is unaffected and the sorbent does not expand on adsorption, which has a positive effect on the mechanical stability over multiple cycles. Many different adsorption materials exist, such as different types of zeolites, silicagel, activated carbon, activated alumina, aluminum phosphates, and metal-organic frameworks (MOFs). Some of these are in wide use (zeolites, silicagel), while others are still largely in the research and development phase (MOFs).

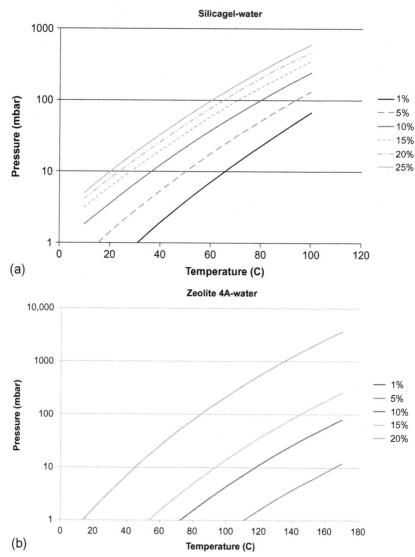

FIGURE 6.6 *p-T* curves for adsorption in solids: (a) silicagel-water and (b) zeolite 4A-water (data from Restuccia et al., 1999). Percentage gives weight percentage of water adsorbed.

    The performance of adsorption materials is mainly determined by their pore size distribution and their polarity. Adsorption materials like activated carbon, activated alumina, or silicagel have a wide range of pore sizes, from small micropores (<2 nm) to large macropores (>50 nm). On the other hand, molecular sieve adsorption materials such as zeolites have a well-defined pore width, ranging from 0.4 nm for zeolite 4A to 1.0 nm for zeolite 13X. Most sorbents are

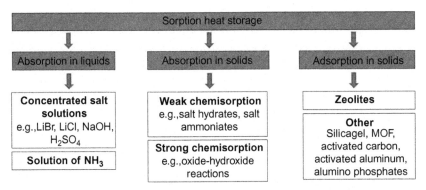

FIGURE 6.7 Classification of sorption materials.

polar, such as activated alumina, silicagel, and zeolites, increasing the sorption enthalpy for polar sorbates such as water. On the other hand, activated carbon is nonpolar, resulting in relatively low sorption enthalpy.

Among the different adsorption materials, zeolite is presently the main adsorption material under research for sorption heat storage, because of its relatively high discharge temperature (much higher than silicagel, for example, as can be seen in Figure 6.6). Zeolites consist of mainly of SiO_2 and Al_2O_3, in which part of the Al is replaced by elements such as Na^+, K^+, Ca^{2+}, and so on, increasing the polarity of the zeolites. Typically, they can absorb an amount of water up to about 22–27% of their dry weight, limiting the energy storage density that can be obtained, compared to salt hydrates, for example. On the other hand, zeolites are more stable than most salt hydrates, and also have fast kinetics. Efforts have been carried out to improve the characteristics of zeolite by impregnation with $MgCl_2$ and $CaCl_2$ (Jänchen et al., 2004) and $MgSO_4$ (Hongois et al., 2010).

6.3.3 Sorption in Liquids

The main practical differences between absorption in solids and in liquids is that on one hand liquids have a lower temperature step, but on the other hand they have much better transport properties. Therefore, in liquid systems the sorbent is usually not reacting in the storage vessel, but is pumped through a separate reactor in which the sorbent is charged or discharged, allowing a decoupling between storage and power generation. Typical hygroscopic liquids are solutions of LiBr, LiCl, NaOH, and H_2SO_4. On charging the liquid (desorbing), too high concentration may cause the liquid to crystallize, which may result in blocking of pipes and pumps. On the other hand, low concentrations reduce the energy density, and the temperature effect will become too small to be useful.

Work on an absorption chiller with integrated heat storage using an LiCl solution has been carried out by SERC in Sweden (Bales, 2006), using a special design that allowed partial crystallization of the solution to enhance the effective energy density. N'Tsoukpoue et al. (2010) carried out a systems study for an LiBr solution-based seasonal heat storage system. He concludes that crystallization is probably necessary for the competiveness of the process, to enhance the effective energy storage density, especially because LiBr is expensive and large volumes are required to reach 100% solar fraction. For seasonal heat storage, EMPA in Switzerland focuses on a hygroscopic solution of NaOH, which is much cheaper than LiBr (Weber and Dorer, 2008; Fumey et al., 2014). According to Fumey et al. (2014), the NaOH solution can be charged to a concentration of 50% (limited by crystallization) and discharged down to a concentration of 38% (assuming a 24 °C return temperature and a 2 °C evaporator temperature). For this case, the energy density of the solution is indicated to be three times the energy density of a water vessel over a temperature range of 65 °C. Another approach is followed at the University of Minnesota (Quinnell et al., 2010), using a $CaCl_2$ solution for both thermo-chemical and sensible heat storage. In order to increase the effective energy density on the system level, a single tank is used to store both the diluted and the strong solution, separated by a thermocline. Researchers indicate that the material energy storage density of the closed system is about twice that of sensible storage with water.

6.3.4 Weak Chemisorption—Hydrates

Higher values for the effective energy density per mass can be obtained by absorption in solids. The absorption of water vapor typically has a ΔH per mole of water in the range of 55-80 kJ/mol, similar to polar adsorption materials. However, the fractional water loading can be much higher for absorption in solids, in the range of 50–100% of the dry sorbent mass, increasing the potential energy storage density. On the other hand, the water sorption in the crystal lattice causes expansion and contraction of the material over the sorption-desorption cycles, reducing the mechanical stability.

For seasonal heat storage, several research institutes focus on salt hydrates, because of their relatively low cost, high potential storage density, and high temperature step. A comparison of different materials for thermochemical heat storage was presented by Van Essen (2009), indicating that $CaCl_2$ and $MgCl_2$ hydrates have a better performance and particularly faster hydration kinetics than $MgSO_4$ and $Al_2(SO_4)_3$ hydrates. Research at the Energy Research Centre of the Netherlands (ECN) focused on $MgCl_2$ (Zondag et al., 2013). Unfortunately, instability of $MgCl_2$ hydrates reduced the performance over multiple cycles (Ferchaud et al., 2014). Mauran et al. (2008) presented research on $SrBr_2$. Work on an open sorption system containing a packed bed with $KAl(SO_4)_2 \cdot 12H_2O$ was carried out by CEA in France (Marias et al., 2011). Studies on crystal level into the detailed mechanisms involved in the hydration-dehydration transition were carried out by Ferchaud et al. (2014) and Lan et al. (2014).

6.3.5   Strong Chemisorption—Hydroxides

In hydroxides, a much stronger water bonding occurs than in hydrates. A typical hydroxide reaction is given by $AO + H_2O \rightarrow A(OH)_2$, in which AO is an oxide. ΔS values are still typically about 150 J/mol/K, similar as for hydrates, as expected. However, the ΔH values show a much larger range. Typically, for the formation of hydroxides from oxides, values of about 50 kJ/mol are found for metal oxides, while values of 80-150 kJ/mol are found for the earth alkali metals of group 2 of the periodical system, with an increase in ΔH for heavier molecules. These high values of the reaction enthalpy make these materials suitable for high temperature heat storage for industrial applications.

Hydroxides have been investigated mainly for high temperature heat pump applications (e.g., Kariya and Kato (2014), examining a heat pump based on the reaction $H_2O + CaO \rightarrow Ca(OH)_2$). However, increasingly, these materials are also investigated for industrial heat storage. Research into improving the kinetics of these reactions, and thereby reducing the effective charging temperature, is carried out by Aristov and Shkatulov, showing a significant improvement of the kinetics for $Mg(OH)_2$ on doping with $LiNO_3$ and for $Ca(OH)_2$ on doping with KNO_3 (Shkatulov and Aristov, 2014). At DLR in Germany, tests are carried out on high temperature heat storage based on $CaO + H_2O \leftrightarrow Ca(OH)_2$, with discharge temperatures as high as 550 °C and an energy density of 323 kWh/m^3 or 1.1 GJ/m^3 (Laing and Wörner, 2013).

6.4   SORPTION HEAT STORAGE SYSTEM DESIGNS

6.4.1   Introduction

Thermochemical storage systems are all still in the R&D phase. In commercial applications, sorption materials are mostly used for industrial process applications such as gas cleaning and separation processes, and on a smaller scale for absorption heat pumps and chillers.

6.4.2   Principles of Sorption Systems

Sorption heat storage systems can be classified in two ways, leading to four different types of storage for the case of seasonal sorption heat storage, as shown in Figures 6.8 and 6.9. A first classification distinguishes between packed bed systems and separate reactor systems:

- In packed bed systems, the packed bed of sorbent functions both as storage and as reactor vessel, without transport of the sorbent.
- In separate reactor systems, on charging, the sorbent is taken from the storage tank with discharged material, is flown through a separate reactor in which the charging reaction is taking place, and is finally stored in the storage tank with charged material. On discharging, this process flow is reversed. In this way, storage and power generation are decoupled.

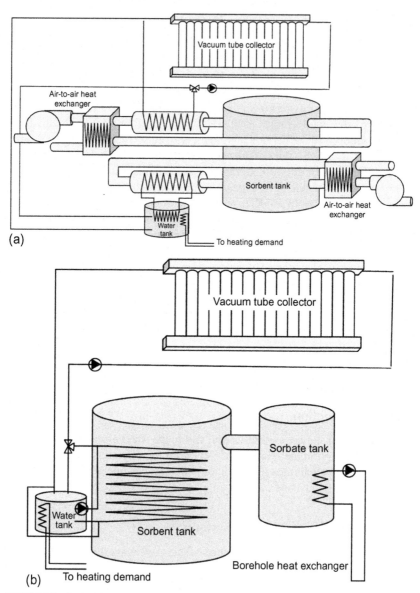

FIGURE 6.8 Packed bed sorption store systems: (a) open sorption system and (b) closed sorption system.

Generally, systems using solid sorption are of the packed bed type (e.g., Zondag et al., 2013), while systems using liquid sorption are of the separate reactor type (e.g., Weber and Dorer, 2008). However, in the literature, a few solid sorption reactors of the separate reactor type are also found, using a granulate flow of sorbent (Kerskes et al., 2011).

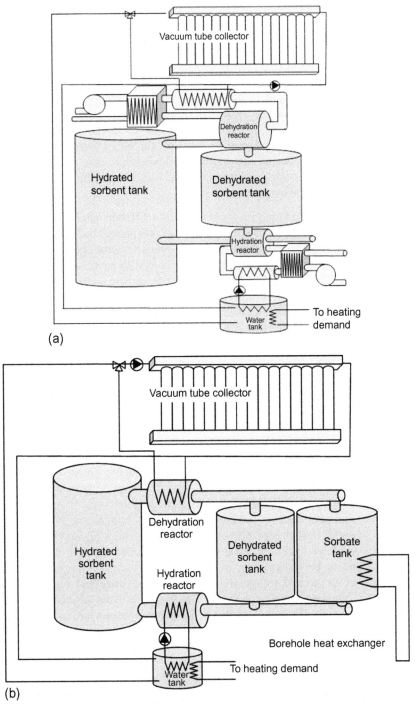

FIGURE 6.9 Separate reactor sorption store systems: (a) open sorption system and (b) closed sorption system.

A second classification is between open systems taking the sorbate from the ambient (e.g., water vapor from the air) and closed systems in which the sorbate remains within the system, and only the evaporation heat is taken from the ambient:

- Open systems are necessarily at ambient pressure and can only use as sorbate components already present in the air, such as water vapor. In an open system, the water vapor is transported as moist air by a fan.
- In closed systems, not only water can be used as sorbate, but also other liquids such as ammonia or methanol. All inert gasses are removed as much as possible by prior evacuation of the system, and the system pressure under operation will be determined by the equilibrium pressure of the sorbate. The water vapor is transported by the difference in vapor pressure between the sorbent and the evaporator.

The storage systems in Figures 6.8 and 6.9 show a different number of tanks for the packed bed open system (one sorbent tank), the packed bed closed system (one sorbent tank and one sorbate tank), the separate reactor open system (two sorbent tanks, one for charged and one for discharged sorbent), and the separate reactor closed system (two sorbent tanks and one sorbate tank). Because multiple tanks increase the total volume of the system, and thereby reduce the effective energy storage density on system level, it is important to keep the design as compact as possible, preferably combining tanks. Furthermore, the closed systems in the figures are connected to a borehole, supplying low temperature evaporation heat to generate the required water vapor. While in a closed system a low temperature heat source for evaporation is inevitable, in an open system, this is not strictly necessary as long as the ambient air provides sufficient water vapor. In open systems, a critical component is the air-to-air heat exchanger that is providing the heat recovery in the charge and discharge air loops. While these loops are drawn in the figure as separate loops (for clarity), in practice the system design is normally such that the same loop is used for both charge and discharge. Finally, in all systems, the sorption heat storage is coupled to a small water tank to deliver peak power; in this way, the sorption system itself can be designed at lower power, which improves its performance.

In the literature, research has been carried out on both closed systems and open systems for heat storage. Closed systems tend to have higher output temperatures and have lower electrical power requirements for fans, while open systems tend to be cheaper (no vacuum tanks or vacuum pumps required) and more robust (no leakage issues), and tend to have better heat transfer characteristics compared to closed packed bed systems (due to the large contact area in the packed bed). It is expected that both types of systems will in time have their own markets and applications where they perform best.

6.4.2.1 Closed System

In a closed system, the maximum temperature rise that can be obtained on absorption can be obtained from the isostere plot. In Figure 6.10, a simplified isostere graph is shown, with curves for the equilibrium vapor pressure of the

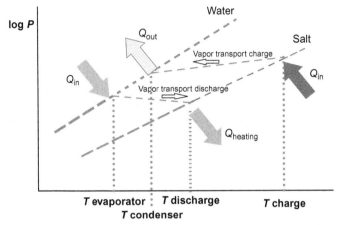

FIGURE 6.10 Temperature levels for absorption and desorption in a thermochemical heat storage using absorption in a solid.

sorbate (in this case water) and the vapor pressure of the sorbent (in this case a salt hydrate). Because the salt hydrate is more hygroscopic than the liquid water, the vapor pressure for the salt hydrate is lower than for water at the same temperature. On absorption, the temperature rise that can be obtained between the evaporator temperature and the sorbent temperature is the temperature difference between these isosteres at the given vapor pressure, corrected for pressure drop in the system. In a closed system, on discharging, the evaporator temperature determines the evaporator vapor pressure and thereby the discharge temperature of the sorbent. On charging, the condenser temperature determines the system vapor pressure and thereby the desorption temperature.

Closed sorption heat storage systems are often found for short-term heat storage applications, but have also been applied for seasonal heat storage. Typically, the system consists of a packed bed with a large heat exchanger area integrated in the bed. The limited heat transfer in the packed bed is often a bottleneck for the performance of these systems. Therefore, the conductivity may be enhanced by impregnating the sorbent into a conductive matrix such as expanded graphite.

Solid adsorption silicagel chillers are commercially available. In addition, several prototypes of solid absorption heat pumps with integrated heat storage have been realized, for example, with water absorption by $SrBr_2$ (Mauran et al., 2008), Na_2S (De Boer et al., 2004), and CaO (Kariya and Kato, 2014), as well as with CO_2 absorption in the system $PbCO_3$-CaO (Kato et al., 1998). In several cases, expanded graphite was used to improve the effective heat transfer through the bed. Mauran et al. (2008) presented research on $SrBr_2$ impregnated in graphite for an air conditioning system application with integrated heat/cold storage; a 1 m^3 prototype was built with 40 kWh cold storage (60 kWh heat storage). In the Modestore project, carried out by AEE Intec in Austria, a closed evacuated sorption storage was built and monitored in a field test. The sorption storage contained a large copper sheet-and-tube spiral as heat exchanger, which was filled with silicagel beads. The silicagel was dried using flat-plate solar thermal collectors. Unfortunately, it had to be concluded that most of the moisture captured by the silicagel did not result in sufficient temperature lift, resulting in a low effective energy density.

Closed liquid sorption differs from closed solid sorption due to the fact that the liquid can easily be transported through the system. This allows separation of the storage of the sorbent (in a vessel with liquid) and the power generation (in a separate sorption reactor). Typically, the principle of a liquid sorption reactor is similar to a chemical absorption column, generating a large contact area between the gas and the fluid by application of dedicated internal structures. Much research has been conducted into liquid sorption systems. Some recent developments have been achieved by Bales (2006) and Weber and Dorer (2008). To enhance the evaporation area, Bales built a prototype in which the liquid sorbent is sprayed over a heat exchanger, thereby generating effective contact between the sorbent and the vapor. Weber and Dorer (2008) indicates a reactor consisting of several trays to improve the contact area.

6.4.2.2   Open System

In an open system, the temperature rise in the bed is limited by the relatively large thermal mass of the inert carrier gas flow (air). Depending on the mass ratio of inert gas over vapor, this may significantly reduce the temperature rise that can be obtained; for a seasonal heat storage system operating at 10 mbar vapor pressure (corresponding to saturated air of 8 °C), the effective temperature rise over the sorption bed is limited to about 25-30 °C, which is much lower than the equilibrium temperature. In order to reach a high output temperature, it is necessary to preheat the airflow into the bed. This can be realized with an air-to-air heat recovery system (Zondag et al., 2013), preheating the incoming airflow with the exhaust airflow that exits the system after heating the load, as shown in Figures 6.8a and 6.9a.

In the Monosorp project, carried out by ITW Stuttgart, an open sorption storage system was developed, based on extruded blocks of zeolite 4A with integrated channels, to reduce pressure drop (Kerskes et al., 2007). During summer, the zeolite was dried with high temperature heat from the solar collector system. During winter, the moist ventilation exhaust air was led through the sorption storage, in which the moisture was captured by the zeolite, heating the exhaust air. This air was subsequently led over an air-to-air heat exchanger, heating the incoming ventilation air. Mette et al. (2012) found that if the hot air for charging the zeolite was pre-dried by means of a sorption wheel, the charging temperature of the zeolite could be reduced from 180 °C down to 130 °C, thereby improving the overall system efficiency (increasing solar collector efficiency and reducing thermal losses).

Salt hydrates have a higher potential energy storage density than adsorption materials like silicagel and zeolite, and in addition are often substantially cheaper. Therefore, salt hydrates are considered an interesting option for seasonal heat storage. In this field, ECN and Eindhoven University of Technology (TU/e) in the Netherlands are involved in both material research and system development. Initially, the focus was on sorption storage using $MgCl_2 \cdot 6H_2O$. A lab scale prototype built at ECN is shown in Figure 6.11. However, in the later European Energy Hub project, due to problems with instability of the $MgCl_2$ hydrate (Ferchaud et al., 2014), ECN built an open sorption prototype based on 200 l zeolite 13X (De Boer et al., 2014). Although the zeolite showed a stable performance, the high required charging temperature led to significant heat losses during charging. Also, the pressure drop in the system required significant auxiliary fan power.

6.5   OVERALL SYSTEM ASPECTS

As previously indicated, the main advantage of sorption heat storage is potentially high energy density, combined with the fact that losses are negligible once the energy is properly stored. This makes sorption heat storage very interesting for compact long-term storage of large amounts of heat. This is reflected in the fact that much of the research on sorption heat storage has focused on seasonal storage of solar heat. Unfortunately, seasonal storage of solar heat is also an area in which the economics of the storage is a difficult issue, because a large amount of storage material is required, which is charged and discharged only once per year. Therefore, this application sets very strong constraints on the cost of the storage material, as well as the system installation as a whole, to keep the cost per amount of energy delivered sufficiently low. In addition, careful design is required to minimize the electrical energy use of the system (borehole pump, fan in open system, solar collector pump, vacuum pump in closed system) to make sure that the effective coefficient of performance (COP) of the system remains sufficiently high, as it should at least be significantly higher than in a conventional heat pump system (e.g., above 10).

A different application in which compactness is important is mobile transport of industrial waste heat. Compared to seasonal heat storage, the sorption material has to undergo many more cycles, affecting both the economics and the requirements for stability of the materials. The use of sorption is particularly interesting for industrial drying, as the sorption material on discharge is absorbing moisture to generate heat. This principle is applied in a project by ZAE in Germany, in which a mobile zeolite sorption heat storage was developed for transport of industrial waste heat from a waste incineration plant (MVA Hamm) to an industrial drying installation for plastics (Krönauer, 2013).

Furthermore, several applications focus on absorption heat pumps or absorption chillers with integrated heat storage. This is an interesting application when mismatch exists between the heat supply and the required heating/chilling.

Finally, because of its loss-free storage of heat, sorption heat storage is also highly suitable for small autonomous applications for one-time use, requiring

FIGURE 6.11 (a) Lab-scaled thermal open energy storage system using $MgCl_2 \cdot 6H_2O$ (Zondag, 2013); (b) Upscaled prototype using zeolite 13X (De Boer et al., 2014).

instantaneous heating or cooling. This has led to the development of applications such as self-heating coffee mugs or self-cooling beer vessels. Although this is a small niche market, it opens new opportunities and a range of applications that can be expanded much further.

6.6 CONCLUSIONS

Sorption heat storage has the potential of very compact storage of heat, as well as very low losses after charging of the storage. Several materials are under investigation, such as hygroscopic NaOH solutions, zeolites, and salt hydrates. Also, different system types are under development, in which the distinction between open systems and closed systems is most prominent. The most challenging aspect presently is to find economically sound applications for sorption heat storage and to develop optimized systems for these applications, both in terms of material selection and also in terms of performance and auxiliary energy use.

REFERENCES

Bales, C., 2006. Solar cooling and storage with the thermo-chemical accumulator. In: Eurosun 2006.

De Boer, R., Haije, W.G., Veldhuis, J.B.J., Smeding, S.F., 2004. Solid sorption cooling with integrated storage: the SWEAT prototype. In: Proceedings of HPC Conference, Cyprus.

De Boer, R., Smeding, S., Zondag, H.A., Krol, G., 2014. Development of a prototype system for seasonal solar heat storage using an open sorption process. In: Eurotherm Seminar 99.

Ferchaud, C.J., Scherpenborg, R.A.A., Scapino, L., Veldhuis, J.B.J., Zondag, H.A., De Boer, R., 2014. Performance overview of different salt hydrates as thermochemical materials for seasonal solar heat storage in a residential environment. In: Eurotherm Seminar 99.

Fumey, B., Weber, R., Gantenbein, P., Daguenet-Frick, X., Williamson, T., Dorer, V., 2014. Parameters effecting economy and heat density of sorption heat storage systems. In: Eurtherm Seminar 99.

Hongois, S., Kuznik, F., Stevens, Ph., Roux, J.J., Radulescu, M., Beaupaire, E., 2010. Thermochemical storage using composite materials: from the material to the system. In: Eurosun 2010.

Jänchen, J., Ackermann, D., Stach, H., Brösicke, W., 2004. Studies of the water adsorption on Zeolites and modified mesoporous materials for seasonal storage of solar heat. Sol. Energy 76, 339–344.

Kariya, J., Kato, Y., 2014. Enhancement reaction performance of chemical heat storage material mixed with expanded graphite for CaO/H2O chemical heat pump. In: Eurotherm Seminar 99.

Kato, Y., Harada, N., Yoshizawa, Y., 1998. Kinetic feasibility of a chemical heat pump for heat utilization of high-temperature processes. Appl. Therm. Eng. 19, 239–254.

Kerskes, H., Sommer, K., Müller-Steinhagen, H., 2007. Monosorp—Integrales Konzept für solarthermischen Gebäudeheizung mit Sorptionswärmespeicher, report ITW.

Kerskes, H., Mette, B., Bertsch, F., Asenbeck, S., Drück, H., 2011. Development of a Thermo-Chemical Energy Storage for Solar Thermal Applications. In: ISES Solar World Congress 2011.

Krönauer, A., 2013. Mobile thermische Energiespeicher zur Nutzung industrieller Abwärme, presentation. In: Berliner Energietage 2013.

Laing, D., Wörner, A., 2013. Wärme speichern mit Salz und Kalk—Faszination der Hochtemperatur-Speicherung. In: Berliner Energietage 2013.

Lan, S., Zondag, H.A., Rindt, C.C.M., 2014. Kinetic study of $Li_2SO_4 \cdot H_2O$ dehydration using microscopy and modeling. In: IHTC-15, Japan.

Marias, F., Tanguy, G., Wyttenbach, J., Rouge, S., Papillon, Ph., 2011. Thermochemical storage: first results of pilot storage with moist air. In: ISES 2011.

Mauran, S., Lahmidi, H., Goetz, V., 2008. Solar heating and cooling by a thermochemical process. First experiments of a prototype storing 60 kW h by a solid/gas reaction. Sol. Energy 82 (2008), 623–636.

Mette, B., Kerskes, H., Drück, H., 2012. New high efficient regeneration process for thermochemical energy stores. In: Innostock, 2012.

N'Tsoukpoue, K.E., Le Pierrès, N., Luo, L., 2010. Theoretical investigation of a long-term solar energy storage based on $LiBr/H_2O$ absorption cycle. In: Eurosun, 2010.

Quinnell, J., Davidson, J., Burch, J., 2010. Liquid calcium chloride solar storage: concept and analysis. In: ASME 2010 4th International Conference on Energy Sustainability.

Restuccia, G., et al., 1999. Performance of sorption systems using new selective water vapour sorbents. In: Proceedings ISHPC, Munich.

Shkatulov, A., Aristov, Y., 2014. Salt-doped Mg(OH)2 and Ca(OH)2 as candidates for middle-temperature heat storage. In: Eurotherm Seminar 99.

Van Essen, V.M., Cot Gores, J., Bleijendaal, L.P.J., Zondag, H.A., Schuitema, R., Van Helden, W.G.J., 2009. Characterisation of salt hydrates for compact seasonal thermochemical storage. In: Effstock 2009.

Weber, R., Dorer, V., 2008. Long-term heat storage with NaOH. Vacuum 82 (7), 708–716.

Zondag, H.A., Kikkert, B., Smeding, S., De Boer, R., Bakker, M., 2013. Prototype thermochemical heat storage with open reactor system. Appl. Energy 109, 360–365.

Chapter 7

Energetic Complementarity with Hydropower and the Possibility of Storage in Batteries and Water Reservoirs

Alexandre Beluco[1], Paulo Kroeff de Souza[1], Flávio Pohlmann Livi[2] and Johann Caux[3]

[1]*Instituto de Pesquisas Hidráulicas (IPH), Universidade Federal do Rio Grande do Sul (UFRGS), Porto Alegre, Brazil*

[2]*Instituto Mensura, Porto Alegre, Brazil*

[3]*École d'Ingénieurs de l'Energie, l'Eau et l'Environnement (Ense3), Institut Polytechnique de Grenoble (INP), Grenoble, France*

Chapter Outline

7.1 INTRODUCTION

Human societies always saw their development strongly tied to energy supplies and the mastering of the processes to exploit them. The growing ability of societies to control and use nature phenomena and resources to improve living conditions, safety, and comfort is always associated to the growth of energy use. Even the quest for more efficient use of energy, in general, just reduces the rate of growth of consumption. This situation today threatens the environment and so poses an immense challenge to humankind as a whole.

Energy, water resources, commodities production, scientific knowledge, and technological expertise are the basis of the development of modern industrial and postindustrial societies.

The recent economic developments implying large-scale industries and inducing large urban concentrations tend to require fairly concentrated energy supplies. Large interconnected systems have been successful in providing these supplies based on thermoelectric plants or hydroelectric generation, wherever there is potential for the latter.

This panorama, however, exerts excessive pressure on primary resources. Fossil fuels are finite and are being consumed at too high rates. A major part of the hydroelectric potentials is already put to use, and most of the remaining potentials are either far away from urban centers or present serious environmental restrictions.

The successive oil crises of the last decades served as incentive for the development of alternative renewable energy sources with lesser environmental impact. Hydroelectric energy has long been an option for micro- and small-scale energy generation. Even in pre-electric times, the water wheel was extensively used as a prime mover. A similar situation is that of wind power, used since ancient times to drive ships and boats, then in windmills and now in wind turbine generators. The direct conversion of solar energy has been a more modern achievement, with heat exchangers and photovoltaic cells. Hydroelectric, wind, and photovoltaic energy sources have reached some technical and economic maturity, becoming the main alternatives in hand.

In specific areas where availability is adequate, geothermal energy is in use. Other types of sources are being studied and shall increase contribution to interconnected systems in great quantities and concentrations such as the oceanic wave, stream, and tide energies.

The main difficulty with systems extensively based in renewable resources is the matching of availability and demand profiles, which are usually quite different. Consumers are too used to burning fossil fuels whenever energy is required. This comfort, available for some centuries now, has deeply marked the relationship between people and energy.

Renewable resources, however, are usually characterized by nonconstant availability profiles. It is usually unlikely that the resource will be available when required, and it is not reasonable nor even feasible to try to concentrate demand when the resource availability is greater.

One way to face source scarcity periods while keeping energy costs down is using two or more different energy sources in hybrid systems, at the expense of higher initial investments. This allows the use of, say, two sources with smaller conversion/storage devices in the system than would be necessary if only one of the sources was used, while still improving supply failure avoidance.

In the last few years, many feasibility studies have been concentrating on systems based on wind and photovoltaic power, with the possible support of diesel-driven generators. Hydropower is frequently considered for support in these types of systems, due to its usually more regular availability. In these systems, energy storage is always fundamental to warrant demand satisfaction during scarcity periods.

A characteristic worth considering is the possible complementarity of availability among the sources considered. A hybrid system based on two different sources may profit from their complementary availability throughout a period of one year. In the season when one source is scarcely available, the other would be mainly in charge of the supply, while in the other half-year the situation would be reversed. It is possible to devise adequate ways of sizing the powers of the conversion devices and the capacities of the energy storage equipment for use under this new conception.

This notion of complementarity opens new perspectives for ideas related to the design of hybrid systems. Given some site or region, the complementarity potentials of the existing resources are usually measurable. Complementarity data are sort of an information support for energy resource managers, useful for the ranking of energy generation projects and/or devising an energy supply strategy for a region. On a lesser scale, for independent entrepreneurs, complementarity may be related to circumstantial opportunities arising from the availability of energy resources or specific equipment, or even from the needs or creativity of the concerned engineer or investor.

This chapter studies energy-complementarity by describing types, establishing the definition of perfect complementarity between two types of sources and presenting a method for the analysis of the performance of a hybrid system. This analysis seeks to evaluate performance as a function of the different degrees of complementarity between the contemplated sources. Some comments on the effect of energy accumulators on the performance of this type of system are included. This chapter also presents some practical cases where feasibility was evaluated for real systems based on partially complementary resources.

7.2   ENERGETIC COMPLEMENTARITY

Complementarity can be seen as the property of one or more energy sources to complement each other's production over a certain space and/or over time intervals. At this point, a better understanding of energetic complementarity and how to evaluate the different degrees of complementarity in time or space is required.

Space-complementarity may exist when the energy availabilities of two or more types of sources complement themselves within a certain region. An example (McVeigh, 1977) is the complementarity between solar and wind resources over the territory of Great Britain, which is scarcely exploited due to the small amounts of energy available.

Time-complementarity may exist when the energy availability of two or more types of energy resources present periods of availability that are interlaced over time in the same region. An example is the complementarity in time between hydro and solar resources throughout the State of Rio Grande do Sul in southern Brazil, characterized by Beluco et al. (2008a,b).

Complementarity may also exist when the availability of energy of only one type of source is considered in different parts of a vast region and over time. As an example, the availability of hydraulic energy over the Brazilian territory (Damázio et al., 1997) may be mentioned, as this was one of the main reasons for interconnecting the south-southeast and the north-northeast electrical energy supply systems. This is an example of complementarity in time and space.

Different energy availabilities may be considered perfectly complementary if the variation of the availability values presents equal periodicity and their respective maximum and minimum values occur in time intervals half a period apart (out of phase). Furthermore, the average values of the availabilities should be the same, as well as the relation between the respective maximum and minimum values of availability of the two sources.

The amounts of energy supplied by generators in a hybrid system are considered complementary in time, even when the availability values are not perfectly complementary. The availabilities may present an imperfect complementarity, and the generators may be sized to supply energy values having the same average throughout the period. Obviously, this is possible only if the available energy is greater than the energy in demand.

Figure 7.1 presents two sinusoidal curves, showing a complementary instance, which will be considered perfect for the purposes of the present work. The two curves present hypothetical availabilities of two sources of energy, expressed in terms of energy or power throughout a year.

Both curves present periods of one year, average values equal to 1, maximum values equal to 1.2, and minimum values equal to 0.8. The minimum of the first curve occurs at 0.75 of the year, whereas the one of the second curve is at 0.25 of the year, or 0.5 year out of phase.

The complementarity between these curves is considered perfect because the minima and the maxima are 0.5 year out of phase, the difference between the maximum and the minimum is 0.5 in both cases, and the average values are equal.

The functions depicted in the figure may represent energy availabilities of two sources or the energy or power supplied by two generators. These functions will be used for the setup of a complementarity index, but they do not adequately describe, in detail, the behavior of solar sources, for example,

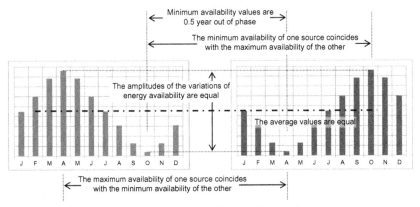

FIGURE 7.1   Two sinusoidal curves showing perfect complementarity.

because these present daily variations have an effect on the detailed behavior of hybrid systems.

The need to evaluate how much two availability functions, which are not perfectly complementary, differ from the situation depicted in Figure 7.1 (considered as a reference) naturally leads to the creation of numerical dimensionless indexes considering the aforementioned characteristics. Indexes varying from 0 to 1 are suggested to evaluate the complementarity in time between, for instance, hydraulic and solar energy availabilities.

7.3   EVALUATION OF COMPLEMENTARITY IN TIME

Among the types of complementarity previously discussed, this section is dedicated to complementarity in time. The article of Beluco et al. (2008a) proposed a dimensionless index evaluating different degrees of complementarity. An index created in this way should be seen as a tool that can be molded according to the decisions to be taken.

The complementarity index, 6, necessarily involves time and is designed to express the degree of complementarity between two energy sources. It is defined according to Equation (1) and includes the evaluation of three components: the phase difference between the energy availability values of the two sources, the relationship between the two average availability values, and the relationship between the amplitudes of variation of the availability functions.

$$\kappa = \kappa_t \, \kappa_e \, \kappa_a \qquad (1)$$

In this equation, κ_t is the partial time-complementarity index reflecting the phase difference, κ_e is the partial energy-complementarity index describing the complementarity between the average values of energy availability, and κ_a is the partial index of amplitude-complementarity taking care of the relation between amplitudes of the variation of the energy availability of the sources.

The partial index κ_t is defined by Equation (2) and measures the effect of the time interval between the minimum values of availability of the two sources of energy. If this interval is exactly half the period, the index will be equal to 1. If the interval is null (i.e., if the availability minima coincide), the index will be equal to 0. Intermediate values are linearly related.

$$\kappa_t = \frac{|d_h - d_s|}{\sqrt{|D_h - d_h||D_s - d_s|}} \tag{2}$$

In this equation, D_h is the number of the day of the maximum value of hydraulic energy availability, d_h is the number of the day of the minimum value of hydraulic energy availability, D_s is the number of the day of the maximum value of solar energy availability, and d_s is the number of the day of the minimum value of solar energy availability. This expression may be rewritten as Equation (3), supposing that the respective $(D-d)$ differences are equal to half a year, which is more practical for estimates.

$$\kappa_t = \frac{|d_h - d_s|}{\sqrt{180 \times 180}} = \frac{|d_h - d_s|}{180} \tag{3}$$

The partial energy-complementarity index κ_e is defined by Equation (4). It evaluates the relationship between the average values of the availability functions. If the average values are equal, the index should equal 1. If those values are different, the index should be smaller and tend to 0 as differences increase. Intermediate values of difference are linearly related to the index.

$$\kappa_e = 1 - \sqrt{\left(\frac{E_h - E_s}{E_h + E_s}\right)^2} \tag{4}$$

In this equation, E_h is the total of the hydraulic energy over the year and E_s is the total solar energy over the same period.

Alternatively, an expression for κ_e may be developed from a coefficient e as defined by Equation (5). This coefficient varies between 0 and 2, being equal to 1 when energies E_h and E_s are equal. When E_h is much bigger than E_s, e tends to 2, whereas if E_h is much smaller than E_s, e tends to 0.

$$e = \frac{2}{1 + \dfrac{E_h}{E_s}} \tag{5}$$

The index κ_e, however, should be $\kappa_e = 0$ if $e = 0$ or $e = 2$ and $\kappa_e = 1$ when $e = 1$. This is obtained by Equation (6), which is equivalent to Equation (4).

$$\kappa_e = 1 - \sqrt{(1 - e)^2} \tag{6}$$

The partial amplitude-complementarity index κ_a is defined by Equation (7) and evaluates the relationship between the values of the differences of the maxima to the minima of the two energy availability functions. If those differences are equal, the index shall be equal to 1. Otherwise, the index shall fall from 1, tending to 0.

$$\kappa_a = \begin{cases} \left[1 - \dfrac{(\delta_h - \delta_s)^2}{(1 - \delta_s)^2} \right], & \text{for } \delta_h \leq \delta_s \\[4mm] \left[\dfrac{(1 - \delta)^2}{(1 - \delta_s)^2 + (\delta_h - \delta_s)^2} \right], & \text{for } \delta_h \geq \delta_s \end{cases} \tag{7}$$

This index is obtained from a suitable manipulation between the δ_h and δ_s, the differences between the respective maximum and minimum values of the energy availabilities. These differences are obtained from Equation (8), where E_{dmax} is the maximum daily energy availability value in a year, E_{dmin} is the minimum daily energy availability value in a year, and E_{dc} is the annual average daily energy consumption.

$$\delta = 1 + \frac{E_{d\,max} - E_{d\,min}}{E_{dc}} \tag{8}$$

This index was created to include the difference between the maximum and the minimum energy availability of the sources in the complementarity evaluation. If one of the sources has no available energy throughout the period of interest, it is impossible to consider it for complementarity purposes. If the two sources have the same difference between maximum and minimum availability, they are ideally amplitude complementary and the index should be equal to 1. In the intermediate cases where the differences are unequal, the index should express it by values between 0 and 1 for less-than-ideal complementarity.

The δ_s difference is always greater than 1 and may be considered constant over quite an extensive area because it represents the availability of solar radiation. The δ_h difference presents a minimum value of 1, in which case the complementarity index should be null. On the other hand, δ_h may present rather large values as compared to δ_s, and in these cases, the index should decay from the maximum value as the δ_h value increases.

For the purpose of the development of the expressions for the κ_a index, described in the next few paragraphs, the value $\delta_s = 2$ was considered. The resulting curve shall be continuous and smooth, it shall present a zero slope for $\delta_h = \delta_s$, and it shall be very nearly symmetrical in the proximity of its peak. It is also desirable to obtain an expression for the quick calculation of the κ_a index. The desired behavior for this function is shown in Figure 7.2.

There is no obvious mathematical expression for a function with those characteristics. The development of an expression for the part of the curve to the right of the maximum can be based on the Agnesi curve, adapted so that its maximum value is 1 and corresponds to an abscissa equal to δ_s, which leads to Equation (9).

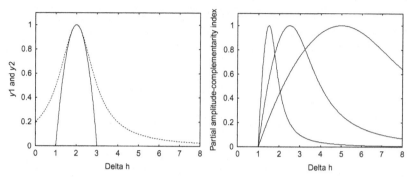

FIGURE 7.2 (a) Superposition of the functions y_1 in solid line and y_2 in dotted line. (b) κ_a index curves for δ_s values of 1.5, 2.5, and 5.

$$y_2 = \frac{1}{1 + (\delta_h - \delta_s)^2} \tag{9}$$

The Agnesi curve presents the desired symmetry characteristics with respect to its maximum for the abscissa δ_h equal to δ_s. The final form for the right side of the function is obtained by slightly modifying Equation (9) to smooth the slopes near the maximum, to ease its connection to the expression to be used for the left part of the curve in Equation (10).

$$y_2 = \frac{(1 - \delta_s)^2}{(1 - \delta_s)^2 + (\delta_h - \delta_s)^2} \tag{10}$$

For the left part of the curve, a quadratic function seems adequate in view of the three contour conditions. The y_1 function, with δ_h as the independent variable, centered in the value of δ_s, is presented in Equation (11).

$$y_1 = 1 - \frac{(\delta_h - \delta_s)^2}{(1 - \delta_s)^2} \tag{11}$$

Figure 7.2(a) shows the superposition of the curves y_1 in solid line and y_2 in dotted line. Figure 7.2(b) shows three different curves' amplitude-complementarity index, for δ_s values of 1.5, 2.5, and 5.

7.4 COMPLEMENTARITY BETWEEN SOLAR ENERGY AND HYDROPOWER

This section is dedicated to complementarity between solar and hydroelectric energies, and discusses the determination of the proposed index in real conditions.

The calculation of complementarity indexes for large regions can provide an important tool for managers of energy resources. For example, the application

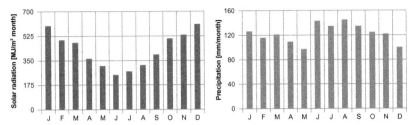

FIGURE 7.3 Average monthly precipitation (left) and mean monthly incident solar radiation (right) for a climatological station in Taquari, Rio Grande do Sul.

of funds for electrification of a region can be decided based on the existing complementarity, because generation systems based on complementary resources may have less installed power.

The climatic conditions in southern Brazil promote the notion that higher rainfall occurs during periods of reduced insolation and, conversely, that lower rainfall occurs during periods of intense sunlight. This notion is the origin of this study, starting with the possible complementarity between solar energy and hydropower. It seeks understanding of its effects on hybrid generation systems based on these resources.

In a real situation, the calculation of the indexes may be carried out through functions adjusted to the monthly average data by the least squares method. Figure 7.3 shows monthly average precipitation data and monthly average of daily incident solar radiation data over a flat horizontal surface, as supplied by a weather station in the state of Rio Grande do Sul (FEPAGRO, 1989).

The solar energy availability in Figure 7.3 is quite similar to the idealized curve of daily availability shown in Figure 7.1. The situation is different for the water availability, a second lesser peak appearing near the position of the valley of the idealized curve.

The monthly average precipitation does not adequately represent the water availability. However, in smaller river basins, flow variations present small phase lag with respect to precipitation variations, and the amplitudes of the variations are also closely related. The water availability curve is lower in the January-May interval. If January is used, because the minimum solar availability is in July, the partial time-complementarity index is equal to 1.00. The partial energy-complementarity index and the partial index of amplitude-complementarity evaluations are not possible using these numbers. However, an expedited evaluation is made to allow the elaboration of the maps that follow.

The determination of the indexes from river flow data may be done in the same way, using adjusted curves over the monthly average data. The use of daily data for water and solar availability would certainly produce better results!

The value of δ_s used for this work is 1.2860. The value for δ_h may vary considerably as a function of the water availability in the plant area and the installed capacity of the hydroelectric generator. Moreover, considering its definition,

δ_h will never be less than 1, which corresponds to the hydroelectric generator always turning out the same power and the corresponding flow presenting a short recurrence time, which implies a high frequency.

The calculation of these indexes for the entire state of Rio Grande do Sul in southern Brazil allows mapping. Beluco et al. (2008a) shows these maps for the partial indexes and the overall index. The systems based on the availabilities with the higher time-complementarity (which can be evaluated from the indexes proposed here) tend to present fewer consumer demand satisfaction failures. So, from this point of view, the complementarity map shows the "best" areas for the use of hydraulic and solar complementary availability.

Figure 7.4 shows the map of complementarity in time for the State of Rio Grande do Sul. This is one of the maps presented by Beluco et al. (2008a). The information that appears in Figure 7.3 was determined from a database of the Agricultural Research Foundation of Rio Grande do Sul (FEPAGRO, 1989, 2000). From this same database, data from 15 weather stations (including Taquari) were used for preparing the map that appears in Figure 7.4.

In areas where the time-complementarity index is less satisfactory, the complementarity condition can be helped by the use of water reservoirs, with the effect of improving the phase difference between the minima to near a half-year, which is the ideal value. Regions with a smaller κ_t will require bigger accumulation volumes for the system to approach the performance levels corresponding

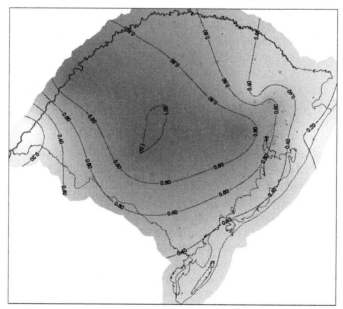

FIGURE 7.4 Time-complementarity between hydraulic and solar energy availability in the state of Rio Grande do Sul, southern Brazil. The different values of complementarity appear in shades of gray, with black corresponding to full complementarity, and white to no complementarity.

to a full complementarity. In this sense, a photovoltaic hydroelectric system can be supported in a more natural way with stored energy in water reservoirs and batteries.

In the same way, small differences in the amplitudes of variation of the hydraulic and solar availabilities also lead to better performing systems. The map of complementarity between amplitudes shows the best-valued areas for the amplitude-complementarity index that are, consequently, the best areas (from this point of view) for consumer demand satisfaction.

In areas with lower values for this index, the generating systems' performance can also be improved by the use of reservoirs. This can help improve the amplitude variation of the hydraulic availability approaching it to the value corresponding to the perfect complementarity, which is the same as the variation of the solar availability. And, as previously mentioned, regions with a smaller index value will need bigger reservoirs for a given level of system performance.

The complementarity maps presented by Beluco et al. (2008a) indicate that about 58% of the area of the state has a κ_t greater than 0.60, corresponding to more than 72 days of phase difference between the minima of the two availability curves, the best values occurring from the southeast to the northwest border. It can also be seen that nearly 50% of the state area presents κ_t values greater than 0.80, that is, phase differences of nearly 50% of the cycle of the hydraulic and solar energies' availability, the extreme values appearing in the center of the state. On the other hand, 4.67% of the state area presents κ values greater than 70%. It is, however, clearly visible that the most adequate area from the time-complementarity point of view is generally not the most adequate area from the amplitude-complementarity point of view. The final complementarity index κ will consequently show intermediate values in these areas, whereas the best values will be near the northwest border.

A precise and reliable evaluation of complementarity in a given place should be based on flow data across a given river section and on incident solar radiation data taken daily. The insight on the hybrid system performance gained through computer simulations and experimental studies allows the evaluation of the effects of the complementarity on the system parameters and may justify deeper and more comprehensive studies for the characterization of the complementarity.

7.5 HYDRO-PV HYBRID SYSTEMS BASED ON COMPLEMENTARY ENERGY RESOURCES

A hybrid hydro-photovoltaic (PV) plant is a generation system based on a hydroelectric plant and a photovoltaic plant operating together to satisfy the demand of an ensemble of consumer loads. The complementarity between the energy sources may then be beneficial for the sizing and the operation of this type of system. The International Centre for Application of Solar Energy

(CASE, 1997) describes the installation of a hydro-PV system in Ban Khun Pae, north of Thailand, comprising 60 120 Wp modules, a 90 kW synchronous generator, batteries with 110 kWh storage capacity, a 40 kW inverter, and a 56 kW diesel backup unit. The hydroelectric plant was preexistent but insufficient to supply the needs of the population of about 90 dwellings. The cost of this system was approximately $170,000. The use of PV modules was considered for supplementing the existing hydroelectric system, but the reference does not discuss the idea of complementarity.

The state of charge of the various storage devices may be managed according to the chosen operation strategy, which shall include the consideration of the effects of complementarity between the energy sources. The chosen operation strategy may contemplate the action of the battery bank to attenuate electromechanical transients in the turbine-generator system, because time constants of electrical transients are much smaller than those of mechanical transients. Van Dijk et al. (1991) suggests an increase in battery life by using 90% of full capacity as the recharge level.

If a direct current (DC) hydroelectric generator is used or the generator current is rectified and the consumer loads accept DC, only a DC bus is necessary. However, because alternating current (AC) hydroelectric generators are more common, two busses are necessary, as depicted in Figure 7.5. In this type of system, the current from the hydro generator is tied to the PV generator, the batteries, and the loads through the DC bus. A regulator may manage the connection and disconnection of generators and loads to the batteries as a function of the state of charge of the latter and define the current level to be supplied by the hydro generator. This configuration is quite usual for systems based on renewable sources, as it is a simple alternative for conditioning the power supplied by induction generators. For this configuration, in case the hydro plant is equipped with induction generators, the rectification avoids the need for a tight

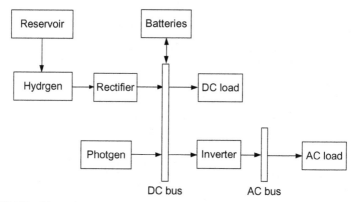

FIGURE 7.5 Schematic diagram of a "series" hydro-PV hybrid power plant with DC and AC buses, with water reservoir and battery energy storage.

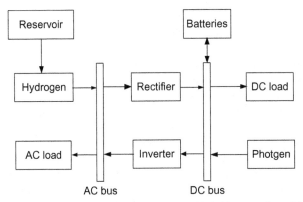

FIGURE 7.6 Schematic diagram of a "parallel" hydro-PV hybrid power plant with DC and AC buses, with water reservoir and battery energy storage.

frequency control. The voltage regulation may be simplified or even omitted, with the excitation being handled simply by a capacitor.

Those systems are suitable for demands of a few kilowatts. A 24 V DC bus for 1-2 kW will handle currents of the order of 40-80 A. On the other hand, a battery bank at 127 V, allowing greater powers with smaller currents and comparable storage capacity, using automotive batteries would have an exaggerated size. Automotive batteries may be replaced with a technical and perhaps economic advantage by lead-acid stationary batteries. Automotive batteries are supposed to work under a certain level of vibration.

The system depicted in Figure 7.6 is much more complex, involving the parallel operation of an AC generator and a power-converting device. In such a system, the AC bus frequency may be easier to regulate if it is the frequency of the hydro generator when this machine is in use. If a synchronous generator is used, the inverter can operate at an approximately constant frequency, equal to that of the generator. If an asynchronous machine is used, the inverter may operate on a variable frequency following the instantaneous values of the generator frequency. In such a configuration, it may be possible that the control of the inverter frequency, in certain conditions, may allow some control of the generator frequency and, consequently, of the bus frequency.

This type of system does not have power limitations. The DC bus currents will certainly be the highest currents supplied by the PV modules. In the AC bus, the voltages can be higher and defined by the hydro generator.

If the hydro and solar energy availabilities are time-complementary, the storage devices may be sized to optimize the operation of the hybrid system. The water reservoir may be sized for a shorter period, implying lower costs and lesser environmental impacts if the hydroelectric generator operates with a photovoltaic generator and a battery bank, provided there is a complementarity between the two energy sources.

It is possible to define an operational strategy for a hydro-PV system as follows: (1) use all the energy provided by the PV generator, polarized by the battery bank voltage (which is determined by its state of charge); (2) operate the battery bank in intermediate charge states (when it has a relatively good capacity for either charge or discharge) to attenuate electromechanical transients due to variations in the sources or in the load; (3) store energy in the water reservoir and in the battery bank, their state of charge being managed as a function of the energy availabilities and energy demand, while considering the possible complementarity of the sources; and (4) operate the hydro generator to address the consumer loads not attended by the PV generator, while considering the states of charge of the storage devices.

7.6 A METHOD OF ANALYSIS

It is necessary to know how a hybrid system based on complementary resources will have its performance influenced by the complementarity of the energy resources. Beluco et al. (2012) propose a method of analysis that has as its starting point the idealization of the available energy. A hybrid system could be simulated with real data, but the results would be subject to the climatological effects. The simulation using idealized data, such as that appearing in Figure 7.1, allows an accurate assessment of the effects of complementarity on the performance of the simulated system.

In this sense, the results of simulations performed with idealized data can be viewed as an upper limit of the simulated system performance. This theoretical limit would be setting an unattainable performance in real conditions. This "maximum performance" would be obtained with the maximum energy that could be available from a certain source if this source were insensitive to random events typical of the environment and insensitive to periods of extreme energy scarcity or extreme availability.

A theoretical limit of performance may be determined from a simulation of a power plant for a year of operation based on functions of energy availability synthesized specifically to this end. These functions may be synthetic series of data obtained from real energy availability data or they can be mathematical functions (such as a sinusoid) with parameters extracted from the series of real data. A further step in the analysis can obviously be the comparison of these results with those obtained with the actual data.

The performance of an energy generation plant may be evaluated by the total of the supply failure times (SFT) observed during a given elapsed time, or by a supply failure index, F, defined as the SFT divided by the corresponding elapsed time. A failure is defined as any situation where the total power supplied by the energy conversion devices is less than the power required by the consumer loads. A failure occurs also if, besides insufficient power, there is no stored energy available for consumption. The elapsed time considered in the analysis may be given in days, weeks, months, seasons, or years.

The failure index is adopted as the performance evaluating parameter and theoretical limit of performance for given operational conditions and energy availabilities. The performance of a power plant will be better as the failure index is smaller and the power or energy required by consumers and effectively supplied by the system is bigger. The upper limit for the failure index is the null index, meaning that all the demand was satisfied, and it may even happen that excessive energy is being offered.

In this chapter, this method was applied in order to obtain some information regarding the effects of complementarity on the performance of hybrid systems based on complementary energy resources. The final part of this section will describe how simulations based on this method were performed. After that, the next section will present some results for the theoretical limit of performance. Then, the following section will present the results obtained for some real cases, simulated with the Homer software.

The hybrid system under study was simulated on the computer with the objective of investigating its performance and subsidizing the establishment of technical bases for a sizing methodology, with special regard for the various degrees of complementarity between the availabilities of the energy resources used. The simulations were made with routines written with MATLAB (MathWorks, 1999). The program simulates the system depicted in Figure 7.6. The values of 0.68 and 0.175 are adopted for the efficiencies of the hydro and PV generators, respectively.

The input to the simulation program consists of information on the yearlong energy availability of the sources used and of the sizes of generators and storage devices. The output is a report on the supply of the demand of the load during the simulation period. Basically, the total of the SFT is evaluated.

The program adopts the per unit (pu) in which the variables are made dimensionless with respect to an outstanding value (typically a reference design value or a nominal value), and it simulates the plant behavior during a two-year time interval without consideration for transient effects. The output file contains hourly records of the system components during the second simulated year, the calculations being repeated every five minutes, with the objective of reducing the influence of the storage devices' initial states on the final results of the simulation.

The total instantaneous power available to the hydroelectric generator is simulated in a theoretical way, using a sinusoidal function. The energy availability represented by this function depends on the location chosen for the hydroelectric plant and may present considerable variation in amplitude and in annual energy availability.

The total instantaneous power available to the PV generator is simulated by three functions. The first one describes the availability during each day; the second one describes the maximum daily availability throughout the year, and the third one, the variation of the duration of the day throughout the year. These functions are not based on the typical equations developed for the description of solar irradiance such as, for example, those presented by Duffie and Beckmann (1980), but they are similar to the function describing hydro availability throughout the year.

The maximum and minimum average values for the total instantaneous power available to the photovoltaic generator were obtained from monthly data as published by FEPAGRO (1989). The pu values of 0.67783 (corresponding to an incident solar radiation of 677.83 W/m^2) and of 0.37108 (corresponding to an incident solar radiation of 371.08 W/m^2) are considered for the respective maximum and minimum instantaneous power available to the PV generator.

The demand profile, p_L, considered in the simulations corresponds to a constant energy consumption throughout the day. Other profiles may be used, such as profiles with a concentration of consumption during the day or with a consumption peak at the end of the afternoon.

Figure 7.7 shows the result of the simulation of a system with total complementarity and the battery bank sized for two days' storage. The horizontal axis is time in years, while in the vertical axis various functions are represented: to the left are the scales for the powers supplied by the generators and consumed by the load; to the right is the state of charge of the batteries (SOC).

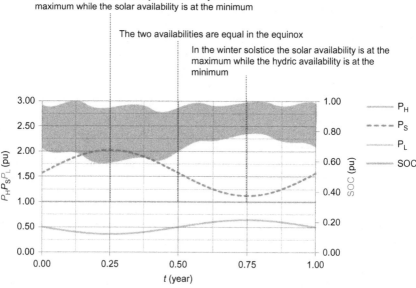

FIGURE 7.7 Results of the simulation of a hybrid hydro-PV system with $\pi_{ad}=1.00$, $\kappa_t=1.00$, $\kappa_e=1.00$ [$\pi_{sh}=1.00$], $\kappa_a=1.00$ [$\pi_{Mm}=1.00$; $\delta_s=1.2860$; $\delta_h=\delta_s$], and $\kappa=1.00$, with $a_p=17.29$. The system is without reservoir, the battery bank having a 24-hour capacity with discharge to 40% and recharge to 100% of full capacity and demand profile being constant. Conventions: SOC: state of charge of batteries; P_H: power generated by the hydro generator; $P_{p\ max\ d}$: maximum daily power from the PV generator; P_L: power supplied to the loads.

The line above, forming a spot shows the SOC. The gray line below represents the hydropower supplied (p_H), the black dashed stands for the maximum daily PV power (p_{Pmaxd}), and the most thin gray is the power consumed by the loads (p_L). It can be seen that there are no shutoffs of the hydroelectric generator and no supply failures.

In the simulated system, the annual energy available for consumption is equal to the annual energy demand, and also the system has a complementarity index equal to 1 ($\kappa = 1.00$), which means perfect complementarity. Consequently, the time-complementarity index, κ_t, the energy-complementarity index, κ_e, and the amplitude-complementarity index, κ_a, are also equal to 1 ($\kappa_t = 1.00$, $\kappa_e = 1.00$, $\kappa_a = 1.00$).

If, for example, the system has loads consuming a constant 600 W of power, the hydroelectric generator should have a minimum 534 W of installed power and the PV generator should use 16.40 m^2 of collector area. The daily consumption is then 600 Wd or 14,400 Wh, and the battery bank should have a capacity for double that (28,800 Wh). This storage capacity may be obtained by 8 150 Ah 24 V batteries.

In Figure 7.7, it can be seen that the maximum hydroelectric power supplied by the generator is equal to 0.89, which, for the example given, means 534 W. It is interesting to note that this power is 11% smaller than the power required by the loads. In small systems such as the one considered in the example, such an advantage is difficult to materialize due to the standardization of equipment commercially available.

The dimensionless area of the PV modules, a_P, is 17.69, as calculated by Equation (12) and indicated in the caption for Figure 7.7, where A_P is the area of the PV modules and $P_{L\,max}$ is the reference value for power. In the example, this area is equivalent to 16.40 m^2 and may be obtained with 16-18 commercial 1 m^2 modules on the market. This area will provide about 1500 Wp of power, where the maximum PV availability is around 2.50 pu.

$$a_P = \frac{A_P}{\left(\dfrac{P_{L\,max}}{1000}\right)} \tag{12}$$

The system of Figure 7.7 was considered the "starting point" for the simulations, a reference for easy comparisons with the results presented. The choice was made based on comparisons among different proportions between the total annual energy available for consumption and the total annual energy required by the loads.

The hybrid system under study, sketched in Figure 7.6, was simulated on the computer with computational routines written with MATLAB, as described in Equation (5). The simulations adopt the pu system to describe the physical quantities, where each quantity is divided by a reference value associated with the dimensions of each component of the system or with the reference values of available energy or demand.

The operational strategy: (1) Use all the energy provided by the PV generator; (2) operate the batteries in intermediate charge states; (3) store energy in the water reservoir and in the battery bank, their state of charge being managed as a function of the energy availability and demand while considering the possible energetic complementarity; and (4) operate the hydro generator to address the consumer loads not attended by the PV generator, while considering the states of charge of the storage devices.

The operation of the hydro generator is obviously different from the usual operation of a hydroelectric power plant, but it should be clear that a system like the one proposed in this work should be focused on better utilization of available power and simplicity of control.

The theoretical limit of performance (Beluco et al., 2013b) is defined as the upper limit for the performance of a power plant corresponding to the performance obtained with the "maximum availability" of the energy resource. This would be the maximum energy that could be available from a certain source if it were insensitive to random events typical of the environment and insensitive to periods of extreme energy scarcity or extreme availability.

The maximum and minimum average values for the total instantaneous available power to the PV generator are obtained from monthly data for the region considered in this study. The pu values of 0.67783 (corresponding to an incident solar radiation of 677.83 W/m^2) and of 0.37108 (corresponding to an incident solar radiation of 371.08 W/m^2) are considered for the respective maximum and minimum instantaneous power available to the PV generator.

The simulated system is designed from the proportions between the total annual available and demanded energies (π_{ad}); the total annual available hydro and solar energies (π_{sh}); and the difference between the maximum and the minimum availabilities (π_{Mm}), as defined by Equation (13). For a constant demand profile, the power required by the loads is considered equal to 1. The available hydro and PV powers are defined as a function of the aforementioned proportions and present sinusoidal variations throughout the year.

$$\pi_{ad} = \frac{E_s + E_h}{E_d} \quad \pi_{sh} = \frac{E_s}{E_h} \quad \pi_{ad} = \frac{\delta_s - 1}{\delta_h - 1} \tag{13}$$

For a constant demand profile, the power required by the loads is considered equal to 1. The available hydro and PV powers are defined as functions of the aforementioned proportions and present sinusoidal variations throughout the year.

Figure 7.7 shows the result of the simulation of a system with total complementarity and the battery bank sized for two days' storage. The horizontal axis is time in years, while in the vertical axis various functions are represented: to the left are the scales for the powers; to the right is the SOC. The line above, forming a spot shows the SOC; the gray line below, the hydroelectric power supplied (p_H); the black dashed line, the maximum daily photovoltaic power (p_P); and the most thin gray, the power consumed by the loads (p_L). It can be seen that there are no shutoffs of the hydroelectric generator and no supply failures.

This system was considered the "starting point" for the simulations, a pattern for easy comparisons with other results. The choice was made after a survey of different combinations of components and their importance for achieving the expected results.

The results presented by Beluco et al. (2012) showed a slight difference in relation to the system of Figure 7.7. Results of that paper clearly show the energy stored in the batteries, with two peaks over a year. The two simulated systems show small differences and, because of that, this result did not appear in Figure 7.7. This behavior of the batteries is being studied, and the results characterizing the transition from the system will be published soon.

The study of the effect of different degrees of complementarity on the performance of the energy system considered is based on the effects of the variations in the energy availability and of different combinations of installed power of hydro and of photovoltaic generator sets, in addition to the effects of the range of water discharges turning the hydroelectric generator.

7.7   EFFECTS OF COMPLEMENTARITY IN TIME

Complementarity can be verified over space and time, as previously mentioned. It is a topic that still has many details to be discovered. This section presents some basic results about the influence of complementarity in time on the performance of a hydroelectric-photovoltaic hybrid system, according to the proposed method of analysis.

7.7.1   Effects of Different Degrees of Complementarity in Time

The system for which simulation results are depicted in Figure 7.7 has a perfect complementarity, with total annual energy available being equal to the total annual energy required by the loads, the battery bank having a 24-hour capacity with discharge to 40% and recharge to 100% of full capacity, and without a reservoir. It does not show any demand satisfaction failures.

The use of idealized data brings out various aspects of the performance of hydro-PV systems running on complementary energy resources. This procedure helps define the applicability of the system and supports the sizing of its components. The configuration of this system is not necessarily a "target configuration" for sizing purposes, but it is the most valuable source of information on the performance of this type of system.

When complementarity is less than perfect, variations in performance as functions of partial indexes being less than 1.00 will be observable. As time-complementarity index κ_t varies, characterizing poorer complementarity, performance degrades and failures may appear, as shown below. A worse time-complementarity, as seen in Figure 7.8, is associated to higher failure

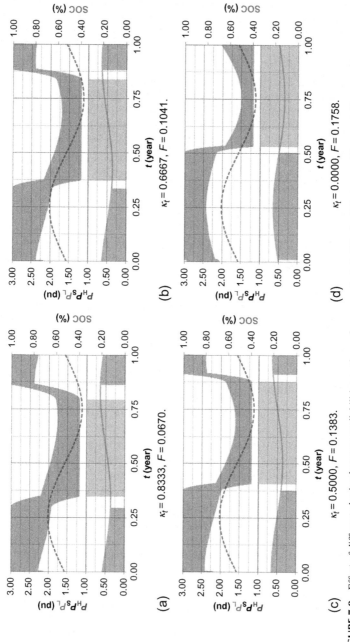

FIGURE 7.8 Effect of different phasing between availabilities on the performance of a hybrid hydro-PV system with $\pi_{ad} = 1.00$, $\kappa_e = 1.00$ [$\pi_{sh} = 1.00$], $\kappa_a = 1.00$ [$\pi_{Mm} = 1.00$], $a_p = 17.29$, without reservoir, with 24-hour-capacity battery bank discharged to 40% and recharged to 100% of maximum capacity and constant demand profile. Phase lags of (a) 150 days [$\kappa_t = 0.8333$], (b) 120 days [$\kappa_t = 0.6667$], (c) 90 days [$\kappa_t = 0.5000$], and (d) no lag [$\kappa_t = 0.000$]. Conventions similar to those adopted in Figure 7.7.

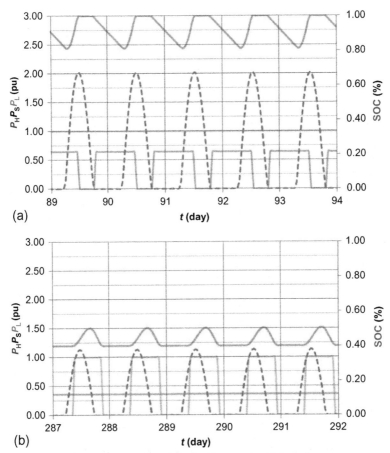

FIGURE 7.9   (a) Details of the period where the disconnection of the hydro generator occurs in Figure 7.8(d) system from the 89th to the 93rd day. (b) Details of the period where supply failures are observed in Figure 7.8(d) system from the 287th to the 291st day. Conventions similar to those adopted in Figure 7.7.

indexes. Values of κ_t equal to 0.8333, 0.6667, 0.5000, and 0.0000, respectively, resulted in F values of 0.0670, 0.1041, 0.1383, and 0.1758.

Figure 7.9(a) details the period between the 89th and 93rd days, showing the shutdown of the hydroelectric generator during the period of intense sunlight. Figure 7.9(b) details the period between the 287th and 291st days, showing the failure in meeting demand due to energy shortages.

These results explain the dark gray, light gray and medium gray spots that appear in the graphs, corresponding respectively to the hourly fluctuations of the SOC, the power supply failures and periods of surplus energy, and consequent shutdown of the hydroelectric plant.

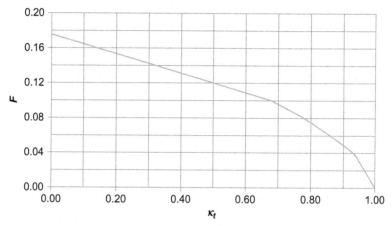

FIGURE 7.10 F as a function of κ_t for results of Figure 7.8.

A reservoir may artificially improve time-complementarity, having the effect of delaying the low hydro availability period. The value of the failure index may imply an initial value for the required reservoir volume, and is strongly related to the conditions of the plant location.

The reservoir design may aim at improving complementarity characteristics, causing imperfect natural complementarity to improve, and to approach a situation leading to a better operation of the hydro-PV hybrid system.

Figure 7.10 summarizes the results of Figure 7.8 and shows the variation of failure index F with the complementarity in time.

7.7.2 Effects of Different Degrees of Energy-Complementarity

The effects of variations in the energy-complementarity index are depicted in Figure 7.11, the graphs being obtained in the same way as in Figure 7.8. The values of the ratio π_{sh}, the index κ_e, and the results for F are listed in the legend of each result. Four results are shown, corresponding to the values of 1.43, 1.67, 0.70, and 0.60 for the ratio π_{sh}, and the values of 0.8275 and 0.7500 for the κ_e index. For reference, the system of Figure 7.7 has unity values for π_{sh} and κ_e.

The values of π_{sh} imply different average energy availabilities. For the first two results, the photovoltaic energy supplied tends to increase, while the supplied hydropower tends to decrease. Because the solar availability presents great daily and yearly variations, when its relative importance increases, so does the failure index. These variations cause corresponding, however subtly amplified variations in the battery-stored energy.

In the other two results, the available photovoltaic power tends to decrease and the hydroelectric power tends to increase. As a growing fraction of the

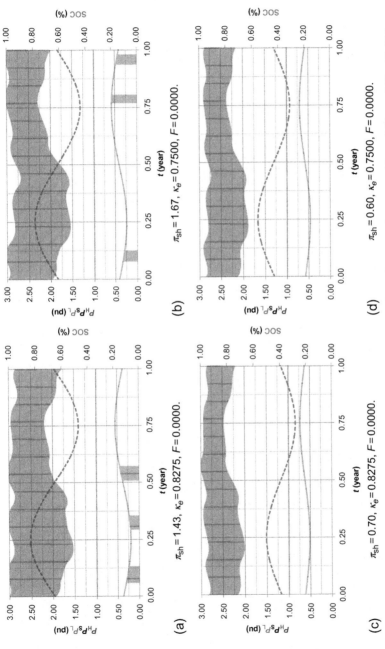

FIGURE 7.11 Effect of the ratio π_{sh} and the energy-complementarity index κ_e on the performance of a system with $\kappa_i = 1.00$, $\kappa_a = 1.00$, with 24-hour-capacity battery bank discharged to 40% and recharged to 100% of maximum capacity. Conventions similar to those adopted in Figure 7.7.

system energy tends to be "firm" in view of its hydro origin, the failure index tends to decrease. One can see from (c) to (d)—chiefly in (d)—an increase in the battery-stored energy in the left side of the graph, as a consequence of the higher hydro availability.

It should, however, be emphasized that the variations of F are quite small, and no changes are seen in the battery-stored energy. In general, the variations in the energy-complementarity index cause variations in the fractions of the total energy supplied by each of the two generators. As the total available energy throughout the year remains constant, no important modifications occur either in the battery behavior or in the system performance.

It can be seen, however, that a bigger contribution by the hydroelectric power component leads to a failure-free system, while a bigger contribution of the energy of photovoltaic modules tends to require a bigger battery bank to avoid the failures. To some extent, this limits the applicability of hybrid hydroelectric photovoltaic generating plants to situations in which the hydroelectric power availability is insufficient.

These results show how a bigger contribution of the energy of photovoltaic modules implies a bigger energy storage capacity. The utilization of a reservoir may become easier to justify in the cases of near perfect time-complementarity, as the suitable effect of the energy-complementarity may be artificially obtained by the accumulation of hydraulic energy. A small increase in storage capacity may be sufficient to accommodate daily and seasonal variations in solar availability.

The small influence of the variations of energy-complementarity on the failure index is quite evident. Failures remain equal to 0.00%, while the failures due to variations in the time-complementarity index may be higher than 15%, and variations due to amplitude-complementarity may be higher than 20%, as seen in the next subsection.

The medium gray spots in Figure 7.11(a) and (b) are due to increased solar availability when the ratio π_{sh} is greater than 1, as previously discussed. However, the failure index is less sensitive to the variations of the partial complementarity index of energy κ_e, and even these small surpluses of energy are always less than 1% of the year.

The graphs in Figure 7.11 also show the connection of the daily variation of the SOC with solar availability. In (c) and (d), when the proportion π_{sh} is less than 1 and water availability becomes larger, this daily variation is reduced.

7.7.3 Effects of Different Degrees of Amplitude-Complementarity

The effect of variations in the amplitude-complementarity index are shown in Figures 7.12 and 7.13, summarized in Figure 7.14. The graphs were made the same way as Figure 7.7. The values of ratio π_{Mm} and index κ_a and the results for F appear listed in the legend of each result. Eight results are shown for the amplitude-complementarity index values of 0.99, 0.95, 0.85, 0.75, corresponding respectively to 1.11, 1.29, 1.63, 1.99 of ratio π_{Mm}, Figure 7.12, and

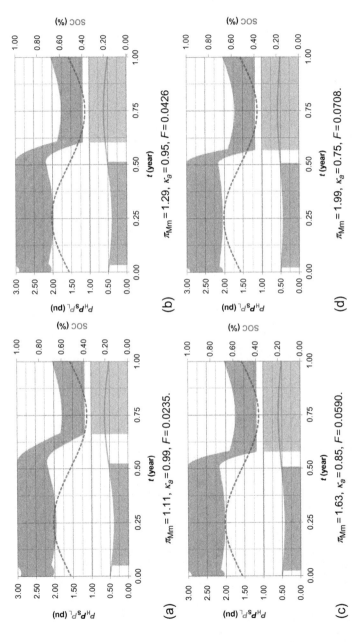

FIGURE 7.12 Effect of the ratio π_{Mm} and the amplitude-complementarity index κ_a on the performance of a system with $\kappa_t = 1.00$, $\kappa_e = 1.00$, with 24-hour-capacity battery bank discharged to 40% and recharged to 100% of maximum capacity. Conventions similar to those adopted in Figure 7.7.

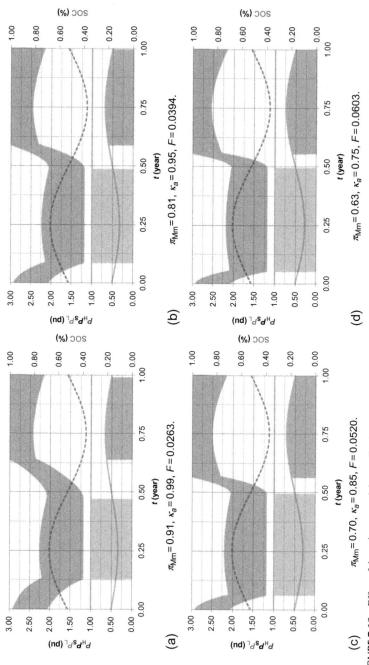

FIGURE 7.13 Effect of the ratio π_{Mm} and the amplitude-complementarity index κ_a on the performance of a system with $\kappa_i = 1.00$, $\kappa_e = 1.00$, with 24-hour-capacity battery bank discharged to 40% and recharged to 100% of maximum capacity. Conventions similar to those adopted in Figure 7.7.

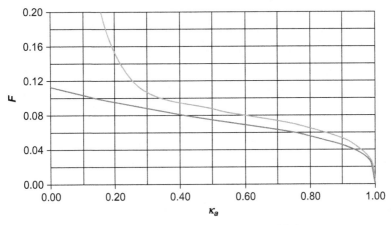

FIGURE 7.14 F as a function of κ_a for results of Figures 7.12 and 7.13.

corresponding respectively to 0.91, 0.81, 0.70, 0.63 of ratio π_{Mm}, Figure 7.13. For reference, Figure 7.7 system has unity values for the ratio π_{Mm} and the amplitude-complementarity index κ_a.

Importantly, the simulated systems with different ranges of energy availability, as reflected in different values for the ratio π_{Mm}, do not involve differences in the gap between the maximum availability (because $\kappa_t = 1.00$) or differences in energy availability (because $\kappa_e = 1.00$).

Different values of π_{Mm} do not affect the average hydro availability but represent the difference between its maximum and minimum values. In the four initially shown results (Figure 7.12), there is a reduction of hydro availability during the second semester, concentrating a growing number of failures in this period as well as keeping a correspondingly lower battery charge. As part of the hydro energy is transferred to the second semester, frequent reductions or shutdowns of the hydroelectric power will be observed.

In the subsequent results (Figure 7.13), there is an increase in hydro energy in the second semester, causing the reduction or shutdown of the hydropower, while failures and lower battery storage levels occur in the first semester. It is interesting to observe how the solar availability modulates the supplied hydroelectric power and the variations in battery-stored energy.

Failures occur in these cases always at the end of the period of sunlight, by depletion of energy stored in batteries. Similarly, the hydroelectric generator is disconnected at the end of the period of sunlight, because the batteries are fully charged.

It shall be emphasized that the failure levels are intermediate between the values caused by time-complementarity variations (which go as high as 15% in the worst situations) and those due to energy-complementarity variations (always zero).

Failures in Figure 7.12 are somewhat larger than in Figure 7.13, and continue growing as complementarity decreases. The greatest failures occur in the second semester because the minimum availability of hydro energy occurs in the first half, and this depletes the energy stored in batteries.

A reservoir with capacity for one week is sufficient to reduce failures to nearly zero in systems with more than 75% complementarity. Systems with less than 75% complementarity will require larger reservoirs. The system of Figure 7.13(d) requires a water reservoir with a capacity for 15 days to avoid failures.

The use of real data, with daily variations of availability instead of the idealized values considered in the simulations, will change these results, requiring more storage capacity to avoid failures. The difference is the result of climatological and meteorological phenomena.

A failure-reducing behavior as a function of the κ_a is to be expected as this index approaches its maximum value. Figure 7.14 shows the behavior of the failure index F due to variation in the partial index of amplitude-complementarity.

Systems with π_{Mm} less than unity show failure slightly higher than those with π_{Mm} above unity. If a smaller battery bank is used, the curve is shifted up. The reverse is verified for bigger batteries. The shape of these lines is obviously related to the definition of the amplitude-complementarity index and must undergo changes with real data.

The use of idealized data brings out various aspects of the performance of hybrid hydroelectric-photovoltaic systems running on complementary energy resources. This procedure helps define the applicability of the system and supports the sizing of its components.

A better time-complementarity is associated to smaller failure indexes. A reservoir may artificially improve time-complementarity, having the effect of delaying the low hydro availability period. The value of the failure index may supply an initial value for the required water reservoir volume, and is strongly related to the conditions of the plant location.

A bigger contribution of hydroelectric origin leads to failure reduction, while an increased photovoltaic contribution implies a higher energy accumulation capacity. This result points to the applicability of hydroelectric-photovoltaic systems when hydro availability is insufficient to supply consumer demand.

A better amplitude-complementarity is associated to lower values of the failure index. A water reservoir may artificially improve amplitude-complementarity, transferring water from one semester to be used in the other, approaching the annual hydro availability distribution tending to the unity value for the index. But this, obviously, involves high costs.

The water reservoir design may aim at improving complementarity characteristics, causing imperfect natural complementarity to improve and to approach a situation leading to a better operation of the hybrid system.

Operation strategy and the design of the cheapest components should consider the investment objective of the reservoir, seeking the best possible use for the stored water.

All these comments are valid for other types of energy storage. This work considers the ways that are most obviously associated with hydropower plants and with photovoltaic modules.

7.8   SOME REAL HYBRID SYSTEMS WITH PARTIAL COMPLEMENTARITY

It is interesting to compare the results of the previous section with results of real systems. In this section, some results of case studies are presented, for which feasibility studies were performed with Homer [Lambert et al., 2005; Lilienthal et al., 2004]. These case studies were originally seeking optimal combinations of the components of the systems to meet consumers. Here, issues related to complementarity are discussed. Three case studies will be reviewed.

Beluco et al. (2013a) sought an optimal solution for power generation during peak hours for a recycled paper factory in southern Brazil. The load is 328 kW of installed capacity, with daily consumption of 975 kWh. The study considered two possible alternatives for power generation during peak hours, both with diesel support and possible connection to the grid.

One alternative considered the restoration of an old plant with 240 kW. The other considered a plant to be undertaken with a power of 427 kW. The two alternatives considered installing PV modules on the roof of the factory building with an area of about 1 ha. The company already had three diesel generators, two with 150 kW and one with 30 kW.

For the first alternative, the index of complementarity in time is equal to 0.98, the complementary energy is equal to 0.82, and complementarity between amplitudes is 0.78. For the second alternative, the index of complementarity in time is equal to 0.92, the complementary energy is equal to 0.74, and complementarity between amplitudes is 0.89.

Observing the annual availabilities of the two resources, the high values of complementarity in time, 0.98 and 0.92, respectively, are evident. Figure 7.15 shows the water availability for the two alternatives, and Figure 7.16 shows the solar availability. Figure 7.17 shows the behavior of the system, obtained by simulations with Homer, for a period of one year.

This system does not use batteries, inserting excess energy in the grid and recovering energy from there when needed. This simulation was performed with low cost for diesel oil, and so the diesel generator set is fired over the hydroelectric plant at precisely the time of year with greater water availability. Building a reservoir, if possible, could allow the use of complementarity in favor of the system, possibly eliminating the need for diesel support.

Silva et al. (2012) studied the case of a middle-class family intending to install a PV-wind-diesel hybrid system that gives them independence from

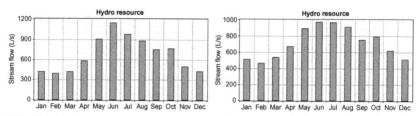

FIGURE 7.15 Water availability for the two alternatives of the case study presented by Beluco et al. (2013a).

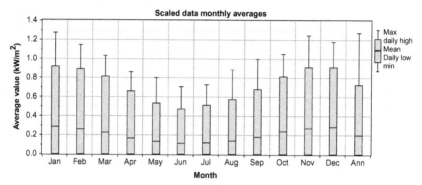

FIGURE 7.16 Solar availability for the case study presented by Beluco et al. (2013a).

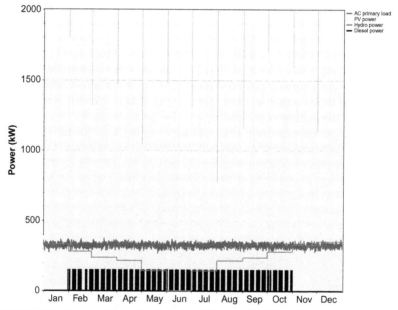

FIGURE 7.17 Behavior of the system studied by Beluco et al. (2013a), with complementarity in time equal to 0.98, as simulated by Homer (HomerEnergy, 2009). The figure shows energy consumption (medium gray), solar power (very light gray), hydropower (light gray), and diesel support (in black).

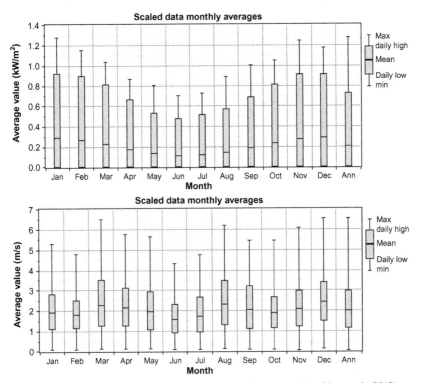

FIGURE 7.18 Hydro and solar resources for the case study presented by Silva et al. (2012).

the grid. The study focuses on the costs to be achieved by the PV modules so that solutions can count on photovoltaics. The house is located in the city of Porto Alegre in southern Brazil, has a total area of about 160 square meters, and has a consumer load of 2.6 kW peak, with a daily consumption of 10 kWh. Water heating represents a load of 0.625 kW peak and a daily consumption of 15 kWh.

The system has a small wind turbine, a diesel gen set, PV modules, and the possibility of connection to the grid, plus a small battery bank. Figure 7.18 shows the solar and wind energy availability for this hybrid system, indicating null complementarity. Figure 7.19 shows the results provided by Homer to this system for a period of one year.

The comparison between solar and wind availability is always difficult because of the variability of wind. Periods of minimum availability of the two resources match in June. If there was complementarity, optimal combinations of components that include PV modules would be obtained with current pricing.

These results are different from the previous ones because climatological effects and a variable demand profile make it difficult to analyze the results. The lack of complementarity and the inconstancy of the winds lead to daily and constant use of the batteries. The power drawn by consumers has many variations, but due to the demand profile, it is not constant. This system does not fail.

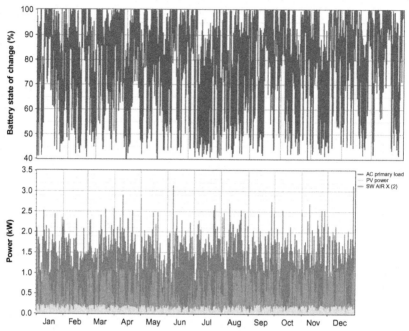

FIGURE 7.19 Behavior of the system studied by Silva et al. (2012), with no complementarity in time. Top: The SOC of the batteries. Bottom: Energy consumption (medium gray), PV (very light gray), and wind power (light gray).

7.9 EFFECTS OF ENERGY STORAGE

The simulations presented here consider only the energy storage in batteries. The study of hydro-PV hybrid systems opens the way for the possibility of accumulation of water in a reservoir. Several results show excess water in one semester and lack of energy in the other semester. A reservoir sized for one or two weeks of consumer demand can solve the problem.

This topic is still the subject of research projects, and the use of a reservoir in combination with batteries can be a complex problem of optimization of resources available for energy storage. The costs of a reservoir and environmental concerns should make its use impractical, but its consideration in studies of complementarity can shed new light on the analysis of its viability.

Just as an example, the system shown in Figure 7.8(d) was simulated again, now with a reservoir included in the system: a water reservoir with a capacity equivalent to 20 days of energy consumption. Its use was linked to low stored energy in the batteries and the imminent failure of the power supply to consumers.

Figure 7.20 shows the results of this simulation. The left graph was assembled the same as the previous ones, indicating the general state of charge of the two storage devices. The graph to the right details the SOC in medium gray line

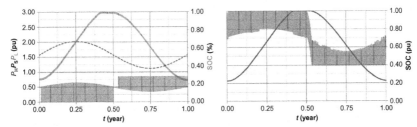

FIGURE 7.20   Simulation of Figure 7.8(d) system with a reservoir with a capacity to supply electricity to consumers for 20 days. Conventions are similar to those adopted in Figure 7.7, with the exception of the state of charge of storage devices. On the left, the general state of charge is shown in medium gray. On the right, the state of charge of the batteries is shown in medium gray and the state of charge of the reservoir is shown in dark gray.

forming a more widespread spot, and the state of charge of the reservoir in dark gray. The main result is that faults are reduced to zero, as was to be expected!

As the capacity of the reservoir is much larger than the capacity of the batteries, it is observed that the first is almost not influenced by the second. The reservoir is filled during the months of highest water availability and emptied when necessary. It is also observed that the hydroelectric plant always operates at full capacity in the second half.

ACKNOWLEDGMENTS

This work was developed as part of research activities on renewable energy developed at the Instituto de Pesquisas Hidráulicas (IPH), Universidade Federal do Rio Grande do Sul (UFRGS). This work was partly funded by CNPq.

REFERENCES

Beluco, A., Souza, P.K., Krenzinger, A., 2008a. A dimensionless index evaluating the time complementarity between solar and hydraulic energies. Renew. Energy 33, 2157–2165.

Beluco, A., Souza, P.K., Krenzinger, A., 2008b. PV hydro hybrid systems (in portuguese). IEEE Lat. Am. Trans. 6 (7), 626–636.

Beluco, A., Souza, P.K., Krenzinger, A., 2012. A method to evaluate the effect of complementarity in time between hydro and solar energy on the performance of hybrid hydro PV generating plants. Renew. Energy 45, 24–30.

Beluco, A., Colvara, C.P., Teixeira, L.E., Beluco, A., 2013a. Feasibility study for power generation during peak hours with a hybrid system in a recycled paper mill. Comput. Water Energy Environ. Eng. 2 (2), 43–53.

Beluco, A., Souza, P.K., Krenzinger, A., 2013b. Influence of different degrees of complementarity of solar and hydro energy availability on the performance of hybrid hydro PV generating plants. Energy Power Eng. 5 (4), 332–342.

CASE, International Center for Application of Solar Energy, 1997. Project Summary Sheet n.5, Micro Hydro Solar Diesel Hybrid RAPS System, Perth, Australia.

Damázio, J.M., Costa, F.S., Ghirardi, A.O., 1997. Análise de complementaridades hidrológicas a nível continental na América do Sul. Rev. Bras. Recur. Hidr. 2 (2), 143–156.

Duffie, J.A., Beckmann, W.A., 1980. Solar Engineering of Thermal Processes. John Wiley & Sons, New York, USA.

FEPAGRO, Fundação Estadual de Pesquisas Agropecuárias, 1989. Agroclimatic Atlas of the State of Rio Grande do Sul (in portuguese), Porto Alegre, 3v.

FEPAGRO, Fundação Estadual de Pesquisas Agropecuárias, 2000. Data series used for preparation of 'Agroclimatic Atlas of the State of Rio Grande do Sul', personally obtained, unpublished data.

HomerEnergy, 2009. Software HOMER, version 2.68 beta, The Micropower Opyimization Model. Available at, www.homerenergy.com.

Lambert, T.W., Gilman, P., Lilienthal, P.D., 2005. Micropower system modeling with Homer. In: Farret, F.A., Simões, M.G. (Eds.), Integration of Alternative Sources of Energy. John Wiley & Sons. ISBN 0471712329, pp. 379–418.

Lilienthal, P.D., Lambert, T.W., Gilman, P., 2004. Computer modeling of renewable power systems. In: Cleveland, C.J. (Ed.), In: Encyclopedia of Energy, vol. 1. Elsevier, pp. 633–647, NREL Report CH-710-36771.

MathWorks (1999) MATLAB, version 5.3, 1 CD ROM.

McVeigh, J.C., 1977. Energia Solar. Cetop, Lisboa, Portugal.

Silva, J.S., Cardoso, A.R., Beluco, A, 2012. Consequences of reducing the costs of PV modules on a PV wind diesel hybrid system with limited sizing components. Int. J. Photoenergy 2012, 1–7. Article ID: 384153.

Van Dijk, V.A.P., Alsema, E.A., Albers, R.A.W., Degner, T., Gabler, H., Wiemken, E., 1991. Autonomous photovoltaic system with auxiliary source for battery charging. In: 10th European Photovoltaic Solar Energy Conference, Lisboa, Portugal, pp. 728–773.

Chapter 8

Revitalization of Hydro Energy: A New Approach for Storing Solar Energy

Z. Glasnovic and K. Margeta
University of Zagreb, Zagreb, Croatia

Chapter Outline

8.1 INTRODUCTION

Energy storage is crucial for increasing the share of energy from renewable energy source (RES) generators. This especially applies to energy from the sun and wind, which is characterized by intermittence.

Solar Energy Storage. http://dx.doi.org/10.1016/B978-0-12-409540-3.00008-6

189

However, the problem with the present energy storage technologies is that they can be stored for a relatively small amount of energy, and in that sense a relatively short period of time to balance intermittent energy from RES generators (i.e., hourly, daily up to a maximum weekly value).

Figure 8.1 shows energy storage system ratings (Margeta and Glasnovic, 2011), while logarithmic scale (b) is converted in linear scale (a) gives insight into the possibilities of storing energy by the different technologies. Pumped storage hydroelectricity (PSH) technology is the solely technology that can significantly increase the share of RES generators in electric power systems (EPS). But precisely because of these qualities (i.e., storing the largest amount of energy), it is possible to balance summer surpluses and winter shortages of solar energy (SE). Thus, the essential characteristic of PSH technology is its storage of large quantities of energy, which allows balancing on an annual basis when weather conditions repeat. In that sense, this technology is suitable for a significant increase in the participation of RES generators in EPS.

When it is revealed that this technology is very acceptable in economic terms (average price of 0.5 $/W; see Deane et al., 2010), the advantages compared to other storage technologies become more significant.

Conventional ways of using PSH technology from EPS include the fact that energy from power plants (mostly nuclear power) is stored in the PSH technology at night (pumping water from the lower to the upper reservoir) when the consumption of EPS is the lowest, or when there is surplus energy in power plants that has nowhere to go. The stored energy is then released to hydroelectric power plants during the day, when energy consumption in EPS is highest.

In that mode of operation of PSH, one tube is enough to connect the lower and upper reservoirs. During the night the pipe is used to pump water to the upper reservoir, and during the day the water is discharged through the tube to hydroelectric turbines and generators. The Powerhouse (PH) shown on Figure 8.2 (Anagnostopoulos and Papantonis, 2008), has two function (motor-pump pumped water from the lower to the upper reservoir and

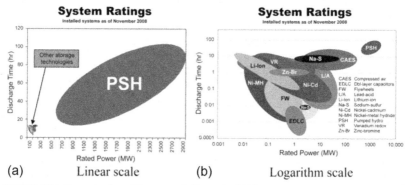

FIGURE 8.1 System ratings (Margeta and Glasnovic, 2011).

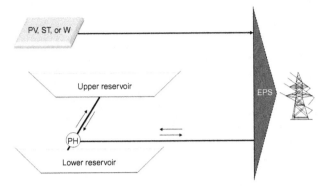

FIGURE 8.2 The conventional way of connecting PV, ST, and W generator with EPS.

turbine-generator produce electricity when water is released from the upper to the lower reservoir).

On the other hand, use of the current RES generators and hydro energy comes down to the fact that RES energy from photovoltaic (PV), solar thermal (ST) or wind (W) (as significant RES generators with intermittent energy) connects directly to the EPS. In this sense, RES generator sends energy to the EPS when it produces energy, while PSH sends energy to the EPS in periods when there is not enough energy from RES (Anagnostopoulos and Papantonis, 2007), Figure 8.2.

8.2 AN INNOVATIVE SOLUTION: INTEGRATION OF A SOLAR-HYDRO SYSTEM

In order to ensure the power supply from SE generators to consumers throughout the year, it is necessary to provide energy storage that can balance the seasonal surpluses and shortages of SE.

However, a different way of connecting PSH technology and generators is required. Unlike the parallel connection of RES and PSH shown in Figure 8.2, Figure 8.3 shows a series connection of SE and PSH, which are provided in the patent solutions of Glasnovic and Margeta (2009, 2013a,b) and paper of Glasnovic and Margeta (2011). In that compound, the energy from the SE generator is first submitted to PSH, where it is stored in the form of gravitational potential energy and supplied to the generators when it is necessary for consumers. This course seeks two PSH pipelines: one for pumping water (with pumped storage) to the upper reservoir, and the other for water discharge toward hydroelectric power plants.

This can ensure a continuous supply to consumers throughout the year. Thus, the new role of PSH consists in the fact that it can be so integrated with the generator as to ensure the continuity of supply to customers throughout the

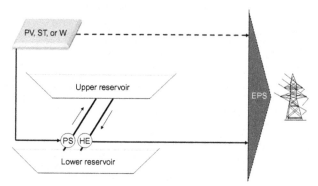

FIGURE 8.3 Old and new ways of using SE generator and PSH technology.

year, because this is the period in which the weather conditions are repetitive. This connection is very suitable for isolated consumers (on islands and other isolated locations); such a system allows their sustainable energy supply.

PSH technology in general is a natural way of storing energy because lakes already exist in nature. These reservoirs have relatively little impact on the environment in which they can and do fit nicely, and they can be multifunctional.

Unlike the conventional storage hydroelectric power plant being built along the river and having a significant negative impact on the environment, PSH can be built practically in all locations where there is a height difference.

If the integrated SE-PSH systems are connected to the EPS (series connection of SE-PSH-EPS), there are no shocks to the EPS and in that case the SE-PSH system allows for management in EPS (energy that is stored in the upper reservoir of PSH may be through transmission lines used by the consumers to whom this energy at this time need). Direct connection RES generator with EPS is only for cases of surplus energy from RES generators, which is then logically delivered RES energy directly to the EPS. PSH systems can be opened and closed, mainly depending on the size of the PSH (smaller systems can be closed, while the larger ones are generally open).

Particularly interesting systems make PSH with seawater because they do not need the lower reservoir (which is basically introduced by the sea), but only the upper reservoir. However, although the hydro energy is essentially *mature technology*, PSH with seawater is new technology. In fact, the only PSH that has ever been constructed is in Okinawa, Japan, and it started to operate in 1998 (Fujihara et al., 1998). The main problems of this technology are reflected in corrosion protection (turbine, for which austenite stainless steel is used; penstock, for which fiberglass reinforcement plastic is used, etc.). In addition, as the PSH technology interacts directly with the sea, it is obvious that in such technologies, one must take into account the reduction of the influence on marine life.

Because of all these problems related to PSH with seawater, there is great potential for further development of these technologies and their integration with SE generators. This is particularly true because of the growing migration of the world population toward the coasts (Black et al., 2011).

8.3 GEOSYNTHETICS AS A PREREQUISITE FOR HYDRO ENERGY STORAGE

Significant improvement of PSH technology and its wider use, especially in semipermeable and permeable soils, could be to ensure the use of impermeable geomembranes and other geosynthetics. This technology could be a very important element in achieving a total renewable electricity scenario (Glasnovic and Margeta, 2011).

Geosynthetics can also be used for *rain harvesting* technology. In this way, they can significantly reduce the required size of the SE-PSH system because all rainfall comes into the upper reservoir, and after that converts into energy.

In this sense, geomembrane technology could enable the revitalization of well-known hydro technology. Namely, hydro energy storage (PSH technology) can be built virtually everywhere with a height difference and with extremely low environmental impact.

8.3.1 Application of Geomembranes in Water/Energy Storage of PSH Technology

A very important polymeric material, geomembranes, which belong to the group of geosynthetics (e.g., geotextiles, geogrids, geonets, geosynthetic clay liners, geopipe, geofoam, geocomposites), can significantly contribute to the further development of PSH technology.

In this subsection, the most commonly used geomembranes are described, and their characteristics for technological applications of water storage are analyzed. For these purposes, geomembranes are used in combination with other geosynthetics to ensure the safety and durability of water reservoirs, and thus ensure sustainable energy supply.

Table 8.1 presents some types of polymeric materials used for the technological preparation and production of geomembranes; such a final product can be later used for various purposes (Koerner, 2005).

Geomembranes are made from relatively thin continuous polymeric sheets, and their primary function is liquid and/or gas containment.

Due to good physical-chemical-mechanical and environmental characteristics of the geomembranes, their production and sales are rapidly increasing worldwide, as shown in Figure 8.4 (Polyethylene, 2014).

Geomembranes have been used in some of the largest hydroelectric and dam reservoir projects in the world. Hydroelectric power plants can require very large reservoirs. Water reservoirs are often lined with geomembranes to avoid

TABLE 8.1 Important Type of Geomembranes in Current Use

Type of Geomembranes	Acronym
High density polyethylene	HDPE
Linear low density polyethylene (or Low density polyethylene)	LLDPE (LDPE)
Ethylene propylene diene terpolymer—nonreinforced (or Ethylene propylene diene terpolymer—reinforced)	EPDM (EPDM-R)
Polyvinyl chloride	PVC
Flexible polypropylene—nonreinforced (or Flexible polypropylene—reinforced)	fPP (R-fPP)
Chlorosulfonated polyethylene—reinforced	R-CSPE

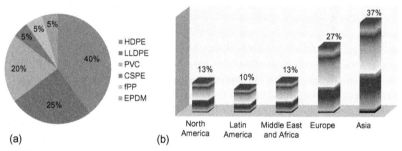

FIGURE 8.4 Geomembranes: (a) on the world market; (b) global HDPE, LLDPE, and LDPE demand in 2013.

leakage and erosion (Figure 8.5). High-density polyethylene (HDPE) and linear low-density polyethylene (LLDPE) geomembranes are effective and affordable options in reservoirs for water storage.

Geomembranes can provide an impermeable barrier for water retention that is superior to concrete or asphaltic lining in both cost and lower permeability. They are relatively impermeable in the range 1×10^{-12} to 1×10^{-15} m/s, which is three to six orders of magnitude lower than the typical clay liner.

Critically important for the proper design of geomembrane-lined side slopes of reservoir is the soil-to-geomembrane shear strength. Geomembranes can also be made from the impregnation of geotextiles with asphalt, elastomer, or polymer sprays, or as multilayered bitumen geocomposites.

The basic use of geomembranes should be their retention of liquids/solids over a long time, and they need to have high endurance properties. One of the best additives for plastics to resist ultraviolet (UV) degradation is carbon black. Black geomembranes will get very hot when exposed to sunlight because

FIGURE 8.5   Application of geosynthetics for water storage of PSH technology.

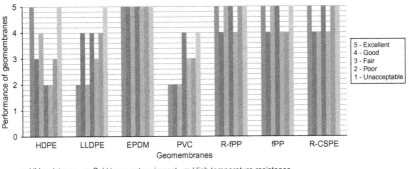

FIGURE 8.6   The most important performance of geomembranes.

they absorb sunlight radiation. They will have more expansion/contraction due to temperature variations. New, innovative technologies enable the implementation of the different geomembranes (*white geomembranes*). In addition to having esthetic advantages, the reflective liner will also reflect solar radiation and reduce the temperature of the geomembrane.

Another innovative solution is an *electrically conductive geomembrane* (in which electrically reactive carbon is incorporated) that, once installed, can be rigorously tested for spark ruptures and leaking through the geomembrane.

For any technological application, geomembranes must have certain properties in order to satisfy demanding requirements for their safe use (Figure 8.6).

In order to maintain their function, geomembranes' test methods and standards (e.g., ASTM and ISO) are available or are being developed by standards-settings organizations around the world.

There are important factors to consider when deciding which type of geomembrane to use in a water storage application. Table 8.2 presents the physical, chemical, mechanical, and other properties for the selected geomembranes.

The geomembrane properties are most involved with resistance or susceptibility to tear, puncture, and impact damage in thickness. For this reason, many agencies require a minimum under any circumstance (e.g., in the United States, minimum thickness is 0.5 mm; in Germany, for similar applications, minimum

TABLE 8.2 Properties of Different Types of Geosynthetics

Properties[a]	Type of Geosynthetics						
	HDPE	LLDPE	EPDM	PVC	fPP-R	fPP	CSPE-R
Physical							
Water tightness	✓✓✓✓	✓✓✓✓	✓✓✓✓	✓✓✓✓	✓✓✓✓	✓✓✓✓	✓✓✓✓
Flexibility	✓	✓✓✓	✓✓✓✓	✓✓✓✓	✓✓	✓✓✓	✓✓
Elasticity	✓	✓	✓✓✓✓	✓	✓	✓✓	✓✓
Tensile strength	✓✓✓✓	✓✓✓	✓✓	✓✓	✓✓✓✓	✓✓✓	✓✓✓✓
Permeability	✓✓✓✓	✓✓✓	✓✓✓	✓✓	✓✓✓	✓✓✓	✓✓✓
Chemical							
Chemical resistance	✓✓✓✓	✓✓✓	✓✓✓	✓✓	✓✓✓	✓✓✓	✓✓✓
Resistance to hydrocarbons	✓✓✓	✓✓	✓	✓✓	✓✓	✓✓	✓
Resistance to plasticizer extraction	✓✓✓✓	✓✓✓✓	✓✓✓✓	✓	✓✓✓✓	✓✓✓✓	✓✓✓✓
Root resistance	✓✓✓✓	✓✓✓✓	✓✓✓✓	✓✓✓	✓✓✓✓	✓✓✓	✓✓✓
Resistance to microbiological attack	✓✓✓✓	✓✓✓	✓✓✓✓	✓✓	✓✓✓✓	✓✓✓	✓✓✓✓
Mechanical							
Dimensional stability	✓	✓	✓✓✓✓	✓✓✓	✓✓✓✓	✓✓✓	✓✓✓✓
Multiaxial strain	✓	✓✓	✓✓✓✓	✓✓✓	✓✓	✓✓✓	✓✓
Resistance to settlements	✓✓	✓✓✓	✓✓✓✓	✓✓✓✓	✓✓	✓✓✓	✓✓

Seamability	✓✓	✓✓✓	✓✓✓✓	✓✓✓	✓✓✓✓	✓✓✓	✓✓✓
Seamability to cold temperatures	✓	✓✓✓✓	✓✓✓	✓	✓✓✓✓	✓	✓✓✓
Seam testing	✓✓✓✓	✓✓✓✓	✓✓✓	✓✓✓✓	✓✓✓✓	✓✓✓✓	✓✓✓
Puncture resistance	✓✓	✓✓✓	✓✓✓	✓✓✓	✓✓✓	✓✓✓	✓✓✓
Surface friction	✓	✓✓✓	✓✓✓	✓✓✓✓	✓✓✓✓	✓	✓✓✓
Slope stability	✓✓	✓✓✓	✓✓✓	✓✓✓✓	✓✓✓✓	✓✓✓✓	✓✓✓
Repairability	✓✓	✓✓✓	✓✓✓	✓✓✓✓	✓✓✓✓	✓✓✓	✓
Service life	✓✓✓✓	✓✓	✓✓✓✓	✓✓✓	✓✓✓✓	✓	✓✓✓✓
Other							
Environmental properties	✓✓✓✓	✓✓✓✓	✓✓✓✓	✓	✓✓✓✓	✓✓✓✓	✓✓✓
Easy of installation	✓✓	✓✓✓	✓✓✓	✓✓✓✓	✓✓✓	✓✓✓	✓✓✓
Details, design and installation	✓	✓✓✓	✓✓✓	✓✓✓	✓✓	✓✓✓	✓✓✓
Conformance to substrate	✓	✓✓✓	✓✓✓	✓✓✓✓	✓✓	✓✓✓	✓✓✓

HDPE, high density polyethylene; LLDPE, linear low density polyethylene; EPDM, ethylene propylene diene terpolymer; PVC, polyvinyl chloride; fPP-R, flexible polypropylene reinforced; fPP, flexible polypropylene; R-CSPE, chlorosulfonated polyethylene reinforced.

[a]✓✓✓✓ *Excellent,* ✓✓✓ *Good,* ✓✓ *Fair,* ✓ *Poor.*

thickness is 0.75 mm). The lifetime of geomembranes is very long—about 25-30 years. HDPE geomembranes can far outlast other engineering materials in comparable situations.

8.3.2 Reservoir Volume

Before selecting the geomembrane type, the desired liquid volume to be contained versus the available land area must be considered. Such calculations are geometric by nature and result in a required depth on the basis of assumed side slope angles. For a square or rectangular section with uniform side slope, the general equation for volume is

$$V = H{\cdot}L{\cdot}W - S{\cdot}H^2{\cdot}L - S{\cdot}H^2{\cdot}W - 2{\cdot}S^2{\cdot}H^3 \,(\text{m}^3) \qquad (8.1)$$

where

V = reservoir volume (m^3)
H = average height (i.e. depth) of the reservoir (m)
W = width at ground surface (m)
L = length at the ground surface (m)
S = slope ratio (horizontal to vertical).

8.3.3 Hydro Energy Storage

Hydro energy that is stored in the reservoir can be calculated according to

$$E_H = \rho{\cdot}g{\cdot}H_{TE}{\cdot}V\,(\text{kWh}) \qquad (8.2)$$

where V (m^3) is water volume in the upper reservoir, H_{TE} (m) is total elevation difference between the lower and upper water levels, g (m/s^2) is gravity acceleration, and ρ (kg/m^3) is water density.

8.4 CONCEPT INTEGRATION OF THE SE-PSH SYSTEM

A concept-integrated SE-PSH system that uses natural resources (i.e., sun, gravity, seawater (in the case of PSH with seawater), and precipitation) is shown in Figure 8.7.

As shown, SE in PV and ST generators is converted into electrical energy, which is supplied to the pump station (PS), by which the water is pumped from the lower to the upper reservoir, which serves as water/energy storage. In the same reservoir, *rain harvesting* technology brings water from rainfall.

When there is demand for energy, the water is discharged from the upper to the lower reservoir, and the hydroelectricity to generate electricity is delivered to a local or regional EPS.

In this way, the possibility of a continuous energy supply to local consumers (i.e., SE-PSH) may enable EPS to give energy continuously.

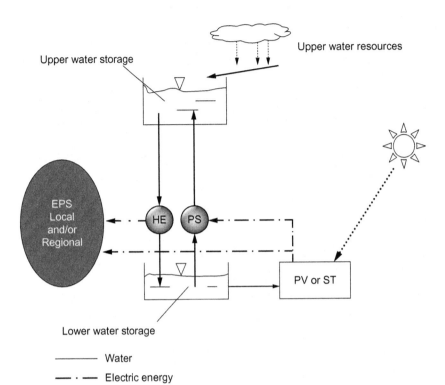

FIGURE 8.7   Concept of SE-PSH system.

8.5   OPTIMIZATION MODEL OF SE-PSH SYSTEM

It is particularly important to determine the required size of the SE-PSH system and it is possible to optimize a model that takes into account all relevant parameters of the system (local climate, terrain configuration, mode of energy consumption, etc.).

8.5.1   Water Balance

The upper reservoir basically represents daily and seasonal energy storage that can balance summer surpluses and winter shortages of SE. However, its charge and discharge dynamics (water volume change from one time period $i-1$ into the other i) are dependent on the inputs and outputs of water in the reservoir. Therefore, the dynamics for time stage i can be expressed by the water balance equation

$$V_{(i)} = V_{(i-1)} + V_{ART(SE)(i)} + V_{RH(i)} + V_{NAT(i)} + R_{(i)} - V_{EV(i)} - V_{HE(i)}, \quad (8.3)$$

providing that

$$V_{MIN} \leq V_{(i)} \leq V_{MAX}, \quad (8.4)$$

where $V_{(i)}$ and $V_{(i-1)}$ are water volumes in the upper reservoir in time steps i and $i-1$, respectively; $V_{NAT(i)}$ is the volume of natural watercourses flowing into the upper reservoir (rivers, precipitation, etc.); $V_{ART(SE)(i)}$ is the volume of artificial watercourses created by the pumping system (PS), which is generated by the energy from the solar generator; $R_{(i)}$ is the volume of rainfall falling into the upper reservoir; $V_{RH(i)}$ is the volume of *rain harvesting* technology that comes into the upper reservoir; $V_{EV(i)}$ is the volume of evaporation from the surface of the upper reservoir; $V_{HE(i)}$ is the volume of water discharged to the turbines to produce electric energy; and V_{MIN} and V_{MAX} are the minimum and maximum water volume in the upper reservoir, respectively.

8.5.2 Water Storage of PSH as Energy Storage of SE Generator

According to Glasnovic et al. (2013), interrelations between the value of the optimal nominal electric power of the SE generator and the working volume of the PSH upper reservoir, V_0, can be calculated from the following expression:

$$P_{el(NOM)} = -\psi \cdot Ln(V_0) + \psi^* \tag{8.5}$$

where Ψ and Ψ^* are parameters on the basis of location characteristics and technological features, and V_0 is the operating volume of the upper reservoir.

Therefore, after determining the optimal values of the nominal electric power of the ST generator, V_0 can be determined from the expression

$$V_0 = e^{\frac{\psi^* - P_{el(NOM)}}{\psi}}, \tag{8.6}$$

From the operating volume V_0 and reserve volume V_r of the upper reservoir

$$V_0 = V_0 + V_r, \tag{8.7}$$

the required total size of the upper reservoir of PSH can be easily determined:

$$V_0 = e^{\frac{\psi^* - P_{el(NOM)}}{\psi}} + V_r \tag{8.8}$$

8.5.3 Model Formulation

Considering the fact that this is a multistage process in terms of time and that the objective and constraint functions are nonlinear in relation to decision variables, the most appropriate model is based on dynamic programming that uses the simulation-optimization model. Optimization is performed by dynamic programming, which is well adapted to the characteristics of the problem, while the necessary constraints are solved by simulation.

The elements of the mathematical simulation-optimization model are shown in Figure 8.8. The system inputs are controlled and uncontrolled (natural), while outputs are also uncontrolled. They are variable in time according to the state of

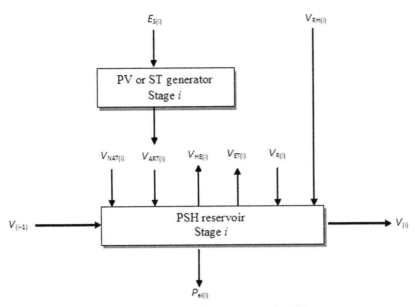

FIGURE 8.8 Mathematical simulation-optimization model of SE-PSH system.

the system and its surroundings. This means that the processes within the system can be represented in relation to time, introducing time stage i.

According to Glasnovic et al. (2013), the general formula of nominal electric power of the solar generator (PV or ST) can be calculated:

$$P_{el(SE)(i)} = \frac{\rho g H_{TE(i)}}{\eta_{SE} \eta_{PSI} R_{coll(i)} E_{S(i)}} V_{ART(SE)(i)} \, (kW) \tag{8.9}$$

where ρ is the water density ($\rho = 1000$ kg/m³) and g is the gravitational constant ($g = 9.81$ m/s²); $H_{TE(i)}$ is the average total head (water level difference in upper and lower storage plus hydrodynamic losses in the pumping system (m)); η_{SE} is the efficiency of the collector field of solar power plants; η_{PSI} is the efficiency of the pumping system and inverter; R_{coll} is the ratio of mean daily irradiation of tilted plane of *tracking parabolic collectors* and horizontal plane (for PV generator $R_{coll} = 1$); $E_{S(i)}$ is the mean daily radiation on the horizontal plane (terrestrial radiation); $V_{ART(SE)(i)}$ is the artificial water inflow, the daily water quantity pumped from the lower into the upper reservoir by the solar generator; and i is the time stage (increment).

Thus, power consumption of the integrated solar-hydro system, for a given manometer height and efficiency, depends on the ratio of the amount of water that solar power can pump and mean daily irradiation on the horizontal plane. This means that higher consumer energy consumption will require greater amounts of water to be pumped into the upper reservoir $V_{ART(SE)(i)}$, and that the specific size of the solar radiation $E_{S(i)}$ (with certain specific local conditions) would be proportional to the needs and higher rated power of the solar generator $P_{el(SE)(i)}$.

The system state has already been described by the corresponding balance Equation (8.1), while oscillations of the upper reservoir level are shown in Equation (8.2). This type of constraint also entered the initial conditions and requirement that the reservoir be full at the beginning of the observed period:

$$V_1 = V_{MAX} \tag{8.10}$$

In that sense, the period with maximum available of solar energy should be taken as the starting point of the calculation. In view of the condition of sustainability of the whole system, the reservoir should also be full at the end of the observed period:

$$V_N = V_{MAX} \tag{8.11}$$

where V_N is reservoir volume in the last time stage.

The decision variable constraint is linked to the motor/pump unit capacity (i.e., the appertaining pipeline), and can be expressed as follows:

$$0 \le V_{ART(SE)(i)} \le V_{ART(MAX)} \tag{8.12}$$

where $V_{ART(MAX)}$ is the maximum amount of water that can be pumped by the motor/pump unit (m^3) in a certain period of time.

As Eck and Zarza (2006) showed that solar collectors produce thermal power for direct nominal irradiance values above 250 W/m^2, it is necessary to introduce the constraint of minimum average daily solar irradiation of about 2 kWh/m^2 day, which can be expressed as (Glasnovic et al., 2013)

$$\text{If } E_{S(i)} \le 2 \text{ kWh/m}^2/\text{day} \Rightarrow P_{el(ST)(i)} = 0 \tag{8.13}$$

In addition to these constraints, it is necessary even to introduce a limit to the amount of water that can lead to the upper reservoir of PSH technology *rain harvesting* $V_{RH(i)}$, which is governed by the surface that provides ground/land around the PSH technology as follows:

$$0 \le V_{RH(i)} \le V_{ART(MAX)} \tag{8.14}$$

where $V_{ART(R)(MAX)}$ is the maximum amount of water that may accumulate geosynthetics at a location along the upper reservoir of PSH.

The general model for determining the optimal power of the solar PV or ST generator (i.e., PV or ST power plant), when integrated with the PSH, can be written with recursive formulas of the optimization process by dynamic programming: (Glasnovic et al., 2013)

$$f_i\left(V_{ART(SE)(i)}\right) = \text{MIN}\left\{ \underset{V_{ART(SE)(i)}}{\text{MAX}} \left[P_{el(SE)i}\left(V_{ART(i)}\right), f_{i-1}(V_{i-1}) \right] \right\} \tag{8.15}$$

with constraints and time step i:

(a) $V_{(i)} = V_{(i-1)} + V_{ART(SE)(i)} + V_{RH(i)}V_{NAT(i)} + R_{(i)} - V_{EV(i)} - V_{HE(i)}$

(b) $P_{el(SE)(i)} = \dfrac{\rho g H_{TE(i)}}{2\eta_{SE}\eta_{PSI}R_{coll(i)}E_{S(i)}} V_{ART(SE)(i)}$

(c) $\left[\text{If } E_{S(i)} \leq 2 \text{ KWh/m}^2/\text{day} \Rightarrow P_{el(ST)(i)} = 0\right]$

(d) $0 \leq V_{RH(i)} \leq V_{RH(MAX)}$ (8.16)

(e) $V_{MIN} \leq V_{(i)} \leq V_{MAX}$

(f) $V_1 = V_N = V_{MAX}$

(g) $0 \leq V_{ART(SE)(i)} \leq V_{ART(MAX)}$

$i = 1, 2, \ldots\ldots, N$

where $f_i(V_{(i)})$ is the optimal return function per state variable $V_{(i)}$ in time stage i, and $f_{(i-1)}(V_{(i-1)})$ is the optimal return function per state variable $V_{(i-1)}$ in time stage $i-1$. In Equation (8.16), (a) is actually Equation (8.1); (b) is Equation (8.7); (c) basically represents the constraint that applies only to the ST power plant (not to the PV power plant), where it operates on irradiated SE of 2 kW/m²/day (Glasnovic et al., 2011); (d) limits the amount of water that can give *rain harvesting* technology $V_{RH(i)}$ between the minimum (zero) and maximum $V_{RH(MAX)}$ value; (e) is the state variable constraint (water volume in the upper reservoir) between the maximum V_{MAX} and minimum V_{MIN} value; (f) is the condition that the first time stage V_1 equals the last V_N; that is, at the beginning and the end of the period, the reservoir volume is maximum V_{MAX} (keeping in mind that the calculation should not be done as usual in calculating the storage hydro power plant (i.e., from the beginning until the end of the year, when there is a rainy period when the hydro reservoirs are full), but calculation should start from July 1 and finish with the same date, because at that time there is maximum SE and hydro reservoirs will be full); and (g) is the control variable constraint, i is time stage (increment), and N is the number of time steps.

8.6 IMPACT GEOSYNTHETICS AND DYNAMIC CHARGING AND DISCHARGING OF PSH SYSTEM

8.6.1 Ratio P, V_0, and Surface of Geosynthetics

The relationship between the PV and ST generators and working volume of the upper reservoir is given in Equations (8.3) and (8.4). However, if geosynthetics are involved in system "rain harvesting" technology, then this relationship changes, as shown in Figure 8.9 for ST and PV generators.

This can be explained by the fact that the capture of solar energy by ST power generator is weaker in winter (for several months, ST generator practically does not work because there is a lower threshold of solar radiation, i.e., 250 W/m², Eck and Zarza, 2006), and the energy produced by the ST generator is more stochastic than the energy produced by the PV generator.

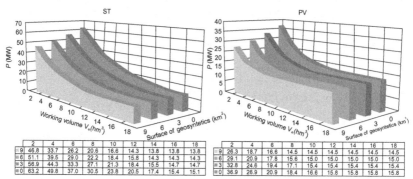

	2	4	6	8	10	12	14	16	18
9	46.8	33.7	26.2	20.6	16.6	14.3	13.8	13.8	13.8
6	51.1	39.5	29.0	22.2	18.4	15.8	14.3	14.3	14.3
3	56.9	44.3	33.3	27.1	21.3	18.4	15.5	14.7	14.7
0	63.2	49.8	37.0	30.5	23.8	20.5	17.4	15.4	15.1

	2	4	6	8	10	12	14	16	18
9	26.3	18.7	16.6	14.5	14.5	14.5	14.5	14.5	14.5
6	29.1	20.9	17.8	15.6	15.0	15.0	15.0	15.0	15.0
3	32.8	24.6	19.4	17.1	15.4	15.4	15.4	15.4	15.4
0	36.9	26.9	20.9	18.4	16.6	15.8	15.8	15.8	15.8

FIGURE 8.9 The ratio of power SE (ST or PV) generators and working volume for different surfaces geosynthetics (0, 3, 6, and 9 km^2).

8.6.2 The Dynamics of Charging and Discharging the PSH System

It is particularly interesting to observe the dynamics of charging and discharging the upper reservoir in the case of different working volumes V_0, which are shown in Figure 8.10. For this purpose, the observed changes in the working volume are from 10% to 90% in steps of 10% for both cases, respectively, and for PV-PSH and for ST-PSH systems; see Figure 8.10.

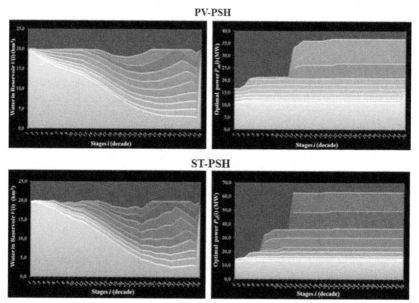

FIGURE 8.10 Dynamics of charge level of the upper reservoir during the year and the corresponding optimal power for PV-PSH and ST-PSH systems.

As shown, PV-PSH systems have a smoother transition from one time step to another compared to ST-PSH systems, which can also be explained by greater stochastic energy production from the ST generator. On the right side of Figure 8.10, the trends of increasing optimal nominal power generators are visible. As expected, the optimal nominal power ST generators are larger than the necessary PV power generators for the same reason that ST generators are inferior to PV generators in winter.

8.7   CONCLUSION

It is evident that at the present stage of PSH technology, it has advantages over other storage technologies, primarily because it can be stored up to the largest amount of energy, with relatively little cost and very little impact on the environment.

Therefore, PSH technology could play an important role in integrating the generators, which should connect the serial (SE-PSH-EPS), thus ensuring continuous energy supply throughout the year. Specifically, the fact that the weather conditions are repetitive throughout the year (annual cycles), as well as the fact that PSH technology can balance summer surpluses and winter shortages of energy, provide PSH technology with a significant advantage in terms of integration with solar generators.

But the factor that could have a significant impact on the greater use of integrated SE-PSH systems is geosynthetics, or new technology geomembranes that could create conditions for the construction of a reservoir on virtually all locations where there is a height difference. Geomembranes can retain water/ seawater over a long time, and they have high endurance properties. Using carbon *black additives, geomembranes* become resistant to UV radiation. Innovative technologies are presented with *white geomembranes,* which can reflect solar radiation and reduce their temperature. Another innovative solution is an *electrically conductive geomembrane* (in which electrically reactive carbon is incorporated) that, once installed, can be rigorously tested for spark ruptures and leaking through the geomembrane.

The new area of application of integrated SE-PSH systems could be along the coast, or PSH with seawater that does not need the lower reservoir; thus, these systems can be expensive. Because the world population is migrating to the shores of the sea, the systems with PSH could be increasingly applied.

If geosynthetics is used as *rain harvesting* technology, then the entire SE-PSH system can significantly reduce the power, which could also have a significant impact on reducing investment in such a system.

REFERENCES

Anagnostopoulos, J.S., Papantonis, D.E., 2007. Pumping station design for a pumped-storage wind-hydro power plant. Energy Convers. Manage. 48, 3009–3017.

Anagnostopoulos, J.S., Papantonis, D.E., 2008. Simulation and size optimization of a pumped-storage power plant for the recovery of wind-farms rejected energy. Renew. Energy 33, 1685–1694.

Black, R., Kniveton, D., Schmidt-Verkerk, K., 2011. Migration and climate change: towards an integrated assessment of sensitivity. Environ. Plan. 43, 431–450.

Deane, J.P., Ó.Gallachóir, B.P., McKeogh, E.J., 2010. Techno-economic review of existing and new pumped hydro energy storage plant. Renew. Sust. Energy Rev. 14, 1293–1302.

Eck, M., Zarza, E., 2006. Saturated process with direct steam generating parabolic troughs, sol. Energy 80, 1424–1433.

Fujihara, T., Imano, H., Oshima, K., 1998. Development of pump turbine for seawater pumped-storage power plant. Hitachi Rev. 47, 199–202.

Glasnovic, Z., Margeta, J., 2009. Solar hydro electric power plant. WIPO, WO/2009/118572.

Glasnovic, Z., Margeta, J., 2011. Vision of total renewable electricity scenario. Renew. Sust. Energy Rev. 15, 1873–1884.

Glasnovic, Z., Margeta, J., 2013a. Solar thermal hydro electric power plant with direct pumping system, WIPO, WO/2013/011333.

Glasnovic, Z., Margeta, J., 2013b. Solar thermal hydroelectric power plant, Espacenet, HRP20110544 (A2).

Glasnovic, Z., Rogosic, M., Margeta, J., 2011. A model for optimal sizing of solar thermal hydroelectric power plant. Sol. Energy 85, 794–807.

Glasnovic, Z., Margeta, K., Omerbegovic, V., 2013. Artificial water inflow created by solar energy for continuous green energy production. Water Resour. Manag. 27, 2303–2323.

Koerner, M.R., 2005. Designing-With-Geosynthetics, fifth ed. Prentice Hall, New Jersey, USA.

Margeta, J., Glasnovic, Z., 2011. Hybrid RES-HEP systems development. Water Resour. Manag. 25, 2219–2239.

Polyethylene (HDPE, LLDPE and LDPE): 2014 World Market Outlook and Forecast up to 2018 http://mcgroup.co.uk/researches/polyethylene-hdpe-lldpe-and-ldpe.

Part II

Economic Assessment of Solar Storage

Chapter 9

Photovoltaics and Storage Plants: Efficient Capacities in a System View

B. Böcker, B. Steffen and C. Weber

Chair for Management Sciences and Energy Economics, University Duisburg-Essen, Essen, Germany

Funded by the Federal Ministry for Economic Affairs and Energy under Grant 03ESP415.

Chapter Contents

9.1 ENERGY OUTLOOK

A salient feature of electricity systems is the requirement to instantaneously balance supply and demand. Traditionally, seasonality and other fluctuations on the demand side were answered by controllable thermal power plants, in some countries complemented by a very limited capacity of indirect electricity storage (almost always pumped-hydro storage (PHS) plants). To mitigate climate change, though, governments recently increased their efforts to reduce greenhouse gas emissions, most prominently by increasing the share of renewable energy sources (RES) in power generation. By 2020, for instance, Australia targets to achieve a share of 20% (up from 11% in 2011), France of 27% (up from 12%), and Spain of 38% (up from 30%) (REN 21, 2013).

Solar Energy Storage. http://dx.doi.org/10.1016/B978-0-12-409540-3.00009-8

Particularly ambitious targets have been set in Germany. The share of RES in power generation shall be ramped up in a narrowly defined corridor, reaching 80% by 2050. In addition, national carbon dioxide (CO_2) emission reduction targets have been fixed, aiming for a reduction of 80-95% by 2050 (compared to 1990). The bulk share of emission reductions thereby has to be achieved in the power generation sector. Given the natural constraints for hydropower and the cropland needs of biomass, additional renewable capacity can primarily be wind power or photovoltaic systems. Given their intermittent load, though, wind and photovoltaics cannot be controlled to follow the system load. To illustrate the magnitude of intermittencies, Figure 9.1 shows the feed-in from German photovoltaics during a summer week in 2013.

The pronounced day-night pattern makes it very natural to consider storage plants to smooth the curve. And the larger the share of renewables, the higher the need of storage technologies (or other backup solutions) to make sure electricity is available when needed.

While many storage technologies are still characterized by such high costs that a widespread commercialization is expected only in the medium term, PHS has been commercially operated for nearly a century, and interest in new sites has recently gained momentum (Steffen, 2012). The renewed interest is fueled precisely by growing shares of intermittent renewables. The typical setup of PHS in lower mountain ranges allows for storage cycles of 6-10 h, which fits well to the day-night pattern of photovoltaic systems as shown in Figure 9.1.

To allow for efficient capital investment, PHS plants are utility scale, with a typical pump-turbine capacity starting from 200 MW. This is a major difference compared to the photovoltaic capacity distributed in many small sites. In Germany, for instance, many installations are smaller than 10 kW. Consequently, the economics of photovoltaics and PHS shall not be looked at considering single plants, but must be analyzed in a system context. This is even more the case because PHS is clearly not the only possible technology to complement renewables. Lithium-ion (Li-ion) batteries are strongly considered (and covered

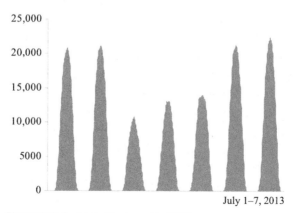

FIGURE 9.1   Photovoltaic feed-in to German grids (MW).

in this chapter), and also the remaining share of conventional generation provides flexibility.

To analyze the efficient use of storage in a system context, Steffen and Weber (2013) proposed a capacity-planning model based on peak-load-pricing theory. It has been applied to study the efficient use of PHS in a power system with a high share of renewables. Recently, Böcker and Weber (2015) extended the model to explicitly model reservoir size restrictions and their impact on efficient capacities—an aspect that is especially relevant for power systems with very high shares of wind and photovoltaics (beyond the 40-60% that have been under study by Steffen and Weber, 2013). While Böcker and Weber (2015) focus on analytical solutions, this chapter takes a more applied perspective, by numerically optimizing generation capacities and power generation in a system with very high shares of wind and photovoltaics, storage, and some thermal power plants. Taking Germany as a case study, the CO_2 reduction target (-80% by 2050) is assumed as fixed and the plant park for a 2040 horizon optimized accordingly, revealing how photovoltaics and PHS complement each other in a power system with very high shares of renewables.

The remainder of this chapter is structured as follows: First, the model to evaluate the role of photovoltaics and storage plants in a system context is described. Second, the regulatory and technical framework is sketched together with the input parameter and scenario assumptions. Third, results of the reference case and a series of sensitivities are presented. Finally, the chapter concludes with a short summary and outlook.

9.2 STORAGE PLANTS IN A SYSTEM VIEW

We aim to derive the efficient portfolio consisting of generation and storage technologies and capable of meeting the energy demand in each hour at the lowest system costs under different technical and political constraints. This is a typical optimization problem. Because we are focusing on a rather distant year in the future, a green field approach is taken, neglecting existing plants.

System costs are described by the sum of investment and operational costs. The objective function minimizing these costs may be written as follows:

$$\min \sum_{u_G} c_{\text{inv},K,G}(u_G) \cdot K_G(u_G) + \sum_t \sum_{u_G} c_{\text{op},G}(u_G) \cdot y_{G,C}(t, u_G)$$

$$+ \sum_{u_S} (c_{\text{inv},K,S}(u_S) + c_{\text{inv},V,S}(u_S) \cdot V_S(u_S)) \quad (9.1)$$

Generation technologies u_G (including conventional technologies u_G^C and renewable technologies u_G^R) are characterized by investment costs and operational costs. Investment costs are computed by multiplying specific investment cost $c_{\text{inv},K,G}(u_G)$ with the installed capacity $K_G(u_G) > 0$. Because we consider a single (representative) year in the future that does not cover the whole technical lifetime of the plants, investment costs are annualized. In the case of conventional power plants, operational costs $c_{op,G}(u_G)$ are dominated by fuel prices and accrue during power generation $y_{G,C}(t, u_G) > 0$. The operational costs for renewable power generation can be neglected and are assumed as zero. Operational costs are assumed to be constant over the observation period $T(t \in T)$.

For storage technologies u_S in contrast to other generation technologies, a distinction can be made between investment costs for installed capacity $K_S(u_S)$ (measured in MW) and storage volume $V_S(u_S)$ (measured in MWh). The specific investment costs are given by $c_{inv,K,S}(u_S)$ for the storage capacity and $c_{inv,V,S}(u_S)$ for the storage volume.

The main constraint in this system model is the balance of supply and demand $D(t)$ in each hour. Besides storage plants, the model distinguishes between flexible generation (mostly conventional technologies) $y_{G,C}$ and fluctuating feed-in by renewable technologies. The latter is assumed to be proportional to the installed capacities $K_G(u_G^R)$ and to the feed-in pattern $y_{G,R}(t, u_G^R) \geq 0$ (taken from July 2011 to June 2012). Storage plants can provide additional supply $y_{S,dc}(t, u_S) \geq 0$ when discharging and increase demand $y_{S,ch}(t, u_S) \geq 0$ when charging:

$$\sum_{u_G^C \in u_G} y_{G,C}\left(t, u_G^C\right) + \sum_{u_G^R \in u_G} y_{G,R}\left(t, u_G^R\right) \cdot K_G\left(u_G^R\right) + \sum_{u_S} y_{S,dc}(t, u_S)$$

$$= D(t) + S(t) + \sum_{u_S} y_{S,ch}(t, u_S) \perp \lambda^D(t) \geq 0 \qquad (9.2)$$

Demand is assumed as inelastic until power prices reach the value of lost load (VoLL). The VoLL is modeled by a virtual "conventional plant" with investment costs of zero and operational costs corresponding to the VoLL. Curtailment of RES $S(t) \geq 0$ is included as an option to obtain an economically efficient energy system. It will only be selected when it is cheaper to dump excess production than to construct and operate additional storage capacities.

Conventional power generation $y_{G,C}(t, u_G^C)$ is limited to its installed capacity $K_G(u_G^C)$. Additionally, renewable energy technologies and storage volume (especially PHS) are limited by available sites in Germany, and corresponding constraints are taken into account.

The consideration of storage plants requires additional constraints in the model. First, storage-filling levels are subject to a continuity constraint, with the next storage level $L_S(t+1, u_S)$ being calculated by adding the charged amount $y_{S,ch}(t, u_S) \cdot \eta_{S,ch}(u_S)$ and subtracting the discharged amount of energy $y_{S,dc}(t, u_S) \cdot \eta_{S,dc}^{-1}(u_S)$ to the current storage level. Efficiencies of storage plants are thereby split in a charging $\eta_{S,ch}$ and a discharging $\eta_{S,dc}$ efficiency. The rate of self-discharge $sd(u_S)$ depends on the average storage level in the current time step.

$$L_S(t+1, u_S) = L_S(t, u_S) + y_{S,ch}(t, u_S) \cdot \eta_{S,ch}(u_S) - y_{S,dc}(t, u_S) \cdot \eta_{S,dc}^{-1}(u_S)$$

$$- 0.5(L_S(t+1, u_S) + L_S(t, u_S)) \cdot sd(u_S) \perp \lambda_L^V(t, u_S) \qquad (9.3)$$

Additional restrictions require that storage plants be empty at the beginning and that their levels during operation be limited by the storage volumes. Cycle stability is a potentially restrictive factor for the lifetime of batteries. It is typically defined as number of full cycles during technical lifetime. Under the assumption of similar storage usage during the lifetime, this is converted into a maximum number CS of full cycle equivalents per year. This can be written as a linear constraint by limiting the sum of charging to the storage:

$$\sum_t y_{S,ch}(t, u_S) \cdot \eta_{S,ch}(u_S) \leq V_S(u_S) \cdot CS(u_S) \qquad (9.4)$$

Charging $y_{S,dc}(t, u_S)$ and discharging $y_{S,chc}(t, u_S)$ are furthermore limited to the installed storage capacity $K_S(u_S)$.

Political objectives are described in two additional constraints. First, the minimal share of RES in power supply f_{RE} is considered through

$$f_{RE} \cdot \sum_t D(t) \leq \sum \left[\left(\sum_{u_G^R \in u_G} y_{G,R}(t, u_G^R) \cdot K_G(u_G^R) \right) \right.$$

$$\left. + \left(\sum_{u_S} y_{S,dc}(t, u_S) - \sum_{u_S} y_{S,ch}(t, u_S) \right) - S(t) \right] \perp \lambda^{f_{RE}} \geq 0 \qquad (9.5)$$

RES power generation used effectively for demand coverage is determined as the sum of RES feed-in minus the curtailed energy and the storage losses. Under the long-term scenarios considered here, storage charging is typically done using renewable excess production rather than conventional power. Consequently, all storage losses are attributable to the usage of renewables and should be deduced from the renewable feed-in.

Second, the political objective of limiting CO_2 emissions is considered using an emission-bound f_{CO_2} in t_{CO_2}. CO_2 emissions are proportional to the conventional power generation $Y_{G,C}(t, u_G^C)$ weighted with the specific CO_2 emission coefficient $e_{CO_2}(u_G^C)$.

$$\sum_t \sum_{u_G^C \in u_G} y_{G,C}(t, u_G^C) \cdot e_{CO_2}(u_G^C) \leq f_{CO_2} \perp \lambda^{f_{CO_2}} \geq 0 \qquad (9.6)$$

Additional costs for the distribution of power (grids) and grid constraints are not taken into account in this model. Furthermore, the energy supply of conventional technologies is not subject to additional restrictions like unavailability or ramp constraints.

9.3 REFERENCE CASE

9.3.1 Scenario Assumptions and Parameters

The following investigation is based on assumptions for 2040 and a planning period of one year with hourly time steps Δt. The optimal portfolio consists of up to nine different technologies. Besides four conventional technologies (lignite, hard coal, CCGT[1], and OCGT[2]) and three RES (wind onshore, wind offshore, and photovoltaics), two storage technologies (PHS and Li-ion) are taken into account.

Table 9.1 provides an overview of the main input parameters.

Hard coal and especially lignite have high investment and low operational costs and are therefore typical base load technologies. The two gas technologies (CCGT and OCGT) are rather peak technologies due to low investment and high operational costs. CO_2 emission prices are not included in the fuel prices because a quantity bound is specified for CO_2 emissions. The specific CO_2

TABLE 9.1 Technology Input Parameters

	Capacity Costs (k€/MW)	Volume Costs (k€/ MWh)	Technical Lifetime (years)	Efficiency (%)	Operational Costs (€/MWh)
Lignite	1500	0	40	49	8.2
Hard coal	1200	0	40	51	23.9
CCGT	700	0	30	62	50.5
OCGT	400	0	25	41	76.3
Wind offshore	1600	0	20	100	0
Wind onshore	1200	0	20	100	0
Photovoltaics	800	0	25	100	0
PHS	840	20	50	80	0
Li-ion	100	150	20	90	0

Source: Based on data by IEA (2013), ISE (2013), RWTH Aachen (2013, 2014), own analyses.

---

1. Combined-cycle gas turbine.
2. Open-cycle gas turbine.

emission rates in t_{CO_2}/MWh_{el} are 0.74 for lignite, 0.67 for hard coal, 0.32 for CCGT, and 0.49 for OCGT.

Wind onshore, wind offshore, and photovoltaics are established renewable technologies with high additional potential. All these RES are characterized by supply-dependent feed-in and high fluctuations. Given the meteorological conditions, full load hours for wind offshore (3500) are almost twice as high as for typical wind onshore locations (1800), and photovoltaics can provide only 900 full load hours (assumed according to AEE, 2013). Full load hours as well as the investment costs are highly dependent on the exact location. In Germany, realistic sites are limited to about 198 GW for wind onshore, 54 GW for wind offshore, and 275 GW for photovoltaics (assumed according to IWES, 2013).

Similar to base load and peak load technologies, a distinction can be made between storage technologies for short-term and long-term balancing. PHS has rather high investment costs for pump-turbine capacity and low volume costs for the reservoir size, hence typically the reservoir volume will provide several hours of full load when discharging. In comparison to that, Li-ion has low capacity costs for power electrics but high volume costs for the reactive substances and therefore advantages for short balancing between load and supply (i.e., low full-load hours). Besides the storage cycle efficiency, the cycle stability and self-discharge rate are important restrictions when it comes to evaluating the potential role of storage plants in an efficient portfolio.

PHS typically has no limitations in charging and discharging cycles, and also self-discharge may be neglected. Yet the total storage volume is limited by available sites in Germany, where a maximum storage volume of 2 TWh is assumed based on an extrapolation of Voith (2014). For Li-ion batteries, in contrast, site restrictions are less relevant but cycle stability is limited. By 2040, Li-ion batteries are expected to reach 8000 full cycles and a self-discharge rate of approximately 2% per month (iSEA, 2013).

In line with the energy concept of 2010 and the coalition agreement of the current German federal government, the minimum share of RES in power generation is set to 65% for the year 2040, and CO_2 emissions in the electricity sector are limited to 20% of the CO_2 emissions in 1990 (Bundesregierung, 2010). The latter exceeds the official objective of -70% emission reduction until 2040 for the entire economy, yet it is expected that power generation will contribute disproportionate to overall emission reduction. The VoLL is assumed as 10,000€/MWh.

9.3.2 Results

In the reference case, more than 250 GW generation capacities (63 GW conventional and 189 GW RES) and 18 GW storage capacities are needed to cover a peak demand of 90 GW (Table 9.2). Hard coal is not part of the efficient portfolio as a consequence of low lignite operational costs compared to the investment costs. The realistic sites for wind offshore are fully exploited (54 GW).

The demand is covered to 65% by RES and to 35% by conventional technologies, in line with the political objectives. 91.1% of the possible RES feed-in is used

TABLE 9.2 Efficient Portfolio Reference Case

	Capacity	Power Supply
Demand	90.0 GW	540.8 TWh
	Peak Load	*(3.2 GWh cutoff)*
Lignite	3.7 GW	24.0 TWh
Hard coal	0.0 GW	0.0 TWh
CCGT	44.8 GW	162.4 TWh
OCGT	14.9 GW	2.9 TWh
Wind onshore	60.1 GW	97.4 TWh
Wind offshore	54.0 GW	171.9 TWh
Photovoltaics	74.5 GW	62.5 TWh
PHS	14.1 GW*(28.8 h)*	16.3 TWh
Li-ion	4.2 GW*(3.3 h)*	3.4 TWh
Curtailment RES	60.8 GW	8.5 TWh
Load cutoff	1.7 GW	3.0 GWh

directly and 5.4% indirectly with storage in between. In total, 2.3% of the possible RES feed-in is curtailed and another 1.2% is lost through the storage process.

System costs are approximately €43.3 bn: 76% is capital costs for generation technologies, 4% is capital costs for storage technologies, and 20% is operational costs of conventional power plants.

The mean electricity price for an additional unit of demand is 63.2 €/MWh and the shadow price of the CO_2 constraint is 75.6 €/t_{CO_2}. Hence the marginal cost of tightening the CO_2 bound by one ton is about 75 € in this future electricity system, and this would also be the market price for an emission certificate if market-based policy instruments like emission certificates are used to achieve the CO_2 objective. Additionally, the constraint on the RES share is also binding, with shadow value of 46.4 €/MWh. Requiring an additional unit of RES electricity would hence increase the system costs by almost 50 €, other things (notably CO_2 emission levels) being equal. Using again market-based instruments, each RES production unit would be valued at 46 €/MWh in addition to its market value. These costs are, however, not reflected in the electricity price previously indicated, because the RES requirement is handled through a separate constraint. So the RES certificate price instead corresponds to a support payment from the government, which on the other hand receives the revenues of the CO_2 certificate sales.

The optimization model used here can be interpreted in economic terms as partial equilibrium on the electricity and related markets. Using duality theory, it can then be shown also that capital and operating costs of the installed technologies plus the costs of CO_2 certificates plus scarcity rents for limited wind offshore sites are exactly matched by the electricity revenues and the revenues from the RES certificate provision. Note that only direct or indirect RES energy supply that can cover the demand receives the certificate price of 46.4 €/MWh in the system described here (see constraint (9.5)). The curtailed RES feed-in and losses during the storage process have no value and are therefore not supported.

Furthermore, the limitation of RES capacity (according to realistic sites) is a key element in the previous results, because otherwise wind offshore expansion would be unrealistically high. The shadow price of this constraint can be interpreted as the additional willingness to pay for wind offshore investment in a system perspective. In the reference case it is 75,600 €/MW.

9.4 SENSITIVITIES

The role of photovoltaics and storage plants in an efficient power system as discussed previously may be very sensitive with respect to different input parameters. Especially for a long-term perspective such as 2040, the actual parameters may deviate considerably from the expected values used in the reference case. This is particularly true for investment costs. Furthermore, the composition of the efficient portfolio may be considerably affected by variations in the policy objectives.

9.4.1 Investment Cost

Concerning the investment cost of storage technologies, a broad range of assumptions is used in the literature. Moreover, the past has shown that ex-ante

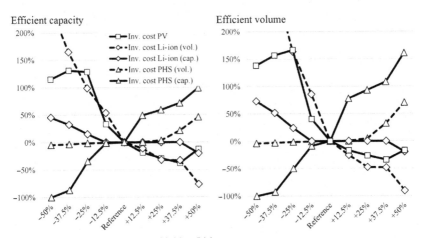

FIGURE 9.2 Investment cost sensitivities: Li-ion.

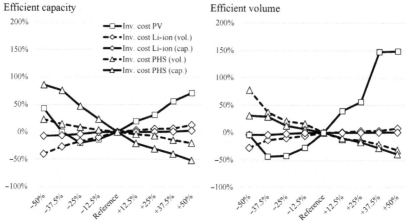

FIGURE 9.3 Investment cost sensitivities: PHS.

studies may be contradicted considerably by later evidence. Therefore, the following sensitivities indicate the impact of variations in the investment costs in a range of ±50%.

Figures 9.2 and 9.3 show the effect of varying investment costs on efficient storage capacity and volume for Li-ion and PHS, respectively. There is a strong impact of the investment costs of storage technologies on both their efficient capacities and volumes.

If storage investment costs (capacity or volume) increase, the corresponding efficient capacity and volume decreases, and vice versa. But cross-price elasticities are also important. For example, if the PHS volume investment cost rises by 50%, the efficient Li-ion capacity increases by nearly 100% and volume by over 150% (see Figure 9.2), while the efficient PHS capacity drops by approximately 50% and capacity by about 40% (see Figure 9.3). In line with actual cost shares for typical plant layouts, sensitivities are strongest with respect to volume cost in the case of Li-ion batteries and with respect to capacity cost in the case of PHS.

Generally, higher investment cost for photovoltaic capacity leads to less Li-ion and more PHS capacity and volume. This means that Li-ion complements photovoltaic feed-in, and PHS the increasing wind share in the case of increasing photovoltaic costs. Also the opposite is mostly true, yet with very low photovoltaic costs (and corresponding high penetration), longer term storage is required for the photovoltaic feed-in. This leads to a shift in storage volumes and capacities from Li-ion to PHS.

The investment cost sensitivity of efficient photovoltaic capacity is shown in Figure 9.4. As expected, variations of photovoltaic investment costs have the strongest influence. A cost reduction of 25% leads to an increase of efficient capacity of 60% (outside the display area). In addition, Figure 9.4 underlines

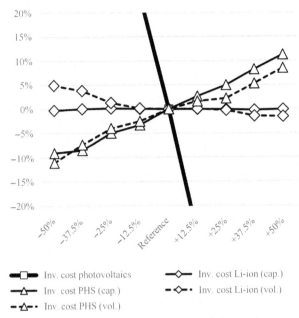

FIGURE 9.4 Investment cost sensitivities: Efficient photovoltaic capacity.

that Li-ion tends to complement photovoltaics, while PHS substitutes for photovoltaics in the scenario under study.

9.4.2 Political Objectives

In the analysis so far, two political objectives have been included (share of RES and emission constraint). Both have the ambition to slow down climate change by reducing emissions and increasing RES, and both lead to an energy system based on renewable energies and with limited direct emissions. Subsequently, we investigate the impact if only a single objective is pursued. Then, an additional investigation describes their interactions.

If the CO_2 restriction is dropped, the total capacities are nearly the same, but there are some shifts between the technologies (see Figure 9.5). In particular, CCGT is replaced by lignite and partly OCGT, because there are no incentives to limit CO_2 emissions. Correspondingly, the CO_2 emissions increase to 138 mt, a 93% increase compared to the reference case. This is accompanied by a slight shift in the renewable portfolio from photovoltaics to onshore wind, inducing also a small increase in the storage capacities. Because renewables lose the benefit of contributing to CO_2 reduction, dumping RES production gets comparatively cheaper and the installations shift to the cheaper wind energy. The resulting increase in the share of unused RES electricity is only partly compensated by the increase in storage capacities.

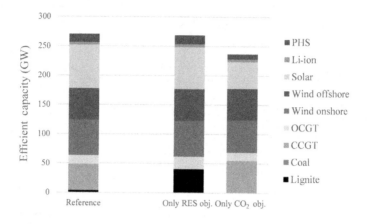

[GW]	Ref.	Only RES	Only CO$_2$
PHS	14.1	15.4	8.9
Li-ion	4.2	4.8	4.5
Solar	74.5	71.9	45.8
Wind offshore	54	54	54
Wind onshore	60.1	61.2	54.8
OCGT	14.9	20.9	14
CCGT	44.8	0.3	54.8
Hard coal	0	0	0
Lignite	3.7	40.2	0.0

FIGURE 9.5 Efficient capacities depending on policy regimes.

In the opposite case, when the RES objective is abandoned but the CO$_2$ objective maintained, lignite drops out of the efficient portfolio and CCGT increases. Correspondingly, less renewable capacities are needed to satisfy the CO$_2$ bound and, consequently, the need for longer term storage via PHS is also reduced. The CO$_2$ price, which indicates the marginal costs of CO$_2$ reduction, increases in this scenario to 181 EUR/t_{CO_2}. This also increases the base electricity price to 104 EUR/MWh. At the same time, system costs drop by about 1 bn€ to 42 bn€. These results seem contradictory at first sight, yet are perfectly consistent with economic efficiency (in the model setting): adding another constraint like the RES requirement will always lead to an increase in system cost, unless the constraint is not binding (i.e., superfluous). On the other hand, this constraint is instrumental for achieving the CO$_2$ target; therefore, the marginal costs attributable only to the CO_2 target drop when the RES constraint is introduced. Because the RES constraint creates additional revenues for the RES producers, the electricity price needed to recover the capital costs decreases. From a consumer perspective, the double objective therefore seems preferable at first sight, yet government revenues are much lower in that case and, consequently, consumers are likely to be charged in other ways to cover government spending.

If climate change mitigation is considered as the primordial objective, the efficient solution hence is the reliance on a single CO_2 constraint. Setting a binding RES objective in addition has the effect of pushing RES, while at the same time keeping low-cost technologies (lignite) longer in the efficient portfolio. These interdependencies between the different objectives are discussed further in the following sensitivity analyses.

The RES policy sensitivity in Figure 9.6 analyzes the effect of a variation of the RES objective between 33% and 98% (sensitivities −50% to +50% compared to the reference case). The RES objective is only binding in the reference case and when the objective gets more restrictive (increasing the minimum share of RES). Up to the +25% sensitivity, CO_2 restriction is binding, and lignite is progressively displaced by CCGTs. After this point, the path is driven only by the renewable energy (RE) policy. Without binding CO_2 constraint, lignite is the most efficient conventional technology, which fades out when RES approaches the 100% scenario. The storage capacity of PHS increases with RES expansion but drops in the last step. This is an effect of limited sites for PHS. With extremely high RES penetrations, it gets more important to store energy for longer periods (increasing the full load hours). In this case, efficient Li-ion capacity raises sharply from 16 GW to 73 GW.

To study the effect of an increasing CO_2 target, Figure 9.7 shows the efficient capacities and generation volumes for power systems with 10% up to 30% of CO_2 emissions compared to 1990 (again −50% to +50% compared to the reference case). With the 30% target, all technologies except hard coal are part

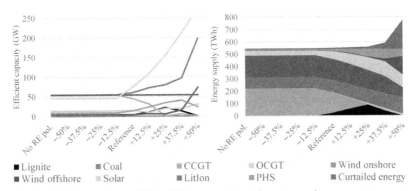

FIGURE 9.6   RES policy sensitivity: Efficient capacity and energy supply.

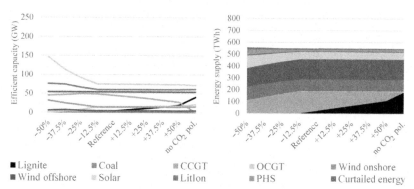

Lignite ■   Coal ■   CCGT ■   OCGT ■   Wind onshore ■
Wind offshore ■   Solar ■   LitIon ■   PHS ■   Curtailed energy ■

FIGURE 9.7 CO_2 policy sensitivity: Efficient capacity and energy supply.

of the efficient generation portfolio. Among storage plants, PHS accounts for more than triple the capacity of Li-ion batteries. Moving from right to left, the additional emission reductions are entirely realized by a coal-gas switch, with generation volumes (and the corresponding capacities) moving from lignite to CCGT, while the renewable capacities and storage remain constant. Only when lignite is entirely replaced by natural gas (at a CO_2 target of about 18% compared to 1990) do wind and solar capacities start to increase further. Most prominently, a sharp rise in photovoltaic capacity can be observed, with capacity almost doubling between 80% and 90% CO_2 emission reduction. This strong rise is complemented by a significant increase of PHS capacity (from 14 GW to 32 GW), while the efficient Li-ion capacity stays roughly the same. Figure 9.7 also indicates that the "curtailed energy" increases only slightly; the additional storage capacity hence allows commercializing most of the additional intermittent feed-in.

Unlike in the RE sensitivity, the CO_2 restriction is binding in all cases. With fewer CO_2 restrictions, lignite will increasingly be part of the efficient portfolio and substitute for CCGT.

9.5   CONCLUSION

Policy makers target large-scale changes to power systems to reduce CO_2 emissions and thereby slow down climate change. To achieve this target, conventional generation technologies using fossil fuels must be reduced, and the use of RES and storage increased. From today's point of view, wind onshore, wind offshore, and photovoltaics will make the largest contribution in an RES-dominated system. With PHS and Li-ion, two proven technologies are available for storage plants, although Li-ion batteries have so far hardly been applied at large scales.

Using an optimization model as presented in this chapter, the main drivers of change in the German energy system can be analyzed. In consideration of the political objectives of a minimum share of RES in the power supply and a limitation in CO_2 emissions, photovoltaics and storage play a key role in the efficient portfolio for 2040. Efficient photovoltaic capacity of 75 GW comes around more or less twice as high as in 2014, and PHS capacity increases from around 6 GW in 2014 to 14 GW in 2040. While today Li-ion is hardly used for power management, the technology could play a role in future power systems (with an efficient capacity over 4 GW, according to the optimization). Sensitivity analyses show the high dependency of storage capacities on investment costs and political objectives.

Analyzing shadow prices of main restrictions allows statements concerning average cost of an additional unit of demand (electricity price), the magnitude of CO_2 prices to achieve the CO_2 target, and required incentives for RES to be competitive in the market. While the limitation to a CO_2 emission reduction target only reduces system cost, the combination with an RES share objective allows for more technologies to be part of the efficient portfolio.

Interpreting the optimization results, it should be noted that the present analysis neglects factors such as cross-border interconnections (increasing flexibility and reducing storage capacity) and grid restrictions (decreasing flexibility and potentially leading to higher storage capacity). In addition, further studies could evaluate power-to-gas as a further technology, especially for extreme cases with close to 100% RES. In sum, though, the present analysis sheds light on important mechanics concerning the efficient interaction of photovoltaic and storage plants in a future power system.

REFERENCES

Agentur für Erneuerbare Energien (AEE), 2013. Studienvergleich: Entwicklung der Volllaststunden von Kraftwerken in Deutschland.

Böcker, B., Weber, C., 2015. Efficient storage operation, capacity and volume—Analytics of the electricity market equilibrium, working paper, University of Duisburg-Essen.

Bundesregierung, 2010. Energiekonzept für eine umweltschonende, zuverlässige und bezahlbare Energieversorgung.

International Energy Agency (IEA), 2013. World Energy Outlook 2013.

Fraunhofer-Institut für Solare Energiesysteme (ISE), 2013. Stromgestehungskosten Erneuerbarer Energien, Studie.

Institut für Stromrichtertechnik und Elektrische Antriebe (iSEA), RWTH Aachen, 2013: Marktanreizprogramm für dezentrale Speicher insbesondere für PV-Strom.

Fraunhofer-Institut für Windenergie und Energiesystemtechnik (IWES), 2013. Energiewirtschaftliche Bedeutung der Offshore-Windenergie für die Energiewende.

REN 21, 2013. Renewables 2013 Global Status Report.

Institut für Elektrische Anlagen und Energiewirtschaft (IAEW), RWTH Aachen, 2014. Genetische Optimierung eines Europäischen Energieversorgungssystems (GENESYS), Eingangsdaten.

Steffen, B., 2012. Prospects for pumped-hydro storage in Germany. Energy Policy 45, 420–429.

Steffen, B., Weber, C., 2013. Efficient storage capacity in power systems with thermal and renewable generation. Energy Econ. 36, 556–567.

Voith, 2014. Die Energiewende erfolgreich gestalten: Mit Pumpspeicherkraftwerken.

Chapter 10

Economics of Solar PV Systems with Storage, in Main Grid and Mini-Grid Settings

Iain MacGill[1] and Muriel Watt[2]

[1]*School of Electrical Engineering and Telecommunications University of NSW, Sydney, NSW, Australia*
[2]*School of PV and Renewable Energy Engineering, University of NSW, Sydney, NSW, Australia*

Chapter Outline

10.1 INTRODUCTION

Recent rapid developments with highly variable and only somewhat predictable renewable energy resources, notably photovoltaics (PV), have highlighted the importance of more efficiently managing the temporal and locational match of supply and demand across power systems. There is, of course, nothing

Solar Energy Storage. http://dx.doi.org/10.1016/B978-0-12-409540-3.00010-4

fundamentally new about this challenge. Load is itself typically highly variable and only somewhat predictable. However, the scale of the challenge is growing, as is interest in options to help manage imbalances.

The potential value of energy storage to assist in managing supply-demand balance has been long appreciated. While there have been a number of technologies such as pumped hydro (covered elsewhere in this book) for storing electrical energy, most storage in large-scale power systems has been in the primary energy resources of major generation sources—coal, gas, nuclear, and hydro. Until recently, there have been only limited cost-effective distributed energy storage options available. Now, however, there is a growing range of options that might assist in the more effective and efficient management of supply-demand balance in the electricity industry. These include a wide range of energy management opportunities, with loads that have some inherent energy storage and hence dispatchability (e.g., heating and cooling services), as well as a number of new or improved electrical energy battery storage technologies and other emerging opportunities such as electric vehicles.

A key question looking forward then, of course, is which particular energy storage options and applications might add the most value to the electricity industry. This is a more challenging question than might be appreciated. The underlying economics of supply-demand balance within electricity industry operation are very complex due to the unique characteristics of electricity itself, as well as the inherent economics of current generation, and network and load infrastructure. Future uncertainties add to the challenge, as seen with rapid changes in energy supply and demand technologies over recent years, including, of course, PV. This complexity and uncertainty, combined with electricity's role as an essential public good, and the many environmental and social externalities of the industry, mean that current market arrangements in almost all electricity industries around the world don't really reflect all of the underlying industry and broader societal costs and benefits involved.

The growing deployment of highly variable and somewhat unpredictable renewable generation, both large-scale (wind) and small-scale (PV within the distribution system), with their own particular economic characteristics, has added to these challenges. On distribution networks in particular, falling costs for PV and supportive policy measures have seen extraordinary growth in grid-connected deployment over the past decade, fundamentally changing supply-demand dynamics and the role of the grid. The future success of the technology will, however, depend on the value it can contribute toward delivering affordable, reliable, secure, and sustainable energy services to end users.

Future success for energy storage in conjunction with PV will also depend on the value it can add. However, economically valuing energy storage in concert with growing PV is highly complex and uncertain. Some particular challenges arise, perhaps surprisingly, from the wide range of potential benefits storage can bring to the electricity industry. For instance, in different contexts, storage may provide similar services to both generation and loads, and even network equipment. Furthermore, distributed energy storage can potentially play an even more useful role than centralized options, given its location within the distribution network, or even at end user premises (Sue et al., 2014; UBS, 2014; Fuhs, 2014).

An additional complexity for economic analysis is the need to consider private commercial economics for PV and storage deployment by key stakeholders as well as their underlying societal economics. The present disconnect between private and public economics might see societally valuable PV and storage options fail to be deployed, or perhaps see deployment that might be privately cost-effective but doesn't offer high societal value. For instance, the beneficiaries of storage deployment with PV depend on the prevailing regulatory framework and tariff structures (Goel and Watt, 2013).

At present, such analysis may seem premature. Only small amounts of electricity energy storage other than pumped hydro are currently deployed in electricity industries around the world. However, this may now be changing, with growing deployments of smart grid technologies including active load management, and policy efforts supporting electrical battery storage in association with PV deployment in jurisdictions including Germany, California, and Japan. A wide range of industries have also been investigating the value of storage through demonstration projects.

Another key area of existing deployment is battery storage systems for off-grid applications. This storage is already cost-effective, or indeed essential, for many smaller stand-alone systems, as highlighted in several other chapters in this book. Energy storage applications for larger off-grid community-sized mini-grid systems and edge-of-grid applications represent a potentially high-value application that falls between these stand-alone systems and the main electrical networks.

In this chapter, we assess the economics of PV and storage. Both societal and private perspectives are considered, along with a range of potential applications, from household systems to utility projects. Most attention is given to battery storage systems, although smart load management options are also considered. The focus of this review is largely on the Australian National Energy Market (NEM) context for such deployment. While findings from the analysis may therefore have limited applicability to other electricity industry jurisdictions in some regards, the work undertaken by the authors and others offers broader insights.

The rest of this chapter is structured as follows. We first consider electricity industry economics in the broadest context, before considering the particular economics of PV and energy storage as separate technology options. In the next section, we consider the economics of PV and storage deployed together in roles from households to commercial and industrial customers, network-driven applications, and finally utility-scale PV and storage systems. This is followed by a section considering possible future developments for the economics of such systems, including the implications of falling costs, challenges to greater deployment, and even the possibilities of grid defection as end users deploy local storage and PV, perhaps in conjunction with other distributed energy options in order to leave the grid entirely. The chapter concludes with some brief thoughts on what might happen next, and the challenge for industry stakeholders in facilitating PV and energy storage systems to contribute to overall electricity industry outcomes in the most appropriate way.

10.2 ELECTRICITY INDUSTRY ECONOMICS

10.2.1 Electricity Industry

Electricity industry economics are highly complex and uncertain. This is not surprising, given that the industry must match variable and unpredictable supply to similarly variable and unpredictable demand at all times and all locations for an invisible and intangible "commodity." Demand arises from the diverse, changeable, and only partially predictable behavior of energy end users to provide desired energy services of widely varying perceived value. The "commodity" itself flows instantaneously through a specialized grid according to complex network physics (with the exception of site-specific pumped hydro), lacks cost-effective direct electricity storage options, and has specific power quality requirements.

There is a range of existing and potential future generation options to meet this demand, with sometimes very different technical and economic characteristics. Conventionally, there have been significant economies of scale, and generation assets have been large, lumpy, and irreversible. The electrical network itself is highly asset-intensive and shared by all electricity providers and consumers. Given this investment intensity, it can be useful to think of the economic value (net benefits) of electricity as having two key components: an energy (supply) value and a network (delivery) value. These components have both fixed and operational costs. Typically, demand has been widely variable and not responsive to changing supply conditions, meaning that large amounts of generation and network are required to supply only occasional periods of particularly high demand. There are also a range of significant environmental and social "externalities" associated with current electricity provision, such as regional air pollution and climate change emissions from key generation technologies.

The real cost and value of electricity, therefore, varies by time and location within the network, depends in large part on the industry's wider social and environmental impacts, and is subject to a wide range of uncertainties. Electricity industry arrangements invariably only capture some aspects of these underlying economics. In some restructured electricity industries, wholesale electricity prices (and associated futures prices) exhibit the temporal and spatial variability and uncertainty of electricity's energy and network value to some extent (MacGill, 2010). However, commercial arrangements for retail electricity consumers are generally far less reflective of these costs and benefits. Many energy users have only basic meters that report cumulative energy consumption, and pay flat electricity tariffs that combine energy and network costs, which have often been set to achieve broader societal objectives within a framework of overall industry cost recovery.

By comparison, mini-grid system economics may be considerably less complex and uncertain, given a much smaller number of generators and loads, and far simpler networks. Still, full economic modeling of these systems remains challenging. In conclusion, the electricity industry has extremely complex and uncertain economics, within which the specific economics of PV and energy storage, and their combination, must be assessed.

10.2.2 PV Economics

PV economics add further challenges to economic assessments. PV technologies convert a highly cyclical, variable, and somewhat unpredictable solar resource with no inherent storage into electricity through a solid-state device with virtually no energy storage either. PV's generation can often exhibit a useful, and hence economically valuable, correlation with daytime periods of higher demand. Furthermore, it is by far the most scalable of all current generation technologies, ranging from appliance level (Watts) through residential (kWs) onto commercial and industrial (hundreds of kWs) to utility scale (tens to hundreds of MWs). It can also be easily integrated into the built environment. Hence, it can be located close to loads and potentially reduce network peaks and losses. Furthermore, it has no operational regional air polluting greenhouse emissions, uses no water or other consumables in operation, and produces no solid wastes (other than the system components themselves at the end of their useful life) (Oliva et al., 2013).

Still, the variability and uncertainty of PV generation has implications for its energy and network value, given the need to precisely match supply and demand. In general, variable and unpredictable technologies without inherent energy storage have lower energy value than those with some measure of storage in their primary fuel supply or energy conversion process, and some level of dispatchability (MacGill, 2010). This needs to be appropriately weighed against the locational flexibility and wider environmental and social benefits that the technology can provide. Another key factor is the level of PV deployment. Generally, the marginal economic value of PV, and any other generation technology, will decline as its penetration grows, although there are particular circumstances, such as the potential role that sufficient PV deployment might play in delaying or averting lumpy network investments, where the reverse can be true.

The present private commercial arrangements for potential PV developers have only limited alignment with these underlying economics. Utility-scale PV systems in a restructured industry such as the NEM will certainly see time-varying and to a lesser extent location-varying and uncertain wholesale energy prices. As penetrations grow, it is likely that the merit order effect will see times of high PV generation associated with generally lower prices as well. However, the NEM does not properly price the environmental externalities of competing, almost entirely fossil-fuel generation. Instead, there is an effective shadow price provided through the Renewable Energy Target (RET), which provides additional cash flow to eligible renewable generation. In other industries, these broader environmental, social, and economic (investment and jobs) externalities may be priced through feed-in tariffs or capital subsidies for PV.

At the retail market level where household and commercial PV systems are deployed, existing customer electricity tariffs, as noted earlier, don't generally capture underlying electricity economics to any real extent. Instead, the private attractiveness of PV deployment depends on nonreflective electricity tariffs and any other explicit PV support measures. Around 140 countries have implemented policies to support renewable power generation, with many of these targeted toward PV (REN21, 2014). Feed-in tariffs (FiTs), which provide a defined payment for eligible renewable generation, have been the most widely implemented policy mechanism. In Australia, recent support has involved an effective capital grant subsidy provided through the small-scale component of the RET and, for a short time, various state government feed-in tariffs (Watt and MacGill, 2014). Now, however, net metering of PV is near universal and sees customers offsetting their own load, or being paid a market set (low or zero) rate for exported PV generation. In a similar manner, FiTs are being wound back in many other jurisdictions, given falling PV costs and hence attractiveness, and budget pressures.

The economics of PV in mini-grids can be extremely compelling, as noted earlier, and system design will often be based around underlying system economics. However, there are fuel subsidies and unpriced externalities that do not always get appropriately factored into decisions regarding the deployment of PV. The most important factor in PV economics, however, has been the marked fall in system costs over the past decade. These price reductions—more than 75% in Australia over four years (Watt and MacGill, 2014)—have entirely transformed PV economics and driven very widespread uptake, even when explicit PV policy support is being reduced or entirely eliminated.

10.2.3 Energy Storage Within the Electricity Industry

Energy storage has always had an important role in the electricity industry in various forms. In a general sense, energy storage is the storing of some form of energy that can be drawn upon at a later time to perform some useful operation. Given the need to match supply and demand, this is a critical function At present, most energy storage within the electricity industry resides in fuel supplies for the major generation technologies—coal stockpiles, gas reservoirs, and pipelines for fossil-fuel plants, and water reservoirs for hydro generation. Interestingly, the large fossil-fuel plants effectively have an energy storage problem in terms of minimum operating levels, ramping rate constraints, and high startup and shutdown costs and time frames. Early efforts at electricity energy storage, including pumped hydro (discussed elsewhere in this book), were intended to help meet varying demand with somewhat inflexible plants. In recent years, however, such electricity storage was displaced somewhat by fast response gas peaking plants.

Now, of course, electricity industry developments in numerous jurisdictions, including Australia, have raised new challenges yet also opportunities for the electricity industry and its operation (Kind, 2013). One development has been the growing deployment of highly variable and somewhat unpredictable renewable energy, primarily wind and solar. Another challenge is that of increasingly peak demand in the residential (and to a lesser extent commercial) sectors, where average demand has been falling due to a range of reasons including energy efficiency and distributed generation, yet peak demand has not, or certainly not to the same extent. The distribution network assets, and hence associated capital expenditure, required to ensure that projected future peak demands can be met have seen growing network expenditure and hence overall industry costs. Technologies that can assist in better managing such peaks therefore offer potential network value in avoiding or deferring such expenditure.

At the same time, a range of emerging distributed energy storage technologies has seen growing technical and commercial progress. These particularly include a range of battery technologies whose development has been driven by factors including portable electronics and electric vehicles. Others include compressed air storage and flywheels. There is also a range of thermal energy storage options. More generally, there are other emerging technologies such as remotely controlled end user appliances (Commonwealth Scientific and Industrial Research Organisation, 2013) (e.g., "peaksmart" remotely controllable air-conditioning now being deployed in Queensland) that effectively take advantage of the inherent energy storage available with some end user energy services such as heating and cooling. Some of these technologies, particularly the battery and smart load options, are highly scalable from residential to commercial, industrial, and utility applications within the distribution network.

There is nothing new about variable and uncertain electrical loads and unpredictable generation, nor peak demand challenges in terms of both supply balance and network management. Neither is there anything fundamentally new in the use of battery technologies for electricity supply; they have been in use for stand-alone systems in remote areas as well as in uninterruptible power supplies (UPS) for critical end user loads for decades. However, trends with both these electricity challenges and energy storage technology opportunities would seem to have created a potentially valuable role for greater deployment of distributed energy storage in the electricity industry (Roberts, 2009).

Perhaps unsurprising, the economics of storage are very challenging to assess. In part, this reflects the broad range of technologies and activities that provide some measure of energy storage, or equivalent service, including demand management as well as these new electrical energy storage technologies. It also reflects the very wide range of potential value that storage can bring to the electricity industry. Storage can add value at the point of the energy consumer by providing greater customer reliability against possible supply interruptions and more economically efficient demand profiles; at the wholesale level as supply costs and demand value vary across time, from periods of seconds (ancillary services) to days to potentially months; at the network level by reducing peak demands and hence network expenditure; and at the retail market level through changing customer patterns of demand upon the system. There is an important question around which of these roles are most valuable (Sue et al., 2014), and how these values are changing as new consumer technologies enter the market.

Storage also poses particular challenges for commercial arrangements, again in part due to the varied technologies and roles it can play. For example, electric storage is both a load and a generator, depending upon particular circumstances. A particular opportunity would seem to lie in distributed applications because of the potential network and customer value that distributed storage can provide, and the lesser competition in these roles posed by conventional large-scale storage options. However, it faces the same disconnect between commercial incentives and underlying economics faced by PV in retail markets.

For the particular context of the Australian NEM, one study has suggested that, economically, the most significant storage applications currently appear to be for increasing end user reliability and deferring network expenditure, although energy arbitrage in the wholesale electricity market also offers potential value. The value of particular projects is very dependent of course on customer values of reliability and network constraints (Sue et al., 2014). Other work has also highlighted the potential role of storage for wholesale energy arbitrage (Wang et al., 2014). Another identified area for high-value storage applications, particularly relevant to this chapter, is in stand-alone and remote mini-grid applications, largely in partnership with PV, as discussed in the next section.

10.3 PV AND STORAGE APPLICATIONS

There would of course seem to be natural synergies between PV and storage in the delivery of assured, affordable energy services. A highly cyclic, variable, and somewhat unpredictable energy source transformed into electricity through a solid-state device with no storage would seem ideally matched with storage technologies that can ensure delivery of time-varying and uncertain energy service demands that will often not prove a good match with the solar resource.

Australia, as explored previously, would seem to provide a particularly relevant case study, with among the world's highest household penetrations of PV. Some 15% of households now have a PV system (Watt et al., 2014), very large distribution networks with low customer densities by international comparisons (Energy Supply Association of Australia, 2013) and hence high and rising electricity delivery costs, and, by some measures, globally high residential and commercial electricity prices (Energy Users Association of Australia, 2012).

There are some potential direct technical synergies between the two technologies: both PV and battery storage are inherently direct current (DC) technologies requiring inverters to supply the alternating current (AC) power provided by the grid. This suggests some potential to share balance of system costs, which, as PV module prices fall, are a growing component of overall PV system costs (Watt et al., 2014), and are also very significant in battery storage systems. In practice, this is certainly evident in some mini-grid technologies, but the requirements of PV and battery storage systems can be rather different in grid-connected systems (e.g., preferred operating ranges).

There are also potential synergies in terms of ownership and the costs and benefits associated with these two technologies. Of particular interest are the synergies between the technologies—that is, values that accrue from the technologies together that exceed their values separately. As we will explore further, this is a key question, and critically depends on the market and broader regulatory arrangements in which the technologies are deployed. We consider a range of potential contexts next.

10.3.1   Household Systems

As noted earlier, residential PV deployment has generally occurred within the context of immature and somewhat dysfunctional retail market arrangements, featuring significant cross-subsidies across and within customer classes (Outhred and MacGill, 2006). Tariffs are largely consumption based and generally still involve flat rates (c/kWh) for residential and small business customers, although there are growing moves toward time-of-use (TOU) rates as metering infrastructure is upgraded. Small businesses had the same commercial arrangements until recently, when demand components have been increasingly added to tariffs.

Complex and challenging interactions can arise with household PV, depending on the metering and tariff arrangements they face. Gross metering and Feed-in Tariffs (FiTs) for PV provide payments based only on the total generation of the system and do not create a commercial case for storage.

However, net metering arrangements, where less-than-load PV generation during a given period effectively offsets consumption (and the corresponding retail tariffs) while periods of net-PV export see the exported PV generation paid at a different tariff, can create a number of commercial possibilities for storage. If the export and retail tariffs are the same, as seen in many US states at present, then there is no value in moving PV generation (or load) across time. However, there are examples of FiT arrangements where the export FiT is far lower, or far higher than the retail tariff. Under such circumstances, there are potentially significant financial advantages in deploying electricity storage and/or load management.

The commercial value of this will depend greatly on the retail electricity market and PV support arrangements in place. It will also depend greatly on the actual performance of the PV system, household load pattern, and the match between them.

Some work has explored these issues for the Australian National Electricity Market (NEM), given a range of retail electricity tariff options (flat and TOU rates) and different PV net FiT tariffs for households with PV and load management options (Oliva and MacGill, 2014). It found that households can certainly deploy load management to either reduce or increase PV generation exports to increase the financial returns of the PV. The extent of these returns depends on the particular households' PV generation and load profiles.

Other work has modeled the use of battery storage systems in the NEM under a similar context (Goel and Watt, 2013). Lead-acid battery systems were added to a sample of Australian households with PV systems. It was found that at current PV and battery system costs and retail and net FiT PV tariffs, battery storage only improved payback periods for larger PV system sizes (above 4 kW) when significant PV exports would occur at very low net FiT rates.

Another finding of both studies was that there were potentially adverse or beneficial impacts on the network service providers under these circumstances of battery storage or load management. PV systems alone had almost no impact on reducing peak network demand. However, load management to shift evening load into the day or charging the battery system when PV was generating could both reduce PV exports and significantly reduce household peak demand in the evening, and hence necessary network infrastructure. Conversely, load management to shift load from daytime to evening to maximize PV exports could worsen impacts. This work highlights the possibility that private commercial arrangements may work toward improving or worsening underlying industry economics, depending on the price signals proved to customers.

10.3.2 Commercial and Industry PV

Larger retail customers such as commercial and industrial facilities typically have more sophisticated metering and tariff arrangements. In the Australian NEM, they generally have separate regulated network tariffs and competitive retail energy contracts. These network tariffs generally have TOU consumption and peak demand components. Retailer energy tariffs also typically have TOU rates. Generally, connection at higher voltages and larger consumption mean lower energy tariffs that reduce the competitiveness of PV to displace load. However, falling PV costs are changing this equation, and the good match between PV and typical commercial load profiles adds value via peak demand reduction. There are often fewer commercial incentives to greatly oversize systems above typical levels of demand, and prohibitions on export are typical, hence there is less question of PV export rates.

However, there are other opportunities for energy storage—for example, the widespread use of uninterruptible power systems (UPS) on critical business loads that already represent some local level of electricity storage.

As penetrations of renewable energy grow, so will the likely need for additional energy storage, and commercial arrangements will likely have to be changed to better incentivize its deployment.

10.3.3   Distribution Network-Driven PV

Although PV doesn't contribute to network peaks, it doesn't generally play a large role in reducing them, certainly in predominantly residential networks with typically evening peak demands and no tariff incentives to change consumption patterns. There are certainly potential roles for storage to assist in this regard, and storage deployed with PV systems could add further value. However, high PV penetrations in the network can certainly raise some other concerns, which are also seen with other distributed generation technologies. Reverse power flows at times of high PV export in networks designed and operated for unidirectional power flows can cause issues, including high voltages, harmonics, phase imbalance, and challenges for protection systems (Passey et al., 2011). Voltage rise has been identified as a particularly important challenge (Braun et al., 2012).

We are now seeing new policy developments to facilitate greater battery deployment with PV. Germany has recently introduced an energy storage financing program to facilitate the deployment of PV systems with battery storage for residential and commercial customers (Parkinson, 2013; Colthorpe, 2014a,b). Japan now also offers subsidies to support the installation of battery storage along with PV systems (Colthorpe, 2014a; Wilkinson 2014). Such battery systems can of course offer potential value beyond that of managing adverse impacts of PV. However, as noted earlier, commercial arrangements in current retail electricity markets do not necessarily provide appropriate financial incentives for such actions, and there are also a number of technical challenges in effective coordination.

Furthermore, battery storage systems and load management are only one possible means of addressing adverse impacts on distribution networks from PV deployment. Other approaches, including reactive and active power management utilizing the capabilities of PV system inverters, tap changers on transformers, and network reconductoring, can also be used (Braun et al., 2012). Battery storage systems may not be the lowest cost means of addressing adverse impacts in many circumstances (Tant et al., 2013), a common theme in the economics of PV with storage systems. Battery storage systems are highly capable but also currently expensive compared with a range of other options, and because the simpler and cheaper grid inverters must be replaced by more costly bidirectional inverters. It is notable that the uptake of the German support scheme for battery storage to date appears to have been relatively slow due to the less than compelling economics, even with subsidy support, although the complexity of the scheme has also been identified as an issue (Colthorpe, 2014b). With time, the value of such approaches should become clearer.

10.3.4 Utility PV and Storage Systems

It may sometimes be the case that the optimal location of storage in the distribution network isn't at the site of the PV systems themselves, creating a case for their deployment at strategically chosen parts of the network. Australian utility Ergon Energy is starting to deploy storage across its network, with a view to reducing costs of service provision in low-density areas (Ergon Energy, 2014). The general challenges of integrating greater levels of renewable generation (particularly solar) at the system-wide level, and the potential role of energy storage, have received some attention (Denholm and Margolis, 2007; Denholm et al., 2010), and motivate a number of chapters in this book. Several complicating factors for such estimations include the potential mix of other generation and level of demand-side participation, and hence the range of options that can be used to assist in ongoing supply-demand balance.

It should not necessarily be assumed that energy storage will be required. For example, a number of scenarios for 100% renewable electricity generation for the Australian NEM have been undertaken that do not include load management or additional electricity storage systems beyond existing hydropower, yet still achieve the required levels of reliability (Elliston et al., 2013). Instead, a range of technologies, including a number with inherent energy storage in either the fuel source (biogas turbines) or through additional thermal storage (concentrating solar plant with molten salt storage), can suffice. Another study for the Australian NEM undertaken by the Australian Energy Market Operator (AEMO, 2013) found additional value from some level of load management, but concluded that additional electricity energy storage through battery systems was not cost-effective at expected prices.

There has been some research on the potential to integrate energy storage into larger commercial and industrial sites through to utility-scale PV systems. For such applications, private commercial incentives are of key importance, and these depend critically on the market arrangements in place. Key aspects of these include time and locational wholesale pricing, and the management of short-term supply-demand deviations through ancillary services. Some work, for example, has looked at the use of battery storage to smooth PV output (Ellis et al., 2012). There is, of course, the potential for storage to be deployed quite separately from particular renewable energy projects. In idealized market arrangements where all potential values of PV and storage are appropriately priced, energy storage might be considered entirely separately. In practice, there are advantages to be had in collocating the PV and storage, including the ability to share site and network infrastructure. Furthermore, there are still potential mismatches between commercial arrangements and underlying economics that may support local storage—one example is the use of storage to cover ancillary loads at PV plants in the NEM (Wang et al., 2014).

10.3.5 Mini-Grids

Mini-grids represent an interesting and important midway point between stand-alone and major grid electricity systems. They typically serve remote communities that are not economical to connect to large grids due to their isolation, but that have a sufficient density and diversity of end users so that it makes sense to connect them together rather than supply them all with stand-alone systems. The use of a mini-grid also permits the use of generation technologies that might not be feasible or economical at smaller scale, such as multiple diesel gensets or biomass or small hydro facilities.

The International Energy Agency (IEA), among others, foresees a critical role for mini-grids in providing universal energy access in developing countries that still haven't achieved near complete electrification (Yadoo and Cruickshank, 2012). There are also some mini-grid deployments in developed countries for isolated communities, such as those on islands, or those located in particularly harsh and remote locations. Australia has a number of these.

Many of these systems are currently based on diesel gensets, given their relatively low capital costs, operational flexibility, and controllability. However, diesel fuel is expensive and, in many of these locations, difficult and expensive to get to site. The cost of diesel genset electricity is, therefore, generally considerably higher than electricity from large-scale grids, although less than for stand-alone systems. There is growing interest in PV deployment, and a growing view that the savings in diesel fuel can more than offset the capital costs of these PV systems.

To maximize the benefit of PV, therefore, the systems need to maximize the displacement of diesel operation. However, conventional diesel gensets have operational difficulties when continually run at low operating levels. The gensets are easy to start and stop, and any interruption to the PV generation when completely supplying the load will lead to supply interruptions. There is, therefore, considerable interest in electrical energy storage in such systems through battery and other technologies. In Australia, there are PV-diesel-storage systems deployed on King Island and Cape Barren Island (battery systems), and a number of isolated diesel grids in Coral Bay and Marble Bar (flywheel systems).

While there have generally been reasonably attractive economic cases for these systems, there have been some challenges in practice. In part, these reflect ongoing technical challenges with batteries, and to an even greater extent, flywheel technologies. However, there is also the high cost of the energy storage systems that, together with these technical risks, makes the commercial case challenging (Hazelton et al., 2014). Furthermore, there are other options that can offer some level of equivalent supply-demand rebalancing, including low-run diesel gensets and load management, potentially with considerably less technical complexity and costs. The economics of diesel genset operation are also complex and not well incorporated into analysis at present—for example, the costs associated with periods of low-output operation or increased stops and starts.

Interestingly, there may even be a move away from the use of storage systems toward straight PV-diesel systems, given their technical simplicity and rugged performance, while still providing useful diesel savings (Corporation, 2014). This might suggest some level of caution regarding uptake of storage with PV systems on the main grid.

Still, recent work for high-value applications of storage in the Australian context has highlighted what is seen as a material opportunity for supporting fringe and remote electricity systems to assist in mitigating unreliable supply, reduce expensive fuel dependence, and assist in managing supply-demand balance, including ramp-rate control of PV and protection of system stability (Marchment Hill Consulting, 2012). In countries like Australia, it may make increasing technical and economic sense to establish local mini-grids centered in regional or suburbs that are interconnected, at least in the short term, via existing electricity distribution networks.

10.4 POSSIBLE FUTURE DEVELOPMENTS

10.4.1 Evolutionary Opportunities

The future for both PV and storage within the electricity sector falls within the broader context of potentially revolutionary change toward a smarter and cleaner power industry. A key factor here has been the emergence of a range of distributed technologies—generation, storage, and load—that may transform the industry entirely from its present large-scale and highly centralized supply and network infrastructure.

PV has been argued to be the most disruptive technology of these, in large part due to its scalability and wide range of possible end user roles. That would certainly seem to be the case to date (Schleicher-Tappeser, 2012). Should it continue to progress, the need to better match electricity demand with PV's daily cycles and weather-dependent generation will become a growing issue and require progress in demand-side technologies, such as controllable loads and electrical energy storage technologies.

There is considerable expectation of continuing falling battery technology costs (BNEF, 2012; Fuhs, 2014) driven, in large part, by the growth and technology development currently seen in electric vehicles. Indeed, an analogy to the extraordinary price reductions seen with PV is often made. However, there are some different characteristics to the technologies, notably that PV is a semiconductor, not a chemical technology, so that similar progress is not assured. It is perhaps surprising, and hence telling, that the venerable lead-acid battery technology that was first developed more than 100 years ago is still widely used in applications such as conventional car electrics, despite its many apparent disadvantages. Lithium-ion batteries have seen remarkable progress over the past decade, initially driven by mobile information and communications technology (ICT) applications and now mobile power applications such as electric vehicles. Battery costs of a range of technologies are certainly falling; however, major cost reductions will also need to see progress in other related balance of system technologies, estimated by Bronski et al. (2014) to represent around two-thirds of the costs of current battery storage systems, and almost three-quarters of residential system costs. Such progress has been seen with PV balance of system costs, so there is some precedent here.

A range of parallel smart technology developments, falling within the broad term "smart grid" and including smart meters and household Wi-Fi Internet connections, also promise to greatly improve the technical capabilities, while reducing the cost of monitoring and controlling distributed energy technologies. The implications for battery storage systems are mixed—these developments will reduce the costs of integrating these technologies, but also enable potential competition through greater demand-side participation, such as smart scheduling of loads with inherent energy storage.

There is a range of policy developments that might drive rapid progress, in a similar manner to that seen with PV FiT schemes in key countries. Recent German and Japanese policy measures for PV and storage were noted earlier. There is also a range of notable policy developments in the United States, which include (Bronski et al., 2014)

- FERC Orders 755 and 784, which facilitate utility-scale grid storage by defining grid-level use and accounting for frequency regulation and ancillary services that appropriately value the highly advantageous characteristics of fast-response battery storage
- AB 2514, which sets a storage target of 1.3 GW in California by 2020, including provisions to facilitate consumer owned or sited grid-connected storage
- California's Self-Generation Incentive Program, which provides subsidies for various generation and energy storage technologies (in particular, an approximate US$2/Watt credit for energy storage systems is facilitating solar and storage solutions)

Beyond these and perhaps additional efforts, the underlying economics of PV and energy storage will likely be the key factors. It is important to note that Schleicher-Tappeser (2012) argues that progress toward such transformation will depend largely on governance of the regulatory frameworks and the ability of key market players to develop appropriate business models. It is certainly possible that storage will have relatively little impact in countries where the deployment of distributed generation is slowing or where there are adverse changes in these regulatory arrangements. Clearly, there are some changes underway by utilities and their regulators in jurisdictions including the United States and Australia to reduce the attractiveness of distributed generation—notably by reducing incentives for DG, and changing underlying energy user tariffs, with greater fixed charge components. However, this does raise the intriguing possibility of utility and regulator attempts to block progress, driving a radical energy storage transformation through grid defection.

10.4.2 Grid Defection

There is growing discussion of whether new battery storage technologies might rid a growing number of energy users entirely of the need to remain connected to the main grid. The compelling economics of grid connection for almost all customers within areas served by the grid have made connection a given. However, this may be changing. The Rocky Mountain Institute (Bronski et al., 2014) has highlighted five forces driving the increased adoption of off-grid hybrid distributed generation and storage systems, even in areas that are served by the grid:

- Interest in reliability and resilience, notably given what would seem to be growing risks to secure grid supply from extreme weather events and increasingly stressed infrastructure
- Demand for cleaner energy, sometimes in the context of only limited mainstream support for cleaner energy by incumbent large-scale electricity industry players
- Pursuit of better economics, particularly in contexts where the economics of grid supply are marginal, including for remote, currently fringe-of-grid electricity provision
- Utility and grid frustration, a growing issue with some incumbent utility players seemingly directly impeding the uptake of innovative distributed energy options
- Regulatory changes by policy makers struggling in many cases to keep up with the rate of change being seen with distributed resources and often under pressure from incumbent players seeing their existing business models being damaged

Bronski et al. (2014) consider a scenario where more than 20 million residential customers in parts of the United States with high residential tariffs could "find economic advantage" in PV with storage systems within a decade, given moderate technology improvements in both solar and storage. Related work in the Australian context has also identified potential opportunities for customers to unplug from the grid (Szatow and Moyse, 2014).

Other work, such as that of the CSIRO future grid forum (Commonwealth Scientific and Industrial Research Organisation, 2013), has also explored possible scenarios that include substantial grid defection as distributed generation and storage costs fall, while the incumbent electricity industry fails to respond effectively.

This is not to say that grid defection will necessarily present the most societally economic outcome. After all, the grid represents a substantial public investment over more than a century and provides important opportunities to take advantage of the considerable diversity across generation and demand. However, there are currently considerable disconnects between societal value and private incentives under retail market arrangements, and some of the proposed changes to these might actually make this worse.

Riesz et al. (2014) highlights some possible future implications of this—in particular, the question of whether network connection remains the highest societal value approach for energy services delivery to particular customer classes, or whether distributed solutions disconnected from the grid become a lower societal cost. Given the apparently different drivers between public and private costs and benefits within the electricity industry, and broader motivations for some key stakeholders, it is of course possible to envisage futures where customers remain connected even when disconnection is more economical, and where customers disconnect even when this is not the economically highest value.

Even if disconnection is not the public or private least-cost option, the possibility of grid defection provides a potentially useful discipline on network operators and the regulators who are meant to regulate them for the greatest public good. Inevitable transition issues between the two are also likely.

10.4.3 What happens next?

The short answer, of course, is that nobody knows. The economics of PV and storage seem likely to play a key, although by no means sole, role. Together, the technologies offer the millions of energy users on the grid the opportunity to participate far more actively in delivery of their energy services. The private incentives they see from retail electricity market arrangements and any explicit support policies will likely play an important role in their decision making. Some might argue that the most important task ahead for policy makers is to better align underlying industry economics with such private incentives. Beyond the complexity of this, there are many reasons why there are limits to what may be achieved. Ideally, policy makers will look at how additional policy measures might better align stakeholders with societally valuable outcomes.

Falling PV and storage costs will of course improve both the social and private economics of deployment. However, that is unlikely to drive appropriate change in the absence of broader policy and regulatory efforts.

The International Energy Agency provides a suitably cautious assessment (IEA, 2014), noting that while electricity storage could play a wide range of roles in future energy systems, it is unlikely to be transformative given current costs and performance, in the presence of competing options both on the supply side (notably flexible and highly dispatchable plants) and demand side (including demand-side response and smart-grid options). However, the IEA also notes that electricity storage offers unparalleled modularity, controllability, and responsiveness, while current drivers for electricity storage (including the challenge of integrating growing variable renewable generation but also system planning and end user applications) are highly context-specific and will likely require changes to market and broader regulatory frameworks. It also notes that off-grid applications represent the most economically attractive deployment opportunities, offering sufficient additional value to offset their high costs. Such options provide, of course, an opportunity to better understand the technical and economic performance of such technologies.

For widespread grid deployment, however, the hard work of market and regulatory change, including the institutional and governance frameworks to direct such change (Sue et al., 2014; Passey et al., 2013), should not be underestimated, regardless of underlying economics.

ACKNOWLEDGMENTS

This work has been supported by a range of research projects funded by sources including the Australian Renewable Energy Agency, the CSIRO future grid cluster, and the Cooperative Research Centre for Low Carbon Living.

REFERENCES

Australian Energy Market Operator (AEMO), 2013. 100 Per Cent Renewables Study. Sydney, May.

BNEF, 2012. Electric Vehicle Battery Prices Down 14% Year on Year. Bloomberg New Energy Finance. April 16, http://www.bnef.com/PressReleases/view/210.

Braun, M., et al., 2012. Is the distribution grid ready to accept large-scale photovoltaic deployment? State of the art, progress, and future prospects. Prog. Photovolt. Res. Appl. 20 (6), 681–697.

Bronski, P., et al., 2014. The Economics of Grid Defection: When and Where Distributed Solar Generation Plus Storage Competes with Traditional Utility Service. In: Rocky Mountain Institute/ Homer Energy/CohnReznick Think Energy From, http://www.rmi.org/electricity_grid_ defection.

Colthorpe, A., 2014a. Japan Launches Subsidies for Lithium-Ion Battery Storage. In: PVTECH Retrieved (10.04.14) from, http://www.pv-tech.org/news/japan_launches_subsidies_for_ lithium_ion_battery_storage.

Colthorpe, A., 2014b. Put Up or Shut Up—Time for Storage. In: SolarBusinessfocus Retrieved (10.04.14) from, http://www.solarbusinessfocus.com/articles/put-up-or-shut-up-time-for-storage.

Commonwealth Scientific and Industrial Research Organisation, 2013. Change and Choice: The Future Grid Forum's Analysis of Australia's Potential Electricity Pathways to 2050. In: CSIRO From, http://www.csiro.au/Organisation-Structure/Flagships/Energy-Flagship/ Future-Grid-Forum-brochure.aspx.

P. and W. Corporation, 2014. Solar/Diesel Mini-Grid Handbook. Power and Water Corporation, Darwin, NT.

Denholm, P., Margolis, R.M., 2007. Evaluating the limits of solar photovoltaics (PV) in electric power systems utilizing energy storage and other enabling technologies. Energy Policy 35 (9), 4424–4433.

Denholm, P., et al., 2010. The Role of Energy Storage with Renewable Electricity Generation. NREL/TP-6A2-47187. Golden, CO, National Renewable Energy Laboratory.

Ellis, A., et al., 2012. PV Power Output Smoothing Using Energy Storage. Sandia National Laboratories, Albuquerque, NM, SAND2012-6745.

Elliston, B., et al., 2013. Least cost 100% renewable electricity scenarios in the Australian National Electricity Market. Energy Policy 59, 270–282.

Energy Supply Association of Australia, 2013. Electricity Gas Australia 2013. Retrieved (18.07.14) from, http://www.esaa.com.au/policy/EGA_2013.

Energy Users Association of Australia, 2012. Electricity Prices in Australia: An International Comparison. In: Prepared by Burce Mountain, Director, CME. Retrieved 18/07/2014, 2014, from, http://www.euaa.com.au/wp-content/uploads/2012/03/INTERNATIONAL-ELECTRICITY-PRICE-COMPARISON-19-MARCH-2012.pdf.

Ergon Energy, 2014. Battery Technology on Network an Australian First. In: From, https://www.ergon.com.au/about-us/news-hub/media-releases/regions/general/battery-technology-on-electricity-network-and-australian-first.

Fuhs, M., 2014. Forecast 2030: stored electricity at $0.05/kWh. PV Magazine.

Goel, S., Watt, M., 2013. Evaluating the Benefits of Implementing Existing Storage Technologies with Residential PV Rooftop Systems. In: EUPVSEC 2013Retrieved (18.07.14) from, http://ceem.unsw.edu.au/sites/default/files/documents/6CV.5.33.pdf.

Hazelton, J., et al., 2014. A review of the potential benefits and risks of photovoltaic hybrid mini-grid systems. Renew. Energy 67, 222–229.

International Energy Agency (IEA), 2014. Energy Technology Perspectives. Paris, June. Available at www.iea.org.

Kind, 2013. Disruptive Challenges, Report for the Edison Electric Institute.

MacGill, I., 2010. Electricity market design for facilitating the integration of wind energy: experience and prospects with the Australian National Electricity Market. Energy Policy 38 (7), 3180–3191.

Marchment Hill Consulting, 2012. Energy storage in Australia: commercial opportunities, barriers and policy Options. Report for the Clean Energy Council MHC, Melbourne.

Oliva, H., MacGill, I., 2014. Value of net-FiT PV policies for different electricity industry participants considering demand-side response. Prog. Photovolt. Res. Appl. 22 (7), 838–850.

Oliva, S., et al., 2013. Estimating the net societal value of distributed household PV systems. Solar Energy 100, 9–22.

Outhred, H., MacGill, I., 2006. Electricity Industry Restructuring for Efficiency and Sustainability—Lessons from the Australian Experience. In: 2006 ACEEE Summer Study on Energy Efficiency in Buildings, Asilomar.

Parkinson, G., 2013. Germany Finances Major Push into Home Battery Storage for Solar. Retrieved (10.04.14) fromhttp://reneweconomy.com.au/2013/germany-finances-major-push-into-home-battery-storage-for-solar-58041.

Passey, R., et al., 2011. The potential impacts of grid-connected distributed generation and how to address them: a review of technical and non-technical factors. Energy Policy 39 (10), 6280–6290.

Passey, R., Watt, M., Morris, N., 2013. The Distributed Energy Market: Consumer & Utility Interest, and the Regulatory Requirements'. Australian PV Association, Australia.

REN21, 2014. Renewables 2013 Global Status Report. R. Secretariat. Paris.

Riesz, J., Magnus, H., Gilmore, J., Riedy, C., 2014. Perfect storm or perfect opportunity? Future scenarios for the electricity sector. In: Sioshansi, F.P. (Ed.), Distributed Generation and its Implications for the Utility Industry. Elsevier, Netherlands.

Roberts, B., 2009. Capturing grid power. IEEE Energy and Storage Magazine.

Schleicher-Tappeser, R., 2012. How renewables will change electricity markets in the next five years. Energy Policy 48, 64–75.

Sue, K., et al., 2014. Distributed energy storage in Australia: quantifying potential benefits, exposing institutional challenges. Energy Res. Soc. Sci. 3, 16–29.

Szatow, T., Moyse, D. 2014. What happens when we un-plug: exploring the consumer and market implications of viable, off-grid energy supply. Energy for the People.

Tant, J., et al., 2013. Multiobjective battery storage to improve PV integration in residential distribution grids. IEEE Trans. Sustain. Energy 4 (1), 182–191.

UBS, 2014. Solar+storage, bottom up analysis confirms opportunity. Utilities Sector.

Wang, J., Bruce, A., MacGill, I., 2014. Electric energy storage in the Australian National Electricity Market – evaluation of commercial opportunities with utility scale PV. In: Proc. Asia Pacific Solar Research Conference, Sydney; 12/2014.

Watt, M., MacGill, I., 2014. Grid parity and its implications for energy policy and regulation. Advanced concepts in photovoltaics. In: Nozik, A.J., Conibeer, G., Beard, M.C. (Eds.), In: Energy and Environment Series No. 11Royal Society of Chemistry, Cambridge, UK.

Watt, M., et al., 2014. PV in Australia 2013. From, http://apvi.org.au/wp-content/uploads/2014/07/PV-in-Australia-Report-2013.pdf.

Wilkinson, S., 2014. Japan Energy Storage Subsidies to Spark Market Growth. Retrieved 09/05/2014, 2014, from, http://technology.ihs.com/495095/japan-energy-storage-subsidies-to-spark-market-growth-says-sam-wilkinson-ihs-research-manager-for-energy-storage.

Yadoo, A., Cruickshank, H., 2012. The role for low carbon electrification technologies in poverty reduction and climate change strategies: a focus on renewable energy mini-grids with case studies in Nepal, Peru and Kenya. Energy Policy 42, 591–602.

Part III

Environmental and Social Impacts

Chapter 11

Environmental Issues Associated with Solar Electric and Thermal Systems with Storage

Bent Sørensen
Department of Environmental, Social and Spatial Change, Roskilde University, Roskilde, Denmark

Chapter Outline

11.1 INTRODUCTION

Solar energy systems generally have a complex structure, making it a major task to identify all possible environmental impacts needed for performing a life-cycle analysis and assessment (LCA). On the other hand, such impacts must be estimated and preferably quantified, in order to make a meaningful comparison between solar solutions and other renewable or nonrenewable options. Impacts derive from the solar radiation-collecting device itself, whether converting the radiation into electricity, heat, or both, as well as from the system for mounting the collectors and transporting the energy to load areas, and further from the specific or general system employed to handle the intermittency of the source flow and mismatch between energy production and use, by employing load management, trade between the system considered and other energy systems, or energy storage. One complication is that some of these systems may serve other units in the overall energy systems besides the solar parts, implying a need for a scheme for dividing impacts fairly between segments.

Solar Energy Storage. http://dx.doi.org/10.1016/B978-0-12-409540-3.00011-6

Historically, life-cycle analysis was first made for the collector cells and modules in isolation, and analysis for other components such as framing, power conditioning, and energy storage (the latter being, as mentioned, not dedicated solely to the solar system) were omitted or rather left to be addressed in a total energy system assessment. The early studies of impacts from photovoltaic cell manufacture, such as Boeniger and Briggs (1980), Watt (1993), Sørensen (1993a), Sørensen and Watt (1993), and Moskowitz et al. (1995), found quite sizable environmental and social impacts, but with the expectation that impacts would diminish as the technologies became more mature. At that time, solar technologies were very expensive, not only relative to the prevailing fossil energy costs, but also to the most cost-effective renewable energy solutions, hydro and wind energy. Thus, it was stipulated that a required reduction in material use and energy input in manufacture and in solar panel installation would reduce environmental impacts as well as cost.

Today, solar systems remain fairly expensive despite substantial price drops, but for some applications and locations they are approaching economic viability. It is therefore important to redo the evaluations, which earlier were based on theoretical forecasts of the technology development. The relative impacts of solar, wind, and biofuel options—the three most important renewable energy technologies to consider beyond the nearly exhausted hydro potential—are of great importance for planning a transition to renewable energy through an optimum mix of these options. Although current knowledge makes it clear that the impacts of solar and biofuel options are fairly significant, they remain smaller than those of fossil or nuclear options (Tsoutsos et al., 2005). In addition, the impacts of wind power (as well as its cost at suitable locations) are coming out in many cases as substantially below those of solar solutions (which also depend strongly on the location considered). Although these conclusions are presently emerging, the uncertainty in the assessment is still quite large, and it would seem wise to explore many renewable energy options, both technically and economically, where economic viability is based on the outcome of a full life-cycle evaluation of direct and indirect economic implications.

11.2 SOLAR CELLS

Currently, the most common solar cells installed are crystalline or multicrystalline silicon cells. Amorphous cells seemed promising a while ago, but have not advanced in terms of efficiency or cost. Various photoelectrochemical cells based on dye sensitizing by use of metal complexes or organic dyes have been manufactured but found unreliable and of short average lifetime. The same is true for cells using organic polymers or fullerenes to increase the electron-transfer effective surface area. Scientific journals and magazines contain a flow of articles on new wonder-cells, none of which have so far made it beyond the laboratory. More serious alternatives to silicon in conventional two or more layer cells include CdTe cells and combinations of chemical group 13-15

(III-V in old notation), such as GaAs. Some of these have higher negative impacts than silicon cells, due to smaller global material abundance or to material toxicity.

Manufacture of solar cells involves several chemical and physical processes that may give rise to health and accident risks as well as pollution, and that require input of energy. Ideally, recycling can reduce the environmental impacts, but in return it needs additional energy use. Currently, few schemes for solar panel recycling are in place. Pollution can be reduced by avoiding emissions to air, wastewater, or soil in manufacturing plants, and by proper handling of decommissioned cells. However, experience with solar cell recycling plants and other end-of-life technologies is not yet available.

The manufacturing process has changed since the early solar cells were marketed. Microprocessor scrap is no longer used as silicon raw material, but a less refined stock of "solar-cell-grade" silicon forms the starting point for photovoltaic cell production. This grade of silicon is much less expensive than the microprocessor grade, but more expensive than the former cost of the microprocessor scrap. Multicrystalline cells have today reached a higher market share than monocrystalline cells, and the radiation-to-electricity conversion efficiency has gradually improved. The amorphous cell market is stagnant and the organic cells have left the marketplace after serious lifetime failures. There is still scientific interest in the aforementioned potentially more efficient crystalline cells based on materials other than silicon, eventually using several stacked layers in order to capture more of the range of frequencies present in solar light. In the current marketplace, however, only the CdTe cells have a modest share. They typically have lower efficiency than the silicon cells, but also lower production cost when using thin-film techniques.

11.3 SOLAR ELECTRICITY SYSTEMS

Cells are connected into modules and modules are assembled in system arrays, including DC-to-AC inverters, integrated into facades or roofs of buildings, or mounted on frames placed on marginal land. Current installed system costs for building-integrated solar cells are typically 1.8-3.5 €/W of rated power (IEA, 2014; 1 € is presently about $1.1). At latitudes around 30°, the modules typically produce some 1300 Wh of energy per year and per rated W (and in locations with low turbidity and cloud cover more), which works out to an average price of 0.1-0.2 €/kWh, or about half the price level prevailing less than a decade ago.

Environmental impacts comprise a number caused by the particular production processes employed and are therefore different for different types of cells, in regard to labor intensity, energy input, and the need for special substances with associated impacts (e.g., silane for amorphous and chlorosilane for crystalline Si cells, phosphine for doping, and other sets of toxic chemicals for CdTe or GaAs cells). Recycling of as many of the chemicals used in manufacture as possible is essential, as is recycling of decommissioned solar cells (Sørensen, 1993b; 1997). Further impacts relate to the system assembly, including any framing and support, as well as circuitry and inverters. During the operational phase, there are few impacts other than possibly from detergents used for occasional cleaning of the collector surfaces, but during decommissioning, there may again be substantial impacts related to energy input requirements, as well as to handling of the many peculiar substances embedded in the module structure.

Because of the special need for chemicals known to present dangers such as toxicity, explosion risk, and diseases associated with inhalation of vapors from the processing of photovoltaic (PV) cells, life-cycle analyses were undertaken at an early stage of the solar cell development toward commercialization. However, many of the studies were not complete life-cycle studies, but single-issue analyses, notably of greenhouse gas emissions and energy payback times (an impact that is in principle irrelevant, if the overall full social cost of the system is acceptable). Figures 11.1–11.4 show the life-cycle results for four specific items in the case of multicrystalline Si-cells, taken from an early quantitative investigation pertaining to the 1995 PV development stage (Kuemmel et al., 1997) and aimed toward estimates of future impacts for a fully mature solar manufacturing industry (Sørensen, 2011a), which in the end will use only energy produced by the same kind of cells or other renewable energy sources, for any energy input required during the entire life.

Greenhouse gas emissions and land use were already mentioned in an early study as potentially important impacts (Phéline, 1981), but without quantification. The emission of greenhouse gases shown in Figure 11.1 shows the quantitative effect of continued reduction in quantities of materials and in required energy input to the processes that cells, modules, auxiliaries, mounting, and

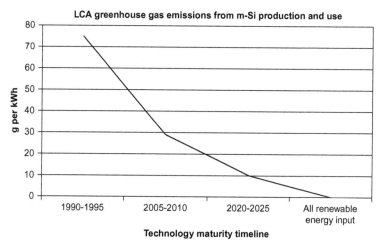

FIGURE 11.1　LCA greenhouse gas emissions from m-Si solar cell production and use. *(Based on Kuemmel et al., 1997.)*

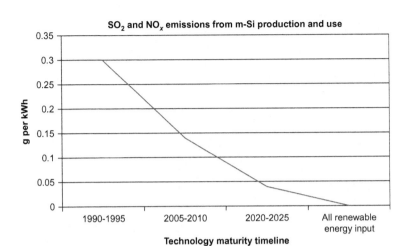

FIGURE 11.2   Air pollution from m-Si solar cell production and use. *(Based on Kuemmel et al., 1997.)*

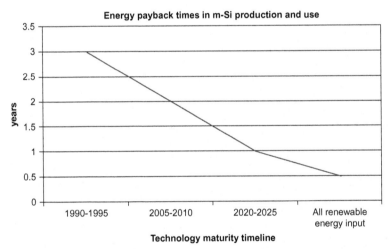

FIGURE 11.3   Energy payback times for m-Si solar cell production and use. *(Based on Kuemmel et al., 1997.)*

other installation tasks required. A similar behavior is found for air pollution, initially entirely from the fossil fuels used, but with time from a mix with more renewable energy, as shown in Figure 11.2. The energy payback period is shown in Figure 11.3, exhibiting again the change in energy input mix, not so much associated with maturity of the industry as with the assumption that mitigation of climate change impacts will lead to a reduced use of fossil energy

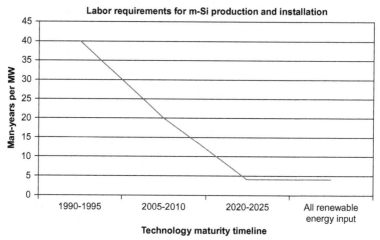

FIGURE 11.4 Labor requirements for m-Si solar cell production and use. *(Based on Kuemmel et al., 1997.)*

sources, as suggested by several political responses to the already observed negative warming impacts. Figure 11.4 shows the labor required by the photovoltaic industry, becoming reduced as the material and energy input diminishes along with the cost reduction primarily sought.

The aforementioned topical investigations are in agreement with the calculations behind Figures 11.1–11.3. For example, Alsema (2000) finds an energy payback period for late 1999 PV cells and projections for 2010 and 2020 cells that are in substantial agreement with those of Figure 11.3, and Fthenakis and Kim (2011) find CO_2, NO_x, and SO_x emissions for cells produced around 2007 that are very similar to those of Figure 11.1. Alsema's CO_2 emissions agree with ours in 1999 and 2010, but are estimated somewhat higher for 2020.

A central problem in LCA is monetizing. On one hand, there are impacts that are difficult to translate into economic terms but that still appear important for deciding among technologies, and on the other hand, even if valuation is possible, it may greatly increase the uncertainty in the estimate. One example is the effect of global warming caused by anthropogenic uses of fossil energy sources. The emissions (Figures 11.1 and 11.2) are indeed very accurate, but their economic consequences are much more uncertain, due to the range of human activities that are influenced by changing climates. Some of these consequences are discussed in the scientific summary reports of the Intergovernmental Panel on Climate Change, IPCC (2014), and additional impact analysis relevant to life-cycle assessment can be found in Sørensen (2011a). Human lives are affected and lost due to extreme events, agricultural setbacks, altered disease patterns, as well as direct impacts of changed temperature variations, in ways that exhibit large geographical differences. For example, as a result of global warming, fewer lives are lost in Siberia due to cold winters, while more lives are lost in Central Africa due to hot summers. This situation influences traditional methods of economic valuation, where the societal cost of a life lost is different in rich and poor countries. These differences are traditionally determined by gross national product (GNP) or by purchasing power parity estimates, but a fully egalitarian valuation with the same value of a life lost everywhere will typically give the monetized climate change impact a two-orders-of-magnitude-higher value than the mere GNP-adjusted figures (Sørensen, 2011a). Figure 11.5 gives the climate change impact of solar cells as estimated when assuming the valuation of a human life made in a major study for the European Union and used globally. One should note that the methodology for life-cycle assessment that I use considers the economic setback for society when losing one of its members and does not enter into an ethical discussion of the value of a human life.

Figure 11.6 shows the results of the analysis regarding the effects of air pollution from the m-Si solar cell life cycle. Here, mortality and morbidity from diseases caused by air pollution are estimated, the latter in terms of workdays or healthy retirement time lost (presented as disability-adjusted shortening of life in years, DALY). Again, the impacts drop to zero if the future energy system is entirely based on renewable energy, as it could well become (Sørensen, 2014).

While Figures 11.1–11.6 give timelines of some important impacts from m-Si solar cell production and use, there are several other impacts that need to be included in a life-cycle assessment, e.g., for comparing different energy solutions (solar, wind, biomass) or choosing the particular technology most beneficial (crystalline or amorphous silicon, CdTe, GaAS, photoelectrochemical

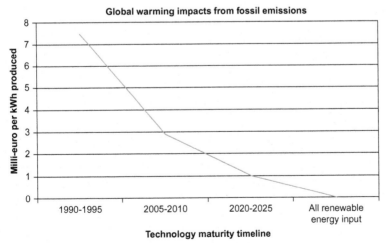

FIGURE 11.5   Global warming impacts from fossil fuel emissions associated with m-Si solar cell production and use. *(Based on Kuemmel et al., 1997.)*

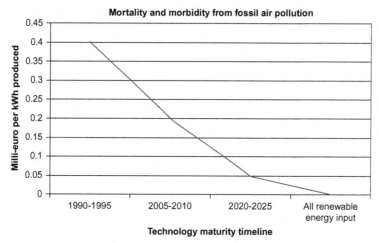

FIGURE 11.6 Mortality and morbidity from fossil air pollution associated with m-Si solar cell production and use. *(Based on Kuemmel et al., 1997.)*

solar cells). A range of these is estimated in Tables 11.1 and 11.2, focusing on technical and economical impacts, respectively. Included are the impacts illustrated in the figures, but taken specifically for the present situation (as of 2015). Land use is set at zero, considering that the solar panels considered are building-integrated, although there may be some additional land used by the manufacturing facility. For dedicated solar cell parks, a land use impact must be attributed (e.g., taken as the cost of purchasing the land area). The last box of impact categories contains those that are hard to quantify, such as resilience and supply security. Even if a monetary value cannot be assigned, such considerations may be important for the political decision of choosing between different energy solutions.

A substantial number of publications give estimates of the range of life-cycle impacts from solar systems, similar to the ones described previously, as regards global warming, energy use, and health impacts, occasionally with an additional focus (e.g., on labor requirements; Stamford and Azapagic, 2012; Masanet et al., 2013). However, labor requirements are an uncertain quantity, partly because they depend on the level of automatization in manufacture, say by use of robot technology, and partly because labor requirements are likely to diminish with maturity of the technology, as needed to reach cost goals. Additional life-cycle studies have been made for large desert installations of solar cell arrays (Ito et al., 2003) and for concentrating solar installations (Whitaker et al., 2013; Corona et al., 2014; Barton, 2014). Such systems are only viable at certain geographical locations with very clean air (low turbidity and hence little scattering of radiation) and high levels of solar radiation.

TABLE 11.1 Life-Cycle Physical Impacts from Building-Integrated m-Si Solar PV Systems

	Impact Type:	Uncertainty
Environmental impacts (m-Si based rooftop systems).	**Emissions (g/kWh)**	
Releases from fossil energy in all steps of cycle	*For year 2015:*	
CO_2	19	L,g,m
SO_2 and NO_x	0.08	L,r,n
Special substances (chlorosilanes, metal particles, etc.)	NQ	
Land use (for building-integrated system)	0	
Social impacts	**Accidents**	
Occupational injuries	$(10^{-9}$ cases/ kWh)	
Silicon provision and cell manufacture	~1	
Panel assembling, mounting, and installing	<1	
Operation	~0	
Decommissioning	<1	
Economic impacts	**Miscellaneous**	
Energy payback time	1.5 (y)	
Labor requirements	12 (man-y/ MW)	
Other impacts		
Supply security (plant availability)	High	
Robustness (technical reliability)	High	
Global issues (nonexploiting)	Compatible	
Decentralization and choice	Good	
Institution building (other than required grid)	Modest	

NA, not applicable; NQ, not quantified; L, M, H: low, medium, or high uncertainty; l, r, g: local, regional, or global impact; n, m, d: near, medium, or distant time frame.

TABLE 11.2 Monetized Assessment of Impacts from m-Si Building-Integrated Solar PV Systems

	Monetized Value 2010 (m€/kWh)[a]	Uncertainty and Ranges
Environmental impacts (m-Si based rooftop systems)		
Releases from fossil energy in all steps of cycle	*For year 2015:*	
Greenhouse effect from fossil emissions[b]	1.9	H,r,n
Land use (for building-integrated system)	0	H,r,n
Visual intrusion (annoyance from reflections, etc.)	NQ	L,l,n
Social impacts		
Mortality and morbidity from fossil-burning air pollution[c]	0.125	H,r,n
Special substances (chlorosilanes, metal particles, etc.)	NQ	
Occupational injuries		
Silicon provision and cell manufacture	0.1	M,l,n
Panel assembling, mounting, and installing	<0.1	L,l,n
Operation	0	L,l,n
Decommissioning	<0.1	L,l,n
Economic impacts		
Direct costs (southern Europe or USA)	100-200	M
Energy payback time	NA	
Labor requirements	NQ	
Benefits from power sold (ignoring intermittency cost)	100-300	
Evaluation of other impacts		
Supply security (plant availability)	NQ	
Robustness (technical reliability)	NQ	

Continued

TABLE 11.2 Monetized Assessment of Impacts from m-Si Building-Integrated Solar PV Systems—Cont'd

	Monetized Value 2010 (m€/kWh)	Uncertainty and Ranges
Global issues (nonexploiting)	NQ	
Decentralization and choice	NQ	
Institution building (other than required grid)	NQ	

NA, not applicable; NQ, not quantified; L, M, H: low, medium, or high uncertainty; l, r, g: local, regional, or global impact; n, m, d: near, medium, or distant time frame.
[a]By 2010, 1€ was approximately 1.1 US$.
[b]Global warming effects over 100 years, as estimated in Sørensen (2011a) using the 2.6 M€ EU valuation of a life lost.
[c]Mortality as for warming-related deaths, morbidity using the 65 k€ EU valuation for a disability-adjusted life shortening of a year (DALY).

11.4   SOLAR ELECTRIC STORAGE

Because of solar source intermittency and more generally due to mismatch between production and energy demand, a full life-cycle analysis should include the additional components and arrangements needed to deal with this intermittency. In this book, the focus is on energy storage, although as mentioned in the introductory chapter, measures for avoiding mismatch by power exchange between different national or international systems, or demand management, may constitute less expensive ways of dealing with intermittency than active energy storage. A practical difficulty in adding storage to the life-cycle assessment is that storage of power for later electricity regeneration is usually a common facility for all electricity units attached to the system, whether solar or based on other energy conversion techniques. Therefore, the impacts have to be distributed between the solar cell systems and other systems, presumably in proportion to how much each system draws on the stores incorporated. This means that an analysis of impacts can only be made when a scenario for the entire power supply system has been established and accepted for the future distribution of investments. Examples of such scenarios and their impacts on intermittency can be found in Sørensen (2014). Here, the analysis will be restricted to studying the impacts for each promising energy storage system in isolation, and it will be left to future energy planners to decide on the solar share and the selection of stores to be used in a particular energy supply system.

The primary energy storage systems relevant to solar power are underground storage of hydrogen or other gases, pumped hydro, batteries, and other reversible electrochemical devices. Magnetic storage at ultralow temperatures, using

superconducting coil charging, is possible but further away from economic viability than the other technologies mentioned. Short-term smoothing by flywheels or capacitors is important for dispersed power generation in isolated
systems, but not very relevant for systems where a large number of electricity
producers are connected to a common transmission grid of sufficient capacity.

Looking first at underground gas stores, there are impacts from establishing
the stores, from producing the gas to be stored (e.g., hydrogen) on the basis of
surplus electricity (by electrolyzers—i.e., fuel cells in reverse mode of operation), and from regeneration of electricity from the stored gas (by gas turbines
or fuel cells). LCA studies have been made for compressed air storage
(Denholm and Kulcinski, 2004; Barnhart and Benson, 2013), often employing
caverns formed in salt domes, i.e., upward salt intrusions reaching shallow depth
below the ground, where excavation can be simplified to water injection and
brine extraction, which over periods of a few years will form a cavity suitable
for gas storage, at much lower cost than drilling. Such geological formations
are frequent in regions having experienced ice age displacements of matter
during progression and withdrawal of ice cover, such as nonmountainous parts
of northern Europe and northern North America. Compressed air storage facilities use compressors and turbines and need heat management, including burners
currently using natural gas for the energy recuperation stage (Sørensen, 2010).
Similar caverns can be used for underground hydrogen storage, sometimes
requiring a cavity liner to reduce pressure losses, and in this case, no auxiliary
energy input for extraction is required. Used as an electricity store, the hydrogen
would, as mentioned, be produced by electrolyzers and electricity regenerated
by turbines or fuel cells, implying that only data for the excavation/flushing of
the cavern can be taken from existing compressed air LCA studies.

Caverns of the kind suitable for hydrogen storage include the salt dome
structures, the even cheaper aquifer upward-bend types, and the more expensive
rock-drilled cavities (see illustration in Chapter 1; Sørensen, 2011b). Both salt
dome and aquifer caverns are currently in use for energy storage (of natural
gas), for example in Denmark, so costs and life-cycle inventories may in principle be derived. Although no LCA has yet been made, but only cost estimates
for future hydrogen stores (Lord et al., 2011; Simón et al., 2014), one may use
some data from the establishment of compressed air caverns in salt domes as a
proxy for the hydrogen stores. Compressed air stores are considered less relevant for solar electricity systems because they are aimed at short-term storage,
such as night to day, but not the weekly effects of changing weather systems that
reduce solar power output in overcast or rainy conditions, or the seasonal variations present in all areas away from the equator. The plant equipment of a
hydrogen store has some similarity to that of a compressed gas store (compressors, gas turbines), but the overriding cost element is electrolyzers. The 2011 US
study by Lord et al. finds a cost of 1.6 US$/kg H_2 for establishing caverns in salt
domes and filling them with compressed hydrogen (two roughly equal shares in
cost), and 0.8 US$/kg H_2 if the caverns are in aquifers. The 2014 EC study by

Simón et al. has the salt dome cavern cost (excavation only) reduced to between 0.1 and 0.3 €/kg H_2 for large installations after the year 2025. To this some 0.7 €/kg H_2 for topside equipment (including compressors) should be added, and another 1.0 €/kg H_2 for the electrolyzers. Finally, there is an estimated 0.6 €/kg H_2 operation and maintenance cost, plus what may have to be paid for the input electricity and its transmission to the storage site.

Because the input electricity in the solar case would be a surplus not able to find other uses, its economic value could be put at zero, whereas the output electricity would have the value that the power market is willing to pay at the moment of regeneration. Alternatively, the hydrogen produced may be sold as a vehicle fuel, an avenue that would be highly profitable if a hydrogen vehicle fleet were already in place. According to the European study of Simón et al. (2014), this would have the potential to be profitable even if the market price were paid to the solar cell power provider of the electrolyzer input electricity.

A large number of studies claim to have calculated life-cycle impacts of hydrogen production pathways (an overview with many references is given in Bhandari et al., 2014; important recent studies are Patyk et al., 2013; Cetinkaya et al., 2012; Dufour et al., 2012). However, the majority of the studies are not full life-cycle analyses, but only estimates of a few impacts such as energy use and the associated greenhouse gas emissions, which in any case is highly dependent on the mix of methods used to generate the energy needed for electrolysis or other avenues of hydrogen production. Furthermore, the results are rarely disaggregated so that the impacts from electrolysis can be sorted out as needed in the present context of storage. For instance, Cetinkaya et al. (2012) include solar PV hydrogen production, but without singling out impacts from electrolysis. Because the largest impacts found in studies disclosing such information are usually from the electricity input to electrolysis, which here is not considered a cost of the electrolysis step (Tables 11.1 and 11.2), the published aggregated data cannot be used. Nearly all the literature studies focus on hydrogen production for delivery to filling stations serving hydrogen-fueled vehicles. This further implies very different conditions from the ones required here, regarding time-wise utilization of the plant as well as compression equipment, where filling station storage is often assumed to be in metal containers at 30 MPa or more (i.e., considerably higher than needed for underground hydrogen storage). Life-cycle analyses for hydrogen storage and subsequent use on the small scale relevant for automotive applications were performed earlier (see Sørensen, 2011b).

A few studies do allow the extraction of data for the electrolysis step separately. Table 11.3 gives the results derived from Patyk et al. (2013) for a high-temperature solid-oxide laboratory-size electrolysis unit (SOEC). Impacts from a solid oxide fuel cell (SOFC), identical to an SOEC except for the reversal of operation, have been studied by Pehnt (2003). The study by Dufour et al. (2012) considers a conventional alkaline electrolysis unit, suggesting lower material requirements than for the SOEC, and it has PV energy input as one

TABLE 11.3 Life-Cycle Physical Impacts from Geological Seasonal Hydrogen Store, Using High-Temperature Electrolyzers to Convert Solar Power into Hydrogen and Back

	Impact Type:	Uncertainty and Ranges
Environmental impacts *construction phase* (*salt dome or aquifer*)	**Emissions (g/rated MJ)[a]**	
Salt dome excavation by flushing		
CO_2 emissions from energy use: natural gas or renewable	<0.2	L,g,m
Water usage	Modest	L,r,n
Land use	Modest	L,l,n
Waste (brine, may be used for salt production)	Low	L,r,n
Alternative: aquifer, accessed by drilling		
CO_2 emissions from natural gas or renewable energy use	<0.1	L,g,m
Land use	Modest	L,l,n
Electrolyzer (high-temperature SOEC) and surface plant		
Energy used in manufacturing: fossil or renewable[b]	0.015 MJ/ rated MJ[a]	L,g,m
Steel[b]	0.015	L,r,n
Nickel oxide[b]	0.004	L,r,n
Zirconia[b]	0.003	L,r,n
Lanthanum[b]	0.001	L,r,n
Solvent used[b]	0.035	L,l,n
Water use (see Figure 11.7)	Modest	L,r,n
Land use (see Figure 11.7)	Modest	L,l,n
Operational phase		
In addition to energy counted under PV system	Small	L,l,n
Social impacts	**Accidents**	
Occupational injuries (based on similar industries)	(10^{-9} cases/ rated MJ)	
Excavation	<1	

Continued

TABLE 11.3 Life-Cycle Physical Impacts from Geological Seasonal Hydrogen Store, Using High-Temperature Electrolyzers to Convert Solar Power into Hydrogen and Back—Cont'd

	Impact Type:	Uncertainty and Ranges
Electrolyzer and other equipment	<1	
Operation	~0	
Decommissioning	<1	
Economic impacts	**Miscellaneous**	
Energy payback time	Small (y)	
Labor requirements	NA	
Other impacts		
Supply security (plant availability)	High	
Robustness (technical reliability)	High	
Global issues (nonexploiting)	Compatible	
Decentralization and choice	Modest	
Institution building (other than required grid)	Modest	

Assumed CO_2 emission from natural gas burning: 115 g/MJ.
NA, not applicable; NQ, not quantified; L, M, H: low, medium, or high uncertainty; l, r, g: local, regional, or global impact; n, m, d: near, medium, or distant time frame.
[a]Rated MJ refers to the capacity of the store, not the production of the electrolyzer.
[b]Based on Patyk et al. (2013) and assuming an electrolyzer life of 10,000 h of operation.

of its case studies, but without separating the impacts from the electrolysis plant. According to Patyk et al., this step is by far the most important for LCA impacts over the operating life of a hydrogen storage facility. It should be kept in mind that an underground hydrogen store may use the same fuel cell in forward and reverse mode for extracting the electric power from hydrogen and converting it to hydrogen for storage. This implies that additional impacts from power regeneration equipment may be taken as negligible, relative to the ones already included. An issue is that both of the studies referred to state the impacts per unit of hydrogen produced. This is meaningless for underground hydrogen stores, because the amount of hydrogen pumped into them depends not only on the renewable energy source of power, but also on the entire electricity and hydrogen usage system, as well as on utility policy, power exchange arrangements, and demand structure and management (Sørensen, 2014). In order to avoid this complication, the numbers given in Table 11.3 are based on the energy rating of the storage itself, and the data pertaining to the electrolysis unit are calculated from those given in the references by assuming an operational life of 10,000 h for the unit (which at present is at the high end of experiences).

The cavern establishment is found to cause very modest impacts, with aquifer drilling even lower than salt dome flushing. The high-temperature electrolyzer consists of a solid-oxide fuel cell in reverse operation, and its fabrication involves a number of impacts related to material and electrolyte losses, as seen in Table 11.3. The possible negative impacts of these depend on schemes for material recycling involved, but most of the impacts are still from the energy required for manufacture of the electrolyzer unit—energy that may be of fossil or renewable origin. The impact assessment presented in Table 11.4 considers both alternatives, reflecting impacts from using currently typical energy mixes as well as future ones in an all-renewable energy system such as the ones discussed by Sørensen (2010, 2011b, 2014). Only global warming impacts are monetized in Table 11.4, because the remaining environmental and social impacts depend on more complex details of the electrolyzer manufacturing process not reflected in the laboratory-scale SOEC data used here. Figure 11.7 relates the various impacts to the specific materials that cause them, indicating the possible need to develop high-temperature fuel cells that do not rely on materials such as lanthanum.

Pumped hydro is another option for solar energy storage. Impacts from hydro installations vary quite considerably, with the establishment of reservoirs as a major contributor (Sørensen, 2010, 2011a). Pumped hydro impacts are similar to those of other hydro installations, as they need an upper reservoir to function. The extra equipment for upward pumping is often a minor source of impacts compared to those of the rest of the facility, but the pumping energy reduces the overall storage cycle efficiency. Ribeiro and Silva (2010) find impacts dominated by energy for dam construction and land use, but in the case of pumped hydro, dams are not necessarily required (Prasad et al., 2013).

Storing electric energy in batteries is another option that has been in use for bulk storage since the early twentieth-century surge in power supply from wind energy (Sørensen, 2011c). Current advanced batteries are declining in price toward a level where they may be contemplated for large-scale storage of renewable energy. A review of their technical status is given by Chen et al. (2009), and specific energy analyses for lithium-ion and other batteries are furnished by Rydh and Sandén (2005), Kushnir and Sandén (2011), and Barnhart and Benson (2013), finding an energy payback time of around 6 months for Li-ion batteries, which is better than that of other battery types included in the study. Extended life-cycle analyses have until now primarily been made for smaller batteries aimed at electric vehicles, such as the 10 kWh lithium-ion battery studied by Zackrisson et al. (2010) and the study of several battery types by McManus (2012). Similar to the Patyk et al. (2013) study for geological stores, they use commercial software for the LCA, which means replacing data not explicitly inserted by the user with generic data, usually derived from

TABLE 11.4 Life-Cycle Environmental and Social Damage from Geological Seasonal Hydrogen Store, Using High-Temperature Electrolyzers to Convert Solar Power into Hydrogen and Back

	Unit	Impact	Uncertainty and Ranges
Environmental impacts from plant construction (cavern, electrolyzer, etc.) and materials used			
Greenhouse effects from fossil emissions (Table 11.3)[a]	kg CO_2-eq./MJ H_2	0.0006 (0)	L,r,n
Monetized greenhouse effects from fossil emissions[b]	m€/MJ H_2	0.00006 (0)	H,g,m
Ozone depletion[a]	kg CFC-11-eq./MJ H_2	0.000000001	L,r,n
Acidification[a]	kg SO_2-eq./MJ H_2	0.000003	L,r,n
Eutrophication[a]	kg NO_x-eq./MJ H_2	0.000001	L,r,n
Photosmog[a]	kg C_2H_4-eq./MJ H_2	0.000002	L,r,n
Water consumption[a]	m^3/MJ H_2	0.00012	L,r,n
Land use[a]	m^2/MJ H_2	0.000013	L,r,n
Visual intrusion		NQ	L,l,n
Social impacts			
Mortality/ morbidity caused by air pollution[c]	m€/MJ H_2	0.000003 (0)	H,r,n
Special toxic substances with human impacts[a]	kg $C_6H_4Cl_{12}$-eq./MJ H_2	0.0008	M,r,m
Occupational injuries		Small	L,l,n
Economic impacts	\$/kg H_2 produced	See text	

The value in parentheses is for the case where renewable energy is used for producing materials and for installation.

NA, not applicable; NQ, not quantified; L, M, H: low, medium, or high uncertainty; l, r, g: local, regional, or global impact; n, m, d: near, medium, or distant time frame.

[a]Plant contribution from Patyk et al. (2013), who give impacts per kg of hydrogen produced without disclosing plant life, and from Table 11.3.

[b]Global warming effects over 100 years, as estimated in Sørensen (2011a) using the 2.6 M€ EU valuation of a life lost.

[c]Mortality as for warming-related deaths, morbidity using the 65 k€ EU valuation for a disability-adjusted life shortening of a year (DALY).

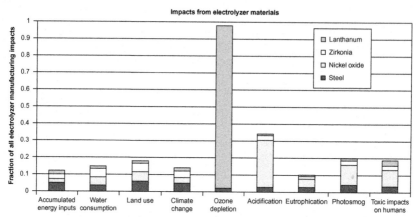

FIGURE 11.7 Impacts imbedded in SOEC electrolyzer materials, as fraction of the total electrolyzer impacts, including those from energy used during manufacture. *(Based on data from Patyk et al., 2013.)*

outdated technology and with assumptions (e.g., on the mix of energy sources used) that may differ greatly from those relevant for the case actually studied. With this reservation, some results for the production of various battery types of 1 MJ capacity are given in Table 11.5.

11.5 SOLAR HEAT SYSTEMS

Solar thermal panels are used to provide heating for individual buildings or for district heating lines. Discussing them in a way is more complicated than discussing solar cell panels, because solar cells are hardly influenced by the remainder of the electricity system they feed into, while the functioning of thermal cells depends critically on the balance of the system, and particularly on the inlet temperature of the fluid being passed through the collectors. As a consequence, dimensioning of the thermal collector, the size of heat store (if present), and the kind of heat distribution system used (typically using air or water) must be done together, in accordance with the characteristics of the solar radiation at the location and the specific mounting of the collector (tilt angle and azimuth angle). Each component will have environmental and social impacts and a generic analysis would treat them separately, leaving the weighing to be done for each particular setting. The discussion here will therefore be confined to analysis of the impacts from the solar thermal panel and, in the following section, from the thermal energy store.

The negative life-cycle impacts associated with solar thermal systems are in any case very small compared with those of present-day heating systems based on fossil fuels such as natural gas. This means that if externalities are included in the price of the systems, the bare cost changes very little. Nielsen and Sørensen

TABLE 11.5 Life-Cycle Environmental and Social Damage from Battery Production

	Unit	Pb Acid	Li-Ion	NiCd	NaS
Environmental impacts from manufacture of various batteries, per MJ battery capacity					
Greenhouse effects from fossil emissions[a]	kg CO_2-eq./MJ	6 (0)	22 (0)	12.5 (0)	2 (0)
Monetized greenhouse effects from fossil emissions[b]	m€/MJ	0.6 (0)	2.2 (0)	1.3 (0)	0.2 (0)
Ozone depletion[a]	kg CFC-11-eq./MJ	0.0000003	0.00044	0.000001	0.0000001
Particulate matter (above 10 μm)[a]	kg PM_{10} eq./MJ	0.025	0.035	0.49	0.009
Ionizing radiation[a]	kg ^{235}U-eq./MJ	1.3	3.7	4.4	0.3
Photochemical oxidants (non-CH_4 volatile organics)[a]	kg NMVOC/MJ	0.035	0.04	0.19	0.007
Water consumption (water not returned to origin)[a]	m^3/MJ	0.085	0.155	0.0024	0.00033
Land use[a]	m^2/MJ	0.25	0.42	0.69	0.09
Visual intrusion		NQ	NQ	NQ	NQ
Social impacts					
Mortality/morbidity caused by air pollution[c]	m€/MJ	0.03 (0)	0.11 (0)	0.07 (0)	0.01 (0)
Special toxic substances with human impacts[a]	kg $C_6H_4Cl_{12}$-eq./MJ	7	4	5	0.3
Occupational injuries		Small	Small	Small	Small

The value in parentheses is for the case where renewable energy is used for producing materials and for installation.
NA, not applicable; NQ, not quantified; L, M, H: low, medium, or high uncertainty; l, r, g: local, regional, or global impact; n, m, d: near, medium, or distant time frame.
[a]From McManus (2012), who also gives impacts per kg of battery material. The uncertainty on impacts is stated as ±20%.
[b]Global warming effects over 100 years, as estimated in Sørensen (2011a) using the 2.6 M€ EU valuation of a life lost.
[c]Mortality as for warming-related deaths, morbidity using the 65 k€ EU valuation for a disability-adjusted life shortening of a year (DALY).

(1998) estimated the average full cost of energy over the system lifetime for solar thermal systems, as it would develop over time of purchasing the hardware, finding costs in an interval of 0.045-0.09 €/kWh for 2015-2020 vintage systems, declining to between 0.03 and 0.07 €/kWh by 2050. The lower figures are for the radiation conditions of southern Europe, the higher ones for high latitudes (50-60°N).

More detailed analysis of LCA impacts from solar thermal collector manufacturing and installation may be found in several recent publications, such as Carnevale et al. (2014) and Lamnatou et al. (2015), again based on commercial software with some European energy mix (primarily based on Switzerland) used for energy input and many impacts taken from databases not specifically derived from current solar thermal installations. This limits the assessment to past installations, and weakens the fact that both studies find solar thermal collector impacts smaller than those of solar cell panels. The current rapid development taking place for all solar cell technologies may make future impacts more similar.

Table 11.6 shows the Lamnatou et al. (2015) results for selected solar thermal systems, omitting the operational impacts that are quantified but smaller than the system installation impacts. The impacts from panel manufacturing are found modest compared with those of transport (manufacturer to building location) and installation, the latter being included in the rightmost column.

11.6 SOLAR HEAT STORAGE

The heat stores used in connection with building-integrated solar thermal systems are currently mostly insulated containers of water, although alternatives do exist, such as gravel and phase-change materials. For communal solar systems used in connection with district heating networks, use of a common solar pond store has been explored. A reasonably complete list of the options may be found in Sørensen (2010).

Oró et al. (2012) have considered the LCA impacts from three storage systems connected to solar concentrator power plants (which, as mentioned, are relevant only for regions of particular solar radiation conditions), using either heat capacity storage in concrete or molten salt, or phase-change materials. Not surprisingly, the simplest system using concrete has the lowest LCA impacts when evaluated for the same rise in storage temperature.

While performing an LCA for the simplest heat capacity stores (e.g., blocks of concrete) is fairly trivial and leads to very small impacts, the more complex storage systems require effort in order to correctly interpret the results. Oró et al. (2012) find that with a prescribed temperature rise, the impacts from the molten salt system are highest, while for the design temperature conditions, the phase-change system has the highest impacts. In any case, the required temperature behavior is determined by the entire system, and it is thus difficult to make a meaningful appraisal by just comparing stores with operating maximum

TABLE 11.6 Life-Cycle Environmental and Social Damage from Solar Thermal Systems

	Unit	Flat-Plate Collector	Evacuated Tube	Flat-Plate System
Environmental impacts from single-dwelling solar thermal unit, per m^2 of collector installed[a]				
Greenhouse effects from fossil emissions[a]	kg CO_2-eq./MJ	10.2 (0)	9.0 (0)	274 (0)
Monetized greenhouse effects from fossil emissions[b]	m€/MJ	1.0 (0)	0.9 (0)	27 (0)
Ozone depletion[a]	kg CFC-11-eq./MJ	0.00001	0.00001	0.00035
Acidification[a]	kg SO_2 eq./MJ	1.0	0.8	2.0
Eutrophication[a]	kg PO_4-eq./MJ	0.7	0.7	1.4
Photochemical smog[a]	kg C_2H_4/MJ	0.05	0.03	1.3
Water consumption	m^3/MJ	NQ	NQ	NQ
Land use (manufacturing plant)	m^2/MJ	NQ	NQ	NQ
Visual intrusion		NQ	NQ	NQ
Social impacts				
Mortality/morbidity caused by air pollution[c]	m€/MJ	0.5 (0)	0.4 (0)	0.12 (0)
Special toxic substances		NQ	NQ	NQ
Occupational injuries		Small	Small	Small

The zero in parentheses would apply if only renewable energy were used at all instances of energy input.

NA, not applicable; NQ, not quantified; L, M, H: low, medium, or high uncertainty; l, r, g: local, regional, or global impact; n, m, d: near, medium, or distant time frame.

[a]From Lamnatou et al. (2015). The system (hot water and space heating) values include Swiss average manufacturing, delivery of system, and installation impacts, but not additional impacts from the operational phase (circulation pumps, etc.).

[b]Global warming effects over 100 years, as estimated in Sørensen (2011a) using the 2.6 M€ EU valuation of a life lost.

[c]Mortality as for warming-related deaths, morbidity using the 65 k€ EU valuation for a disability-adjusted life shortening of a year (DALY).

temperatures more than 300 °C apart. On one hand, the molten salt system stores about twice as much energy per kg as the two other systems do, but on the other hand, salt can cause considerable environmental damage if it is lost from the plant. In contrast, a block of concrete does not seem to invite any accidental escape from the plant.

11.7 COMBINED SYSTEMS

As noted in the review article by Lamnatou et al. (2015), no life-cycle studies seem to have been made for combined power and heat producing solar collectors (PVT systems). Because of the limitations to reasonably south-facing surfaces on detached dwellings or larger buildings, it is in practice rarely possible to install both solar PV and solar thermal panels of sufficient area to supply a reasonable fraction of both the heat and the power demanded within the building. Therefore, the combination consisting of a PV panel with heat tubes on the back side to carry hot water into the building's heat-distribution system, or alternatively a combination with airflow in a confined, transparent space in front of the PV collector plate and connected to a heat distribution system based on warm air, is likely to become the preferred system in the future (Sørensen, 2001).

In general terms, the LCA impacts from a PVT system must be considerably smaller than the sum of impacts from a corresponding PV plus thermal flat-plate system, due to the common collector surface, common framing, and combined installation effort.

11.8 CONCLUSION

The entire area of storage systems to use in supply systems based on renewable energy is in its infancy. We do not know which storage systems will be selected for future energy systems characterized by solar and other renewable energy intermittency, although we do have a catalog of possible candidates. Therefore, it is not so straightforward to perform life-cycle analyses and assessments for the entire future energy system envisaged. Without knowing the mix of renewable energy sources, the detailed layout of the system and its operating conditions, the extent of transmission facilities and arrangements for using them, the future demand structure and its possible modification by various energy management options, and the technical development of the relevant components, it is not possible to describe precise conditions for the use of energy storage and choose between the storage solutions presently known. What can be done is to sketch some scenarios for combining all the mentioned concerns, and then analyze these with respect to environmental and social impacts. The previous examples should be seen in this light.

Despite such large uncertainties, the investigations described do suggest that there exist solar storage options that are neither prohibitively expensive, nor have unacceptable levels of externalities. Underground hydrogen storage in salt domes and aquifers is suggested as one such possibility—one that is strangely absent from the majority of published renewable energy storage work as well as from the policy debates on energy futures that go on in many countries.

REFERENCES

Alsema, E., 2000. Energy pay-back time and CO_2 emissions of PV systems. Prog. Photovoltaic Res. Appl. 8, 17–25.

Barnhart, C., Benson, S., 2013. On the importance of reducing the energetic and materials demands of electrical energy storage. Energy Environ. Sci. 6, 1083–1092.

Barton, N., 2014. Life-cycle assessment for BRRIMS solar power. Renew. Energy 67, 173–177.

Bhandari, R., Trudewind, C., Zapp, P., 2014. Life cycle assessment of hydrogen production via electrolysis—a review. J. Cleaner Prod. 85, 151–163.

Boeniger, M., Briggs, T., 1980. Potential health hazards in the manufacture of photovoltaic solar cells. In: Rom, W., Archer, V. (Eds.), Health Impacts of New Energy Technologies. Ann Arbor Science, Ann Arbor, MI (Chapter 43).

Carnevale, E., Lombardi, L., Zanchi, L., 2014. Life cycle assessment of solar energy systems: comparison of photovoltaic and water thermal heater at domestic scale. Energy 77, 434–446.

Cetinkaya, E., Dincer, I., Naterer, G., 2012. Life cycle assessment of various hydrogen production methods. Int. J. Hydrogen Energy 37, 2071–2080.

Chen, H., Cong, T., Yang, W., Tan, C., Li, Y., Ding, Y., 2009. Progress in electrical energy storage system: a critical review. Prog. Nat. Sci. 19, 291–312.

Corona, B., Miguel, S., Cerrajero, E., 2014. Life cycle assessment of concentrating solar power (CSP) and the influence of hybridising with natural gas. Int. J. Life Cycle Assess. 19, 1264–1275.

Denholm, P., Kulcinski, G., 2004. Life cycle energy requirements and greenhouse gas emissions from large scale energy storage systems. Energy Convers. Manag. 45, 2153–2172.

Dufour, J., Serrano, D., Gálvez, J., González, A., Soria, E., Fierro, J., 2012. Life cycle assessment of alternatives for hydrogen production from renewable and fossil sources. Int. J Hydrogen Energy 37, 1173–1187.

Fthenakis, V., Kim, H., 2011. Photovoltaics: life-cycle analysis. Sol. Energy 85, 1609–1628.

IEA, 2014. Trends 2014 in photovoltaic applications. PVPS Report T1-25:2014. International Energy Agency, Paris.

IPCC, 2014. Climate change 2014. Reports from working groups I-III. Available at: http://www.ipcc.ch.

Ito, M., Kato, K., Sugihara, H., Kichimi, T., Song, J., Kurokawa, K., 2003. A preliminary study on potential for very large-scale photovoltaic power generation (VLS-PV) system in the Gobi desert from economic and environmental viewpoints. Sol. Energ. Mat. Sol. C. 75, 507–517.

Kuemmel, B., Nielsen, S., Sørensen, B., 1997. Life-Cycle Analysis of Energy Systems. Roskilde University Press, Frederiksberg.

Kushnir, D., Sandén, B., 2011. Multi-level energy analysis of emerging technologies: a case study in new materials for lithium ion batteries. J. Cleaner Prod. 19, 1405–1416.

Lamnatou, C., Chemisana, D., Mateus, R., Almeida, M., Silva, S., 2015. Review and perspectives on life cycle analysis of solar technologies with emphasis on building-integrated solar thermal systems. Renew. Energy 75, 833–846.

Lord, A., Kobos, P., Klise, G., Borns, D., 2011. A life cycle cost analysis framework for geological storage of hydrogen: a user's tool. Report SAND2011-6221. Sandia National Laboratories, Albuquerque and Livermore.

Masanet, E., Chang, Y., Gopal, A., Larsen, P., Morrow III, W., 2013. Life-cycle assessment of electric power systems. Annu. Rev. Environ. Resour. 38, 107–136.

McManus, M., 2012. Environmental consequences of the use of batteries in low carbon systems: the impact of battery production. Appl. Energy 93, 288–295.

Moskowitz, P., Steinberger, H., Thumm, W., 1995. Health and environmental hazards of CdTe photovoltaic module production, use and decommissioning. In: Proc. 1994 IEEE First World Conf. on Photovoltaic Energy Conversion (Kona), vol. 1, pp. 115–118.

Nielsen, S., Sørensen, B., 1998. A fair-market scenario for the European energy system. Ch. 3 in: Long-Term Integration of Renewable Energy Sources into the European Energy System (LTI Research Group). Physics-Verlag, Heidelberg, pp. 127–186.

Oró, E., Gil, A., Gracia, A., Boer, D., Cabeza, L., 2012. Comparative life cycle assessment of thermal energy storage systems for solar power plants. Renew. Energy 44, 166–173.

Patyk, A., Bachmann, T., Brisse, A., 2013. Life cycle assessment of H_2 generation with high temperature electrolysis. Int. J Hydrogen Energy 38, 3865–3880.

Pehnt, M., 2003. Life-cycle analysis of fuel cell system components. In: Vielstich, W., Gasteiger, H., Lamm, E. (Eds.), In: Handbook of Fuel Cells, vol. 4. Wiley, Chichester (Chapter 94).

Phéline, J., 1981. Solar energy. Ch. 3 in: El-Hinnawi, E., Biswas, A. (Eds.), Renewable Sources of Energy and the Environment. Tycooly International Press, Dublin, pp. 57–96.

Prasad, A., Jain, K., Gairola, A., 2013. Pumped storage hydropower plants environmental impacts using geomatics techniques: an overview. Int. J. Comput. Appl. 81 (14), 41–48.

Ribeiro, F., Silva, G., 2010. Life-cycle inventory for hydroelectric generation: a Brazilian case study. J. Cleaner Prod. 18 (1), 44–54.

Rydh, C., Sandén, B., 2005. Energy analysis of batteries in photovoltaic systems, parts I and II. Energy Convers. Manag. 46, 1957–1979, 1980-2000.

Simón, J., et al., 2014. Assessment of the potential, the actors and relevant business cases for large scale and long term storage of renewable electricity by hydrogen underground storage in Europe. Final report D 6.3 from the European Commission funded project "HyUnder", Zaragosa.

Sørensen, B., 1993a. Environmental impacts of photovoltaic and wind-based electricity production evaluated on a life-cycle basis. In: Proc. Heat and Mass Transfer in Energy Systems and Environmental Effects, Cancun. Instituto Ingenieria UNAM, Mexico City, pp. 516–521.

Sørensen, B., 1993b. What is life-cycle analysis? In: Sørensen, B. (Ed.), Proc. Expert Workshop on Life-Cycle Analysis of Energy Systems, Paris 1992. OECD, Paris, pp. 21–53.

Sørensen, B., 1997. Impacts of energy use. Ch. 9 in: Diesendorf, M., Hamilton, C. (Eds.), Human Ecology, Human Economy. Allen & Unwin, St. Leonards, pp. 243–266.

Sørensen, B., 2001. Modelling of hybrid PV-thermal systems. In: McNelis, B. et al., (Ed.), In: 17th European PV Solar Energy Conference, Munich, vol. 3. WIP-ETA, Florence, pp. 2531–2534.

Sørensen, B., 2010. Renewable Energy—Its Physics, Engineering, Environmental Impacts, Economics & Planning, fourth ed. Academic Press-Elsevier, Burlington (5th edition currently in preparation).

Sørensen, B., 2011a. Life-Cycle Analysis of Energy Systems—From Methodology to Applications. Royal Society of Chemistry RSC Press, Cambridge.

Sørensen, B., 2011b. Hydrogen and Fuel Cells, second ed. Academic Press-Elsevier, Burlington.

Sørensen, B., 2011c. A History of Energy. Earthscan-Routledge, Cambridge.

Sørensen, B., 2014. Energy Intermittency. CRC Press Taylor & Francis, Baton Rouge.

Sørensen, B., Watt, M., 1993. Life-cycle analysis in the energy field. In: Proc. 5th Int. Conf. "Energex 93" (Seoul), vol. 6. Korea Inst. Energy Research, pp. 66–80.

Stamford, L., Azapagic, A., 2012. Life cycle sustainability assessment of electricity options for the UK. Int. J. Energy Res. 36, 1263–1290.

Tsoutsos, T., Frantzeskaki, N., Gekas, V., 2005. Environmental impacts from the solar energy technologies. Energy Policy 33, 289–296.

Watt, M., 1993. Environmental and health considerations in the production of cells and modules. Centre for Photovoltaic Devices and Systems report 1993/02. University of New South Wales, Sydney.

Whitaker, M., Heath, G., Burkhardt III, J., Turchi, C., 2013. Life cycle assessment of a power tower concentrating solar plant and the impacts of key design alternatives. Environ. Sci. Technol. 13, 5896–5903.

Zackrisson, M., Avellán, L., Orlenius, J., 2010. Life cycle assessment of lithium-ion batteries for plug-in hybrid electric vehicles—critical issues. J. Cleaner Prod. 18, 1519–1529.

Chapter 12

Consumer Perceptions and Acceptance of PV Systems with Energy Storage

Naoya Abe[1], Junichiro Ishio[1], Teppei Katatani[1] and Toshihiro Mukai[2]
[1]*Department of International Development Engineering, Tokyo Institute of Technology, Tokyo, Japan*
[2]*Central Research Institute of Electric Power Industry, Tokyo, Japan*

Chapter Outline

12.1 BACKGROUND

This chapter will illustrate consumer perceptions and acceptance of photovoltaic (PV) systems with energy storage in Japan. Though this dual-system combination is not yet common, the product appears set to grow in the near future. Because energy storage systems (ESS) are not common at the residential level, the significance of ESS for consumers is appropriately understood in conjunction with energy policy—more specifically, renewable energy policy and its related

Solar Energy Storage. http://dx.doi.org/10.1016/B978-0-12-409540-3.00012-8

subsidy schemes. Because Japan experienced a serious nuclear power plant accident in Fukushima on March 11, 2011, Japanese energy and renewable policies cannot be discussed without inquiring how and for what electricity should be generated, though the energy policy itself has not changed dramatically since the Fukushima accident (Kurokawa, 2014). Dr. Kiyoshi Kurokawa was appointed by the National Diet and led the investigation of the nuclear power plant accident, and criticizes the current state of the Japanese energy policy.

As of June 2014, more than 3 years have passed since the Great East Japan Earthquake, the consequent tsunami, and the Fukushima Daiichi nuclear power plant accident. Those severe events resulted in consequences including forced evacuation, "temporary" relocation, and planned outages (rolling blackouts). These inconveniences were further highlighted by the loss of taken-for-granted conveniences such as refrigerators, televisions, and/or air conditioners, among others. Reasonably, those impacts in turn strengthened interest in renewables and related technologies among those who were directly and indirectly affected by the accident. At the same time, however, it could also be the case that some people are no longer vividly conscious of the accident and its impact, simply using electricity as a taken-for-granted service, as before.

The general attitude toward and recognition of energy policy and renewables after March 2011 can be partially confirmed through a survey by NHK (2013). The survey revealed that about 67% of the 2536 respondents thought renewables should be promoted in the coming years. The survey also indicated that if the price of electricity increased due to the reduction of the number of nuclear power plants, then about 50% of the total respondents would not desire that policy. Reasonably, the general attitude toward and perception of the energy policy was influenced not only by the accident, but also by the realistic costs associated with changes to the policy.

Notably, while the number of installed residential PV systems was increasing before the accident in Fukushima, the accident also triggered accelerating sales in Japan, which will be discussed further in Section 12.2. The rapid increase of installed PV generating capacity, however, does not necessarily mean that there is an appropriate understanding of the system and its potential risks. Mukai et al. (2011), for example, found that residential PV owners in a Japanese municipality did not know about its possible failure and its consequent influence on profitability, which is often emphasized by retailers and manufacturers. This is partly due to the fact that residential PV manufacturers emphasize the positive aspects of the system and substantially neglect the negative aspects of their products. The same could be the case for residential ESS, so an investigation into how people think of ESS is worthwhile. Consumers should understand both the positive and negative aspects of energy-saving products.

For individuals looking to reduce their dependency on grid-oriented electricity in Japan, the choices have been relatively limited and the systems have been quite expensive in the past. Among the available options, the residential gas-based fuel-cell system (cogeneration system for electricity and heat, called

"Ene-Farm," to which subsidy has been provided by the Japanese government since 2009) has been the most significant and realistic option. However, its diffusion is limited mainly by its high cost and the space required to house the system; Japanese urban residential areas are extremely densely populated, and each household is restricted in size, meaning that many households cannot spare the space for the equipment.

In April 2013, a Japanese firm called ONE Energy, a joint venture established by three Japanese companies (i.e., ORIX, NEC, and EPCO; http://www.orix.co.jp/grp/en/news/2013/130425_ORIXE.html), began offering to install a lithium-ion (Li-ion) storage battery with 5.53 kWh capacity for detached houses and renting the system to customers, which in turn allows customers to use the system at an affordable price (ORIX, 2013). The combination of residential PV systems and residential electricity storage batteries could be the most plausible option for many Japanese households in the future, if prices continue to fall. However, the general public interest and perception of residential ESS is not well known. This chapter aims to shed light on this aspect.

This chapter consists of the following sections: Section 12.2 offers a brief sketch of Japanese energy policies, including the feed-in tariff (FiT). Section 12.3 briefly outlines the current use of ESS in Japan, followed by the results of a survey of consumers' perceptions and knowledge of renewable technologies, including electricity storage.

12.2   JAPANESE ENERGY POLICY, INCLUDING FiT

A turning point in Japanese energy policy and industry occurred in 2011. Indeed, a law on the FiT for renewable energy sources (RES) went into effect (Ministry of Economy, Trade, and Industry (METI), 2011) after the earthquake and consequent disasters. Furthermore, the Japanese METI has promoted liberalization of the electricity market to improve power exchange across regions as well as to increase renewable energy generation capacity by creating full retail-sector competition and unbundling the generation, transmission, and distribution sectors (METI, 2013a). Notably, the current administration assumes nuclear power as one of the main sources of energy in Japan, which has caused complicated discussions in terms of necessity, safety, and legitimacy.

When we focus on the historical trend of residential-level renewable energy diffusion in Japan, substantial installation started in 1974, with gradual increases in the number of installed systems. Notably, the number of residential PV system owners has increased sharply after 2011, triggered not only by the natural and human-made disasters in Fukushima, but also by economic incentives from the Japanese FiT scheme introduced in 2011. According to statistics provided by Japan's Agency for Natural Resources and Energy, as of March 2011, the accumulated installed generation capacity for the residential sector was about 4700 MW. After the FiT scheme took effect in July 2012, the newly added capacity as of March 2014 is about 2688 MW (approved) and 2276 (in operation), indicating how fast the total national capacity of residential PV systems has expanded (METI, 2014).

In Japan, the "Sunshine Project" in 1974 (Ikeda, 2013) marked the initial development of substantial renewable energy policies. The project, triggered by the oil crisis in 1973, aimed to encourage the research and development of alternative energy technologies. Afterward, in 1980, the Act on the Promotion of Development and Introduction of Alternative Energy set national targets for the amount of power supplied by alternatives, aiming to reduce dependence on oil by developing and introducing new energy sources (Ikeda, 2013) because Japan imported (and still imports) the majority of its energy sources from abroad.

Since the 1990s, RES have gradually increased, beginning mainly as pilot projects in various local municipalities because utility companies voluntarily purchased surplus electricity generated from renewable energy. This was furthered by a grant (subsidy) scheme introduced by the New Energy and Industrial Technology Development Organization (NEDO), a government research and development funding agency in Japan (Ohira, 2006; Suwa and Jupesta, 2012; Kaizuka, 2012). Nevertheless, renewables (i.e., solar PV and wind) experienced lower adoption rates compared to other alternatives to oil, such as nuclear, natural gas, and coal energy. Hence, the Act on Special Measures for the Promotion of New Energy Use was enacted in 1997 to encourage further diffusion of RES to fulfill targets set in 1980.

The Act on Special Measures Concerning New Energy Use by operators of electric utilities, also known as Renewable Portfolio Standards (RPS), took effect in 2003. This law obliges utility companies to produce and/or supply a certain amount of electricity from RES, including hydropower (less than 1000 kW), solar PV, wind, biomass, and geothermal. Practically speaking, the act requires the power company to generate 1.35% of its total electricity from renewables in 2010. When the RPS targets in other countries are considered, such as the United Kingdom (10.4% of total power supply in 2010), Italy (3.05% of total power supply in 2006), Australia (3.8% of total power supply in 2010), and Sweden (16.9% of total power supply in 2010), their figures are more ambitious (Ohira, 2006). Prices charged to utility companies for energy from RES were at the time relatively low (e.g., JPY 24/kWh from solar PV, and approximately 10¥/kWh from wind power), thus the policy was not beneficial for power producers.

The partial FiT scheme, which requires utility companies to purchase electricity at JPY 48/kWh from residential PV systems and JPY 24/kWh from nonresidential PV systems, went into effect in 2009 with the Act on Sophisticated Methods of Energy Supply Structures. These prices are fixed for the next 10 years, though they exclude facilities with a capacity greater than 500 kW. Other than the purchase price (almost double that of the previous scheme), the general structure was the same as that of the subsidy policy for residential PV systems effective between 1994 and 2005 (Myojo and Ohashi, 2012; Kaizuka, 2012).

The Act on Special Measures concerning the Procurement of Renewable Electric Energy by Operators of Electric Utilities (METI, 2010) was enacted in August 2011. This energy policy scheduled cabinet approval on March 11, 2011—before the earthquake occurred, by curious coincidence—and became effective in July 2012 (Esteban and Portugal-Pereira, 2014), establishing the "full" FiT scheme. Table 12.1 describes the tariffs and duration for each

TABLE 12.1 Tariffs (¥/kWh) and Duration (Years) of Renewable Energy Sources in Japan

Energy Source	Tariff	Duration
Solar PV		
10 kW or more (2012→2013→2014)	40→36→32	20
Less than 10 kW (2012→2013→2014)	42→38→37	10
Wind power		
Onshore, 20 kW or more	22	20
Onshore, less than 20 kW	55	20
Offshore	36	20
Small- and medium-scale hydropower		
1-30 MW	24	20
200 kW-1 MW	29	20
Less than 200 kW	34	20
Geothermal power		
15 MW or more	26	15
Less than 15 MW	40	15
Biomass		
Biogas	39	20
Wood-fired power plant		
Timber from forest thinning	32	20
Other woody materials	24	20
Recycled wood	13	20
Waste, excluding woody waste	17	20

Note: FiT requires utility companies to purchase only surplus electricity of solar PV whose capacity is less than 10 kW.
Adapted from "Tariffs and duration of feed-in tariff in Japan" by the Agency of Natural Resources and Energy (2014).

RES as of 2014. The share of electricity from renewables in the total electricity consumed remained at only 1.6% in 2012, implying that a further diffusion of renewable energy in Japan could be expected. It is also notable that the price of solar PV has decreased since the inception of the FiT in Japan.

12.3 ESS IN JAPAN

The massive earthquake, nuclear power plant accident, and consequent black-out focused serious attention on the fundamental role of and high dependency on electricity for those who live not only in the affected areas, but also in other areas, including the greater Tokyo region. Partially because the impact of the nuclear power accident was so serious, many people came to believe that renew-able energy would be the main power generation option pursued in the future. Some also seriously started considering PV systems and other stand-alone tech-nology as a backup source of energy for natural-disaster-related grid power fail-ures. In terms of backup systems, general public awareness of electricity storage could have increased, but the magnitude of awareness is unknown. This part of the book aims to clarify this point.

In Japan, the development of Li-ion and nickel-metal hydride batteries have advanced for use as batteries for plug-in hybrid electric vehicles (PHEVs) and electric vehicles (EVs), which could also supply power in an emergency, to some extent.

Additionally, on the power supply side, some institutions have conducted demonstration tests of bulk energy storage with generally more than 1 megawatt power output. These include sodium-sulfur (NaS) and redox flow batteries, in addition to Li-ion and lead-acid batteries, to compensate for the intermittent nature of renewable energy, which depends on climate conditions, and for load leveling and peak shaving of micro-grids on isolated islands (Ishida et al., 2012; Kawakami et al., 2010; Shigematsu, 2011; Tamaki et al., 2012).

On the demand side, the need for more RES and the usefulness of energy storage as an emergency power supply, realized after the disaster, have raised awareness of ESS for households and businesses. Besides ESS, a succession of new products to address these needs have appeared on the market, including the "smart house," equipped with residential PV systems, energy storage, and home energy management systems (HEMS).

Presently, households mainly use Li-ion and lead-acid batteries as an emer-gency power source and to save power with PV systems. Most of the Li-ion bat-teries have a capacity between 1 and 15 kWh and cost between JPY 700,000 and JPY 2,000,000. Although ESS are initially expensive, one can receive a subsidy for certain Li-ion storage systems of one-third of the expenditure, up to JPY 1 million, as of 2014 (Sustainable open Innovation Initiative, 2013; METI, 2013b). Simultaneously, some municipalities provide grants (approximately JPY 100,000: about $1000 if $1 = JPY 100) for those who purchase energy storage. As another option, PHEVs and EVs generally carry batteries with a

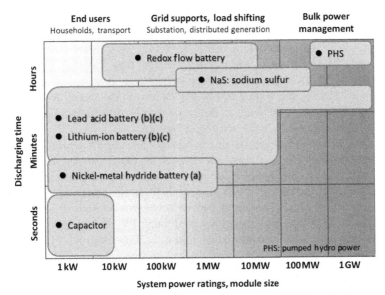

FIGURE 12.1   Battery types displayed by system power ratings and discharge time.

capacity ranging from 5 to 10 kWh, and 15 to 25 kWh, respectively. Their prices range between JPY 3 and 4 million.

Figure 12.1 illustrates various types of batteries based on system power ratings and discharge time. Among them, the general public could recognize the battery storage system through some actual products and familiar devices such as hybrid car batteries (a), automobile starter batteries and uninterruptible power supply (b), EV batteries, portable batteries for tablets and smartphones, and residential electricity storage systems (c).

Toyota (2012) conducted an Internet-based survey of electricity consumers to analyze how the awareness of energy on the demand side changed after the calamity. In the survey, more than half of the 500 total respondents answered the top two in five categories (strongly agree and agree) to the statement that it is necessary to install energy storage for an emergency, and that one prefers to purchase energy storage for one's own house. The results of the survey suggest the need for distributed generation systems.

Goto and Ariu (2012) also investigated consumers' preferences for residential PV systems after the earthquake. They analyzed the purchase intentions for residential PV systems using contingent valuation methods, and calculated the willingness to pay for an emergency power supply by installing energy storage with a PV system. From the 2663 valid responses, consumers placed an estimated value of about JPY 1,180,000 on a half-day supply of electricity from a battery in an emergency. The results indicate a high level of consumer interest in power supply in the event of a calamity. Please note that Goto and Ariu (2012) further discussed why consumers demonstrated a significantly high willingness to pay for the PV and storage technologies, pointing out three main reasons. First, the respondents may have a potentially stronger motivation to mitigate disaster-related risk, because the survey was conducted right after the March 2011 earthquake and tsunami. Second, they may hold exaggerated expectations for the capacity of storage technologies provided at the time of the survey. Finally, the willingness-to-pay options in the survey are relatively higher than in reality, on average. For more details of their discussion, please refer to Goto and Ariu (2012), written in Japanese.

A Japanese company specializing in homes with solar PV and smart homes, Sekisui Chemical Co., Ltd., and its research institute, EnviroLife Research Institute, Co., Ltd., administered surveys in January 2011 (before the disaster), July 2011 (after the disaster), and January 2013 to the tenants of their houses with solar PV systems. The survey carried out before the disaster investigated the changes in attitudes to energy usage through PV, and gained 1226 valid replies from those who moved into their houses in 2009 (Sekisui, 2011a). In July 2011, after the earthquake, they administered a follow-up survey of the respondents from the previous investigation (Sekisui, 2011b), but focused more on the changes in awareness of energy usage due to the disaster. From 716 valid replies, 34% indicated a keen interest in energy storage, 57% had an interest, and the remaining 9% indicated that they had no interest at that time.

In January 2013, the companies administered a similar survey, aiming to investigate changes in consumer awareness, targeting tenants who occupied houses with residential PV systems in 2011 (Sekisui, 2013a). This inquiry also asked about batteries, as in the previous examination. According to the report, from 1097 valid responses, 40% replied that they had a keen interest, 54% had an interest, and the remaining 6% indicated no interest, which suggests an increase in concerns about backup energy, compared to 2011.

In August 2013, the companies conducted another investigation, targeting those living in smart homes equipped with residential PV, energy storage (5.53 or 7.2 kWh), and HEMS from September 2012 to March 2013 (Sekisui, 2013b). This study aimed to understand the actual usage and consumers' satisfaction with the use of energy storage in their homes. In terms of satisfaction related to battery usage, 44% indicated that they were very satisfied, 43% were fairly satisfied, 10% did not say either way, 3% answered somewhat dissatisfied, and there were no answers of very dissatisfied from 115 valid respondents. These consumers answered that the ability to reduce monthly electricity expenses and access to an emergency power supply were the features providing satisfaction with owning energy storage.

Tagashira et al. (2012) investigated consumer intentions to purchase home storage batteries through a survey in November 2011. The survey revealed that consumers recognized the advantages, including access to electricity in an emergency, and reduced electricity costs from using grid-generated power at night rather than during the day. However, they also recognized the requirement to use an electricity plan offering comparatively expensive daytime rates to qualify for less expensive evening rates as a disadvantage—a purchase disincentive. Users reported that they had purchased their storage batteries to gain the stated advantages. In the question asking about the desire to purchase energy storage, outlining some patterns of initial costs and the potential gains over the estimated useful life of the battery, a majority of respondents made it a purchase condition that the initial cost be recoverable.

12.4   CONSUMER PERCEPTION SURVEY: RENEWABLES AND ESS IN JAPAN

12.4.1   Background

It is assumed that people's attitudes toward and perceptions of ESS are influenced by the characteristics of the geographical location in which they live. This is mainly because many local municipalities promote renewables, and many actually provide various types of assistance to promote purchases, notably of PV systems. Indeed, when considering the characteristics of each local area, the ongoing aging and depopulation trend in Japan should be taken into account, because if an area has few or no people, there is not much sense in a discussion about consumer perceptions. In addition, the relative benefits and costs that people associate with these systems would vary, influenced by this demographic trend, making it a trend worth examining.

12.4.2   Objective and Design of the Survey

An Internet-based survey was conducted in March 2014. The groups of people surveyed were residents in three areas: Fukushima, Kochi, and Kanagawa prefectures. These areas were chosen based on careful selection criteria meant to include and represent the current aging and depopulation trends in Japan.

Each of the three prefectures characterizes the current issues in Japan. Fukushima prefecture is well known as the area suffering massive losses from the tsunami and nuclear power plant accident, which still forbids access within a 20 km radius of the plant. It has been 3 years since the disaster, but the area still suffers economically and socially. To accelerate the reconstruction and revitalization of the affected areas, the federal and local governments have launched many initiatives, including the Fukushima Revival Plans (Plan for Revitalization in Fukushima Prefecture), which clearly state that they should abandon nuclear power in favor of robust decentralized generation systems based on RES. Therefore, it is assumed that the residents in Fukushima could have a relatively strong intention and motivation to install various RES-oriented technologies, such as residential PV and ESS. Kochi prefecture is representative of rural depopulated and aging areas in Japan. Except for those prefectures where heavy snowfall is common, Kochi prefecture and Fukushima prefecture have the highest depopulation rate following the disaster, and thus a comparison of the responses from residents in both Kochi and Fukushima could be of interest. Kanagawa prefecture, part of the greater Tokyo region, is representative of Japan's urban areas. While depopulation is creating serious effects in many rural prefectures in Japan, Kanagawa prefecture does not have any depopulating subareas in its jurisdiction, and income levels are generally relatively higher than those of the other two prefectures. These representative areas were selected to investigate the ownership characteristics of storage battery systems in Japan.

In total, 1397 answers were collected: 484 from Fukushima, 477 from Kanagawa, and 436 from Kochi. Among the samples, there are no significant disproportions in the sample attributes. The respondents are between 20 and 69 years old.

12.4.3 Awareness of Storage Battery Systems

Figure 12.2 illustrates the awareness of ESS in Japan. At least 90% of respondents have heard of them; however, the number of people who know what they are is less than half. To better understand these figures, we compared them with knowledge of the other types of residential RES, shown in Figure 12.3.

Compared to the other technologies, awareness of heat pump systems and storage battery systems is relatively low. On the other hand, awareness of PV systems is much higher, with only a few people having never heard of them, and about 80% of respondents knowing what they are.

We can conclude that almost everyone has heard of ESS, but detailed information about what these are has not been shared to the same extent as for PV systems.

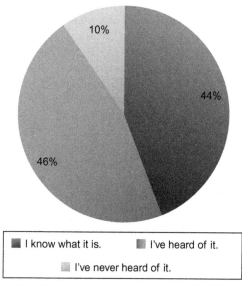

FIGURE 12.2 Awareness of storage battery systems in Japan ($N = 1397$).

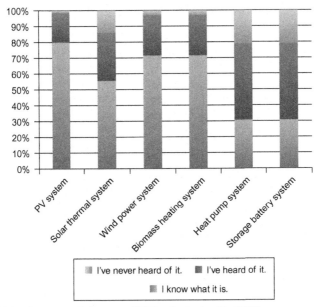

FIGURE 12.3   Awareness of six types of RES technologies ($N = 1397$).

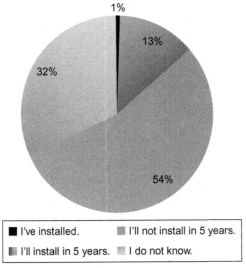

FIGURE 12.4   Installations of "storage battery systems" in Japanese households ($N = 1397$).

12.4.4 Installations of Electricity Storage Systems

Figure 12.4 shows the current ownership and future installation intentions for storage batteries. Only 11 respondents have one installed, and 13% of respondents have plans to install one. However, more than half of respondents do not intend to install a storage battery system within the next 5 years. The rest could not clearly state their future ownership intentions because, as Figure 12.2 shows, they do not know enough about the technology to make an installation decision.

12.4.5 The Relationship Between Electricity Storage Systems and PVs

Generally, residential ESS should provide their owners with electricity during a power outage, while saving the owners money on their monthly utility bills. If a resident installs both a storage system and a PV system, the home electricity supply system is robust, further reducing reliance on peak usage. Some expect synergistic effects from storage batteries and a PV system, and install both in their homes. Figure 12.5 shows the current state of installations and intentions, demonstrating that the owners or future owners of PV systems tend to also install storage battery systems, compared to non-PV owners.

Table 12.2 shows a detailed description of the current installations. The number of respondents who use a storage battery alone is 56, though the number of

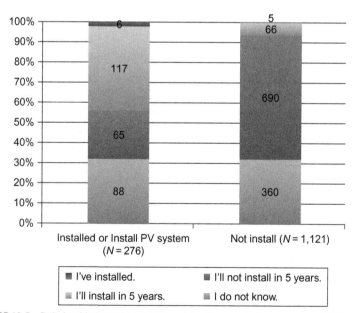

FIGURE 12.5 Relationship between installation state of storage battery systems and PV systems.

TABLE 12.2 Cross Tabulation of Ownership Status of and Attitudes Toward PV Systems and Storage Battery Systems

			PV System			
		Installed	Install Within 5 Years	Will Not Install	I Do Not Know	Total
ESS	Installed	4	2	3	2	11
	Install within 5 years	30	87	53	13	183
	Will not install	37	28	657	33	755
	I do not know	36	52	123	237	448
	Total	107	169	836	285	1397

residents who intend to use both an energy storage and a PV system is 123 out of 1397 respondents. Of the 123 respondents, about 71% (87 respondents) indicated that they will install both systems. This result implies that more people assume that the combination of these systems is something like "bread and butter."

12.4.6   Who Owns/Will Own ESS?

To learn more about the characteristics of storage battery ownership, a cross-tabulation analysis was conducted by comparing storage battery system ownership patterns with age, annual household income, respondents' prefecture, duration of residence, and residence status. Figure 12.6 shows the results.

Storage battery ownership is uniformly distributed across each generation (age) and duration of residence. However, annual household income and residence status show significant differences. Groups whose annual household income is more than JPY 10,000,000 (about $100,000 if $1 = JPY 100) show a slight decrease in purchase intentions, though higher household income may be linked to ESS installations. In addition, only owners of private houses have installed systems as of yet, though some respondents representing other residence status groups are planning to install a system.

In terms of residential area, we could not find significant differences between Fukushima and Kanagawa prefectures; however, people living in Kochi prefecture have a relatively low motivation to install ESS in their homes.

12.5   CONCLUSION

While consumer perceptions of and attitudes toward residential PV systems are positive and growing, they may not be well informed about the true costs when

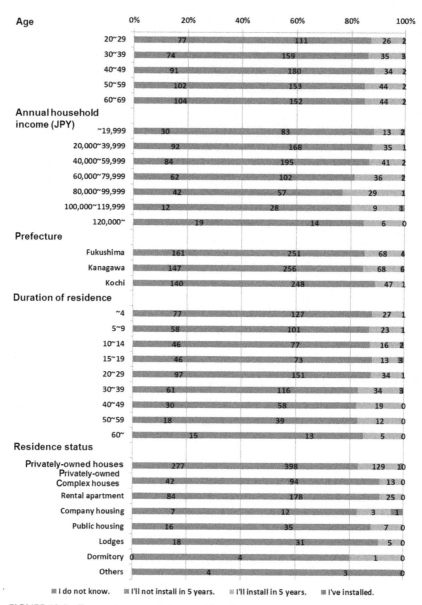

FIGURE 12.6   Energy storage system ownership patterns.

the system fails. This is especially important for system owners. The perceptions and attitudes related to ESS are still in the early stage and are thus weak. We cannot confirm any significant difference based on consumers' ages; however, there are differences based on annual household income and residence status. After all, the system is expensive, so it is appealing for limited groups of people. The results of the surveys in this chapter also indicate that many current residential PV owners are interested in a storage system, and thus the combination of the two systems could become something like "bread and butter" in the future. Given the current FiT scheme in Japan, there is growing criticism pointing out a potential market bubble for residential PV systems. In addition, the financial burden on the general public to maintain the scheme could become socially and politically unacceptable in the near future. Therefore, providing more reliable information about the merits and demerits of electricity storage systems is crucial to shaping consumers' perceptions and attitudes.

REFERENCES

Agency of Natural Resources and Energy, 2014. Tariffs and durations of feed-in tariffs in Japan (in Japanese). Retrieved May 21, 2014, from: Agency of Natural Resources and Energy Website: http://www.enecho.meti.go.jp/category/saving_and_new/saiene/kaitori/kakaku.html.

Esteban, M., Portugal-Pereira, J., 2014. Post-disaster resilience of a 100% renewable energy system in Japan. Energy 68, 756–764.

Goto, H., Ariu, T., 2012. An analysis on consumers' preferences toward home solar power generation system after the Great East Japan Earthquake—focusing on feed-in tariff and emergency use (in Japanese). Central Research Institute of Electric Power Industry Research Reports, Y11029.

Ikeda, T., 2013. The renewable energy development and related promotion and pricing mechanism in Japan. In: Asia-Pacific Economic Cooperation Workshop on Renewable Energy Promotion and Pricing Mechanism, Taipei, Chinese Taipei, September.

Ishida, K., Shimogawa, Y., Sato, Y., Takano, K., Imayoshi, T., Kojima, T., 2012. Demonstration tests of microgrid using renewable energy for small remote islands. In: International Council on Large Electric Systems Session 2012, Paris, France, August.

Kaizuka, I., 2012. Net billing schemes, experience from Japan—evolution to net-export FIT—assigning a fair price to photovoltaic electricity. In: International Energy Agency Photovoltaic Power Systems Programme Workshop, Frankfurt, Germany.

Kawakami, N., Iijima, Y., Sakanaka, Y., Fukuhara, M., Ogawa, K., Bando, M., Matsuda, T., 2010. Development and field experiences of stabilization system using 34 MW NAS batteries for a 51 MW wind farm. In: Proc. ISIE 2010, Bari, Jul. 4-7, 2010, pp. 2371–2376.

Kurokawa, K., 2014. The 3rd anniversary of March 11, Personal blog. Retrieved June 15, 2014, from: http://kiyoshikurokawa.com/en/2014/03/the-3rd-anniversary-of-march-11.html.

Ministry of Economy, Trade and Industry, 2010. 2010 Annual report on energy. (Japan's "Energy White Paper 2010"). Retrieved May 21, 2014, from: http://www.meti.go.jp/english/press/data/pdf/20100615_04a.pdf.

Ministry of Economy, Trade and Industry, 2011. Feed-in tariff scheme for renewable energy. Retrieved May 21, 2014, from: http://www.meti.go.jp/english/policy/energy_environment/renewable/pdf/summary201209.pdf.

Ministry of Economy, Trade and Industry, 2013a. Announcement of a subsidy program to support the installation of lithium-ion battery-based stationary storage systems (in Japanese). Retrieved May 21, 2014, from: http://www.meti.go.jp/press/2013/03/20140317004/20140317004.html.

Ministry of Economy, Trade and Industry, 2013b. Report of the electricity system reform Expert Subcommittee. Retrieved May 21, 2014, from: http://www.meti.go.jp/english/policy/energy_environment/electricity_system_reform/pdf/201302Report_of_Expert_Subcommittee.pdf.

Ministry of Economy, Trade and Industry, 2014. Announcement for current state of renewable introduction (in Japanese). Retrieved July 30, 2014, from: http://www.meti.go.jp/press/2014/06/20140617003/20140617003.html.

Mukai, T., Kawamoto, S., Ueda, Y., Saijo, M., Abe, N., 2011. Residential PV system users' perception of profitability, reliability, and failure risk: an empirical survey in a local Japanese municipality. Energ. Policy 39 (9), 5440–5448.

Myojo, S., Ohashi, H., 2012. Effects of consumer subsidies for renewable energy on industry growth and welfare: Japanese solar energy. In: 39th Annual Conference of European Association for Research in Industrial Economics. Rome, Italy, September.

NHK (Japan Broadcasting Cooperation), 2013. National poll survey on nuclear power and energy in March, 2013 (in Japanese). Retrieved June 15, 2014, from: https://www.nhk.or.jp/bunken/summary/yoron/social/pdf/130523.pdf.

Ohira, T., 2006. Measures to promote renewable energy and technical challenges involved. National Institute of Science and Technology Policy, Science & Technology Trends, 18, 98–114.

ORIX Corporation, Corporate Planning Department, 2013. Launch of Japan's first energy service for households using storage battery rental. Pre-order accepted from April 26, Service to be rolled out progressively from Jun. 1, News Releases. (April 25). Retrieved July 30, 2014, from: http://www.orix.co.jp/grp/en/news/2013/130425_ORIXE.html.

Sekisui Chemical Co., Ltd., EnviroLife Research Institute Co., Ltd., 2011a. Awareness and behavior of the consumers who have solar PV systems. April (in Japanese). Retrieved May 21, 2014, from: http://www.jkk-info.jp/files/topics/13_ext_05_0.pdf.

Sekisui Chemical Co., Ltd., EnviroLife Research Institute Co., Ltd., 2011b. PV system consumers' awareness and behavior change because of the earthquake. August (in Japanese). Retrieved May 21, 2014, from: http://www.jkk-info.jp/files/topics/12_ext_05_0.pdf.

Sekisui Chemical Co., Ltd., EnviroLife Research Institute Co., Ltd., 2013a. Survey of the PV system consumers' awareness in the fiscal year of 2012. March (in Japanese). Retrieved May 21, 2014, from: http://www.jkk-info.jp/files/topics/40_ext_05_0.pdf.

Sekisui Chemical Co., Ltd., EnviroLife Research Institute Co., Ltd., 2013b. Questionnaire survey on actual usage of home energy storages. December (in Japanese). Retrieved May 21, 2014, from: http://www.jkk-info.jp/files/topics/43_ext_05_0.pdf.

Shigematsu, T., 2011. Redox flow battery for energy storage. SEI technology review, 73, 5–13.

Sustainable Open Innovation Initiative, 2013. A subsidy program to support the installation of lithium-ion battery-based stationary storage systems (in Japanese). Retrieved May 21, 2014, from: http://sii.or.jp/lithium_ion25r/.

Suwa, A., Jupesta, J., 2012. Policy innovation for technology diffusion: a case study of Japanese renewable energy public support programs. Sustain. Sci. 7 (2), 185–197. http://dx.doi.org/10.1007/s11625-012-0175-3.

Tagashira, N., Ikeya, T., Tsuchiya, Y., Baba, K., 2012. Public attitude toward vehicle-to-home systems and home-use storage batteries (in Japanese). Central Research Institute of Electric Power Industry Research Reports, Y11021.

Tamaki, M., Takagi, K., Shimada, K., Kawakami, N., Iijima, Y., 2012. Development of PCS for battery system installed in megawatt photovoltaic system. In: In 15th International Power Electronics and Motion Control Conference EPE-PEMC ECCE Europe, Novi Sad, Serbia.

Toyota, S., 2012. Analysis of the consumers' awareness toward energy policy after the earthquake (in Japanese). The review of economics and business management 39, 21–33.

Part IV

Case Studies

Chapter 13

Photovoltaic-Energy Storage Systems for Remote Small Islands

J.K. Kaldellis

Lab of Soft Energy Applications & Environmental Protection, TEI of Piraeus, Athens, Greece

Chapter Outline

Symbols

C_{APS}	cost of keeping the existing thermal power stations as backup stations (€)
c_e	specific energy capacity cost of the ESS (€/kWh)
CF_{grid}	capacity factor of the electrical network under study
CF_{PV}	capacity factor of the PV installation
c_p	specific power cost of the ESS (€/kW)
$C_{PV\text{-}ESS}$	total life-cycle cost of the PV-ESS configuration (in present values) (€)

Solar Energy Storage. http://dx.doi.org/10.1016/B978-0-12-409540-3.00013-X

$c_{PV\text{-}ESS}$	electricity generation cost of the PV-ESS configuration (in present values) (€/kWh)
c_{tps}	current electricity production cost of the existing thermal power stations (€/kWh)
c_w	specific input energy cost (€/kWh)
d_o	energy autonomy of the ESS (hours)
DOD	instantaneous depth of discharge of the ESS
DOD_{bat}	maximum depth of discharge of the batteries
DOD_L	maximum permitted depth of discharge of the ESS
DOD_{PHS}	maximum depth of discharge of the water reservoir
D_p	diameter of the piping system (m)
e	mean annual escalation rate of the produced electricity price
EC	cost of input energy utilized to charge the energy storage system (€)
E_{dir}	energy demand covered directly by the existing power stations (kWh)
E_{ESS}	energy storage capacity of the ESS (kWh)
E_h	average hourly load of the electrical network under study per annum (kW)
E_{in}	ESS energy input (kWh)
E_{out}	ESS energy output (kWh)
E_{PV}	energy production of the PV installation (kWh)
$E_{PV\ dir}$	energy yield of the PV installation absorbed directly by the local network (kWh)
$E_{PV\ min}$	minimum annual energy production of the PV installation (kWh)
E_{stor}	energy demand covered directly by the ESS (kWh)
E_{stor1}	energy contribution of the ESS during daytime (kWh)
E_{t1}	energy demand of the local electricity network during sunlight periods (kWh)
E_{t2}	energy demand of the local electricity network during sunlight absence (kWh)
E_{tot}	energy demand of the local electricity network (kWh)
f	balance of the plant coefficient
FC_{ESS}	fixed M&O cost of the ESS (€)
FC_{PV}	fixed M&O cost of the PV installation (€)
$FC_{PV\text{-}ESS}$	fixed M&O cost of the entire PV-ESS configuration (€)
g_{ESS}	mean annual change of cost for the ESS
g_j	mean annual change of cost for the ESS' major parts to be replaced
g_k	mean annual change of cost for the PV installation's major parts to be replaced
g_{PV}	mean annual increase of cost for the PV installation
H	head (m)
i	capital cost of the local market
IC_{ESS}	initial cost of the ESS (€)

IC_{PV}	initial cost of the PV installation (€)
$IC_{PV\text{-}ESS}$	initial cost of the entire PV-ESS configuration (€)
j_{max}	number of time steps for the period under study
jth	major components of the ESS
k_o	major parts to be replaced during the system's service period for the PV installation
k_s	major parts to be replaced during the system's service period for the ESS
kth	major components of the PV installation
l_j	replacement times for the ESS' major parts (integer number)
l_k	replacement times for the PV installation's major parts (integer number)
L_p	length of the piping system (m)
m_{ESS}	ratio of annual M&O cost to the total initial investment for the ESS
m_{PV}	ratio of annual M&O cost to the total initial investment for the PV installation
n	years of operation for the PV-ESS configuration (years)
N_{APS}	rated power of the existing autonomous power stations (kW)
N_d	mean hourly load demand
N_{ESS}	nominal output power of the ESS (kW)
n_{ESS}	Service period of the ESS (years)
N_{hydro}	total nominal power of the hydroturbines (kW)
N_{in}	maximum input power of the ESS (kW)
n_j	lifetime of the ESS' major parts to be replaced
n_k	lifetime of the PV installation's major parts to be replaced
N_{p1}	peak load demand of the local electricity network during sunlight (kW)
N_{p2}	peak load demand of the local electricity network during the evening (kW)
$N_{p\text{-}grid}$	peak load demand of the local electricity network (kW)
N_{pump}	total nominal power of pumping station
N_{PV}	rated power of the PV installation (kW)
$n_{PV\text{-}ESS}$	lifetime of the entire PV-ESS configuration (years)
Pr	specific price of the PV installation (€/kW)
Q_{bat}	capacity of the battery storage system
r_j	replacement cost coefficient for the ESS' major parts
r_k	replacement cost coefficient for the PV installation's major parts
s	ratio of energy demand during sunlight to energy demand during sunlight absence
SF	safety factor considering the electrical network and the PV installation
T_{amb}	ambient temperature (°C)
t_{ch}	charging period of the ESS
t_{dis}	discharging period of the ESS

t_r	response period of the ESS
t_{st}	storing period of the ESS
$U_{w\text{-}max}$	maximum permitted water speed (m/s)
$VC_{PV\text{-}ESS}$	variable maintenance cost of the entire PV-ESS installation (€)
V_{ss}	volume of upper water reservoir of volume (m³)
V_{wt}	volume of the water storage tank (m³)
w	mean annual escalation rate of the input energy price
W_{des}	total capacity of the reverse osmosis units (m³/h)
x_3	ratio of the PV installation contribution during daytime

Greek Symbols

β	PV panel-tilt angle (degrees)
γ	ratio of state subsidy to the total investment cost
Δt	duration of each time step
δE	energy contribution of the local APS (kWh)
δE_1	energy contribution of the local APS during daytime (kWh)
ε	energy demand ratio covered directly by the ESS
ζ	peak load demand ratio covered by the ESS
η_{bat}	round-trip efficiency of the batteries
η_{cycle}	cycle efficiency during the charging-discharging of an ESS
η_{ESS}	energy transformation efficiency of the ESS (round-trip)
η_{hydro}	efficiency of the hydroturbines
η_p	power efficiency of the ESS
η_{pump}	efficiency of the pumping station
λ	ratio of the maximum ESS input power to the corresponding rated output power
ξ	ratio of the ESS contribution during daytime
ρ_k	mean annual technological progress change for the PV installation's major parts
ρ_j	mean annual change of technological progress for the ESS' major components
ρ_w	water density (kg/m³)
σ	absolute roughness of pipelines (mm)
Y_n	residual value of the PV (€)

Abbreviations

AFC	alkaline fuel cell
APS	autonomous power station
BOS	balance of system
CAES	compressed air energy storage
ESS	energy storage system
FC	fuel cells
FC-HS	hydrogen storage coupled with fuel cells

L/A	lead-acid batteries
Li-ion	lithium-ion
M&O	maintenance and operation
Na-S	sodium-sulfur batteries
Ni-Cd	nickel-cadmium batteries
PCS	power conversion system
PHS	pumped-hydro storage
PV	photovoltaic
PV-ESS	photovoltaic-energy storage system
PV-PHS	photovoltaic-pumped-hydro storage
RES	renewable energy source
SC	supercapacitors
SMES	superconducting magnetic energy storage
VRB	vanadium redox battery
Zn-Br	zinc-bromine

13.1 INTRODUCTION

The ongoing electricity consumption increase, along with the need for environmental protection, has imposed the introduction of applications based on renewable energy sources (RESs). To this end, solar photovoltaic (PV) technology has shown a remarkable growth over the last decade and has established its role as a mature source for electricity generation. Even during a time of economic crisis, PV installations have shown an exponential increase in most European countries, with the current installed PV capacity in the European Union being higher than 80 GW_e, in comparison with the modest 5 GW_e only 5 years ago (EPIA, 2013). At the same time, outside the European Union a rapid development of solar energy systems has also been noticed, with the PV systems' installed global power capacity now reaching 140 GW_e.

In the meantime, there are several remote areas across the globe that rely on electricity grids of small scale (micro-grids), normally employing oil-fired power generation solutions of low energy quality at very high electricity production cost. On the other hand, many of these areas are favored by high-quality wind and solar potential that encourages the development of applications such as wind farms and PV systems. However, the stochastic nature of wind speed and the variable electricity generation of a PV unit, which depends on the amount of solar irradiance during daytime (and not available at night), underline the necessity of collaboration with various energy storage solutions for reducing or even eliminating the contribution of oil power generation in isolated electricity grids.

In this context, this chapter is dedicated to investigating several commercially established or emerging energy storage system (ESS) configurations, such as batteries, fuel cells, flow batteries, compressed air energy storage

(CAES), and pumped-hydro storage (PHS) applications, which may interact with the primary RES (e.g., the sun) and provide a reliable and secure electric power supply into a remote island grid. In addition, special emphasis is given to the methodology applied for determining the optimum PV-ESS combination along with its appropriate dimensions, based on several technical criteria (e.g., operational characteristics of the technology employed, electricity demand profile data, autonomy period, available solar radiation) and economic indexes (e.g., electricity production cost). Finally, a representative case study is evaluated comprising a PV plant—able to satisfy a significant amount of the electricity demand of a remote small-island network—as well as an appropriate ESS that contributes to electricity generation cost reduction while also providing great level and high-quality electricity, without the environmental and macroeconomic impacts of oil-fired power generation.

13.2 THE NEED FOR ENERGY STORAGE IN REMOTE ISLANDS

Storing energy is essential for three main reasons. The first reason is the need to increase the reliability and flexibility of current electricity grids. The second is the need to reduce current costs of electricity generation (especially at peak times), and the third is the accommodation of the projected high solar energy penetration (or any other type of RES) in future energy consumption patterns.

A power system must reliably meet consumer demand for electricity, every moment ensuring a constant dynamic balance between the demand and the amount supplied by the power generation units (Denholm et al., 2010; Fthenakis and Nikolakakis, 2012). Up to now, in autonomous (isolated) island grids, the load demand is usually covered by either low-efficiency heavy oil engines (base load) or high operational cost gas turbines (peak load). The flexibility of a power system to vary its output in order to meet the demand depends on the mix of its generators. Meeting the demand requires a specific number of base load engines designed to run at nearly full capacity for 24 h/day. At the same time, in order to maintain the reliability of the grid, an additional number of generators (spinning reserve) operate at partial load in order to satisfy rapid changes on load demand. On the other hand, peak load generators operate only for a few hours a day or a few hours over a month or year. Usually, peak load generators are gas turbines and diesel generators.

In general, penetration of RES in isolated electricity grids may lead to reduced need for overall system capacity (especially in the case of solar energy, because solar output usually coincides better with load demand than wind energy does), fossil-fuel combustion reduction, and protection of the environment. However, such penetration in the grid normally requires additional costs for maintaining operating reserves (thermal power units) in case of a sudden loss of RES-based energy generation (especially in the case of wind energy). Furthermore, with RES penetration in remote grids, the need for frequency regulation is increased (due to the variable nature of RES) and the requirements for ramping rate of load-following units become larger (Denholm et al., 2010). In this context, several studies have showed that the variable nature of wind speed causes an extra cost of the order of about 5 €/MWh of energy produced by the wind, while the respective cost in the case of solar energy drops significantly when PVs are geographically dispersed over large areas (Fthenakis and Nikolakakis 2012; Mills and Wiser, 2010).

It should be noted that the intermittent nature of solar and wind resources, along with the remarkable fluctuations of daily and seasonal electrical load demand encountered in isolated systems, pose strict penetration limits in RES-based applications (without energy storage) that hardly exceed 15-20% of the overall electricity consumption (Kaldellis et al., 2012; Kapsali et al., 2012). When these levels are exceeded, RES-based electricity generation normally needs to be curtailed, especially in low load-demand periods, in order to avoid grid perturbation (frequency, voltage, reactive power) and grid congestion. In this context, energy storage solutions may smooth the variability of RES-based applications and move their energy production to peak load-demand hours. More specifically, as shown in Figure 13.1—though depending on site-specific climate conditions—the electricity demand curve during a typical day presents one peak at midday and a second peak in the evening. On the other hand, one may notice that the peak load-demand hours may be different during the hours when maximum wind or solar energy generation occurs. In this context, the use of storage units allows part of the electricity produced during off-peak hours to be stored and consumed at high load-demand periods, where the cost of electricity production is very high. In this context, Figure 13.2 shows an example of how energy storage may affect the cost of electricity generation.

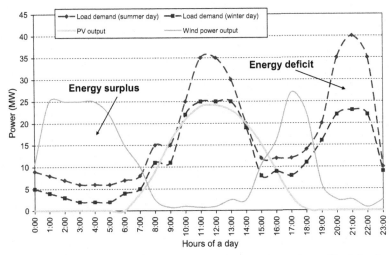

FIGURE 13.1   Daily load-demand fluctuations in a typical island case, along with wind and solar power generation.

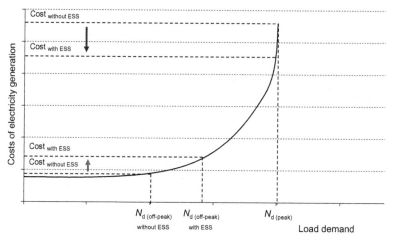

FIGURE 13.2 Large-scale ESS influence on electricity generation prices.

Through "discharging" stored electricity during peak load-demand hours (see x-axis), the electricity generation cost may be significantly lowered ($\text{Cost}_{\text{without ESS}} \rightarrow \text{Cost}_{\text{with ESS}}$) because the use of expensive peak-load power plants is avoided. On the other hand, the electricity generation cost during off-peak hours(see x-axis) ($\text{Cost}_{\text{with ESS}} \rightarrow \text{Cost}_{\text{without ESS}}$) may be characterized by a slight increase (which may become zero in the case where the power supply source is an RES-based plant that provides the ESS with energy free of charge) in order to store cheap electricity. In this way, high-cost electricity generation during peak load-demand hours is (fully or partially) substituted by the output power of the energy storage device.

Based on this, the main benefits stemming from the adoption of ESS may be summarized as the exploitation of otherwise wasted amounts of energy (e.g., rejected amounts of RES-based electricity can be stored), the increased reliability of energy supply (because an extra power source is available), and the improved operation of the power system and existing power units (e.g., operation of conventional units at optimum point, reduction of costs at peak load-demand periods).

13.3 OPERATION MODES OF A TYPICAL ESS

As previously mentioned, a typical ESS stores energy during low energy-demand periods and delivers the stored energy amounts during hours of high energy consumption. A rather simplified profile of the charging and discharging modes for a typical ESS is illustrated in Figure 13.3. According to this chart, the charging period "t_{ch}" of the ESS starts late in the day (late evening/night hours),

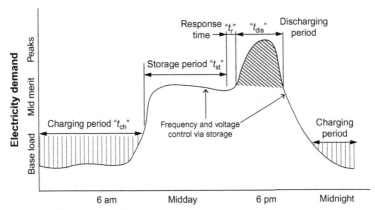

FIGURE 13.3 Energy generation management and frequency-voltage control through energy storage. *(Based on Boyes, 2000; Makansi and Abboud, 2002.)*

when electricity demand is kept at base load levels. As day comes, electrical loads gradually increase and the entire power generation (from electricity generation units: conventional or RES-based) is used to satisfy the demand. In the meantime, management of the power injections used to regulate frequency and voltage are also among the duties of the ESS. Subsequently, depending on the characteristics of the ESS incorporated, certain amounts of stored energy (storing period "t_{st}") after specific response period "t_r" are used to cover parts of the (or the entire) energy deficit (i.e., the extra electrical demand in comparison to either the base load or the mid-merit levels) that appears during peak load-demand hours (discharging period "t_{dis}"). This diurnal cycle is finally completed by the gradual decrease in load demand and the subsequent recharging of the ESS, which has been partly or entirely discharged during the previous discharging period.

Before proceeding to the presentation of the main characteristics and applications of ESS, a short analysis of the energy flows in a typical ESS is necessary. In this context, Figure 13.4 presents the Sankey diagram of a typical ESS (e.g., a battery system). As shown, input energy in the ESS during the charging period is reduced due to distribution and conversion losses. Distribution losses are noticed when energy is delivered from the energy source (e.g., PV system) to the ESS. On the other hand, conversion losses (usually comprising the most critical losses) occur during the conversion of electrical energy to the form of energy required to charge the ESS. Other losses include self-discharge or idling losses, and take place during the standby or off-duty mode of the system. Thus, based on all of this, one may define the cycle efficiency during the charging-discharging of an ESS as the ratio of the ESS energy output to the ESS energy input; i.e.,

$$\eta_{cycle} = \frac{E_{out}}{E_{in}} \qquad (13.1)$$

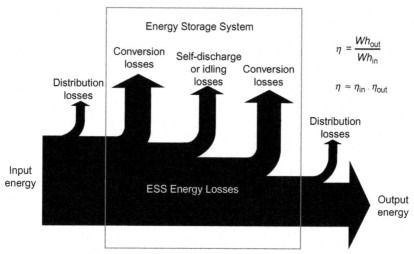

FIGURE 13.4   Energy flows in an ESS. *(Based on Denholm and Kulcinski, 2003; Zafirakis, 2010.)*

13.4   AVAILABLE ENERGY STORAGE TECHNIQUES

A widely accepted differentiation divides the storage systems into those described by high-power provisions that are able to confront power quality issues (e.g., flywheels, supercapacitors (SC), superconducting magnetic energy storage (SMES)), and into those presenting high-energy capacity rates that are able to deal with energy management applications (i.e., PHS, CAES, hydrogen storage coupled with fuel cells (FC-HS), high energy density batteries, and flow batteries). A more detailed classification is presented in Table 13.1.

Based on the data presented in Table 13.1, one may classify energy storage technologies into three major categories: power quality, bridging power, and energy management. The main difference between these categories is the operation timescale and the power covered by each ESS. In this context, power quality applications refer to the provision of considerable power within periods of seconds, to ensure that power disturbances are eliminated. Energy management encompasses the applications concerned with harmonizing the energy generation and demand profiles (i.e., for demand to coincide with generation via the implementation of storage), and requires the use of bulk ESS with considerable storage capacity, storage duration, and discharge time (see also Figure 13.5a and b). On the other hand, bridging power is less demanding as far as capacity requirements are concerned, and involves applications where the discharge duration is kept within a timescale of minutes.

Figure 13.5a provides a classification of ESS in terms of energy storage capacity and discharge time (i.e., the period over which the ESS discharges at its rated power; Electricity Storage Association (ESA), 2009). Systems found on the upper-right side of the chart (where discharge time and energy storage

TABLE 13.1 Categories for Various Electricity Storage Technologies

Categories	Applications	Operation Timescale	Technologies
Power quality	Frequency regulation, voltage stability	Seconds to minutes	Flywheels, supercapacitors, superconducting magnetic storage
Bridging power	Contingency reserves, ramping	Minutes to ~1 h	Nickel-cadmium batteries, lead-acid batteries, high energy-density batteries
Energy management	Load following, capacity, transmission and distribution deferral	Hours to days	PHS, CAES, high energy-density batteries

Based on Data from Fthenakis and Nikolakakis, 2012

capacity are considerable), such as PHS, CAES, and FC-HS, are ideal for the applications of commodity storage, rapid reserve, and area control frequency responsive reserve. By contrast, systems found in the lower-left side of the chart (where the power-to-energy ratio is high and the discharge time requirements are low), such as flywheels, SCs, and SMES, are suitable for power quality/reliability and transmission system stability applications (Zafirakis, 2010). Batteries cover a wide range of applications, from power quality to the early stages of energy management, with flow batteries being more appropriate for transmission and distribution deferral. Note that in the chart, a concentrated point of view is provided, considering battery technology as a whole. More information about the performance of actual systems may be obtained from the updated database of the ESA (2009); see also Figure 13.5b.

In this context, advanced sodium-sulfur (Na-S) batteries comprise the battery technology with the highest discharge time, not influenced by rated power output, while the opposite is true for the discharge time of lead-acid (L/A) batteries, nickel-cadmium (Ni-Cd) batteries, and lithium-ion (Li-ion) batteries. Na-S and L/A demonstrate similar power outputs (up to a scale of tens to hundreds of megawatts), with Li-ion showing the most moderate available power among battery systems (up to hundreds of kilowatts). Ni-Cd batteries, on the other hand, cover a wide range of power, from a few kilowatts to tens of megawatts.

Finally, the discharge times of vanadium redox batteries (VRBs) and zinc-bromine (Zn-Br) are not really affected by variation of power outputs, while VRBs extends their power range back to the scale of a few kilowatts as well, in the interstage between customer energy management and power quality

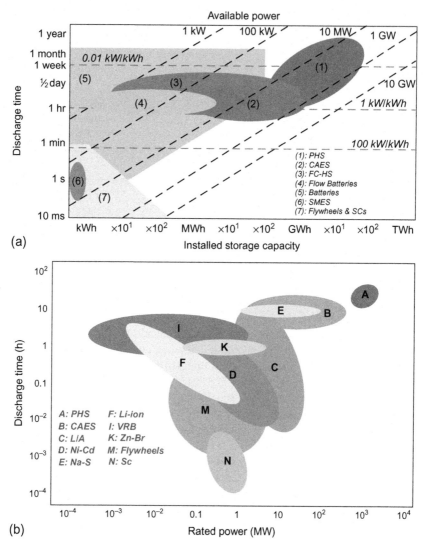

FIGURE 13.5 Comparison of storage systems in terms of discharge time, energy capacity ratings (a), and rated power (b) (PHS, pumped-hydro storage; CAES, compressed air energy storage; FC-HS, hydrogen storage coupled with fuel cells; SMES, superconducting magnetic energy storage; L/A, lead-acid batteries; Ni-Cd, nickel-cadmium batteries; Na-S, advanced sodium-sulfur batteries; Li-ion, lithium-ion batteries; VRB, vanadium redox battery; Zn-Br, zinc-bromine; SC, supercapacitor) *(Based on data from ESA, 2009; Sauer, 2006; Zafirakis, 2010.)*

applications. Furthermore, SCs apply in conditions of high-rated power (even at the megawatt scale) and minimum discharge time (in a scale of seconds), with flywheels having the ability to satisfy high-power applications for short duration (high-power flywheels) and considerable time applications at moderate power output (long-duration flywheels) (Zafirakis, 2010).

In the following, the PHS technology is described in brief, because this technology is the one adopted in the proposed analysis (Chapter 7) concerning a small island (Tilos island) of Aegean Archipelago.

13.4.1 Pumped-Hydro Storage

PHS historically has been used to balance load on a system, enabling large nuclear and thermal power stations to operate at peak efficiencies. These systems are considered the most mature large-scale (>100 MW) commercial energy-storage technology (Figure 13.6), with more than 300 units installed around the world that have total power that exceeds 100 GW (ESA, 2014), covering approximately 3% of the current global power generation. In Europe, installed capacity reaches approximately 45 GW. Next are Japan and the United States, with 25 and 22 GW, respectively.

PHS technology is characterized by high technical and financial reliability (Anagnostopoulos and Papantonis, 2007; Benitez et al., 2008; Kapsali et al., 2012; Kaldellis et al., 2010a), is completely friendly to the environment, and is easy to install in areas where local topography and water availability allow. Nowadays, the classic use of PHS is load balancing with power units that operate under acceptable efficiencies and cheap fuel in energy systems that present periodic weakness to cover the demand. In such systems, pumping of water into an elevated (upper) storage reservoir occurs mainly during the periods of low demand, by either absorbing "cheap" energy from thermal power stations or from any other power generation unit (e.g., wind and/or solar power plant;

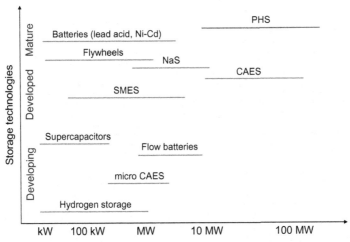

FIGURE 13.6 Power capacity vs. technological maturity of various energy storage methods. *(Based on data from ESA, 2014.)*

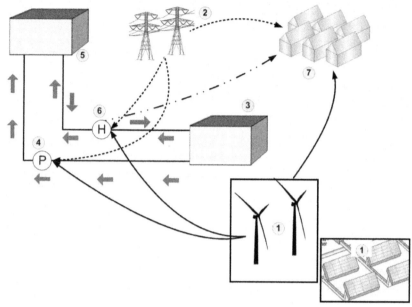

FIGURE 13.7 Pumped-hydro storage configuration. (1) PV park and/or wind park, (2) electricity grid, (3) lower reservoir, (4) pumping system, (5) upper reservoir, (6) reversible hydroelectric machines, and (7) electricity consumption.

see Figure 13.7). Subsequently, during peak demand hours, water is released from the upper reservoir, and hydroturbines operate to "feed" a connected electric generator. As a result, the system is able to cover an existing power deficit by using the appropriate amount of previously stored energy.

In another version, hydroturbines may be replaced by reversible hydraulic machines working either way (in pumping and turbine mode; see Figure 13.7). In that case a double piping system is unnecessary, thus significantly eliminating the initial and the maintenance cost of the project. It is estimated that a double piping system can increase the initial cost of the plant up to 25%. However, it is not always certain that the extra profit of the flexibility provided by the simultaneous modes of storage and electricity generation offsets the additional initial and operation costs; thus, design and selection of the type of piping system is a matter to be considered on a case-by-case basis (Kapsali et al., 2012).

The typical efficiency of pumped storage hydropower mostly ranges between 65% and 80% (Papantonis, 1995), while the maximum depth of discharge is up to 95% without affecting the considerable service period ($\sim$50 years). Such systems can provide energy balancing, stability, storage capacity, and ancillary services such as frequency control, spinning reserves, and so on. This is due to the ability of PHS plants, just as conventional hydroelectric plants, to respond to electrical load changes within seconds (ESA, 2014).

The idea of a PHS plant that consumes renewable energy, although not so common yet, may now be considered realistic (Jaramillo et al., 2004; Katsaprakakis et al., 2008), at least for island systems, providing long-term solutions with several economic and social benefits (Kapsali and Kaldellis, 2010; Katsaprakakis et al., 2012; Bueno and Carta, 2006). By installing such systems, further penetration of renewable energy may be achieved in autonomous island networks, utilizing the amount of power produced by wind and solar power plants that is otherwise impossible to absorb (Kaldellis, 2002; Kaldellis et al., 2004). Furthermore, the introduction of these systems may significantly contribute to the reduction of the considerable oil-fuel quantity consumption, as they may substitute one or more thermal power stations.

13.5   ESS SIZING

In the following subsections, a methodology is presented concerning the determination of the main dimensions of an energy production system, incorporating RES-based power generation (using PVs) able to meet the electricity demand of a remote island network as well as an appropriate storage facility that contributes to the local community energy autonomy for a desired time period (Kaldellis, 2008). Note that any application of the current methodology is based on the exploitation of the maximum available solar potential of the area under investigation.

For a complete energy analysis of a typical remote island network, one needs to first define (with 1 or more years of time-series data) the annual electrical load-demand profile and the available local solar radiation potential. At this point, an interesting relationship between the annual load-demand distribution and the respective solar radiation profile of a typical remote island case is worth mentioning. As shown in Figure 13.8, in most cases, the highest amount of solar radiation is available when increased electricity demand is encountered (i.e., in summer months).

13.5.1   Main Components of the PV-ESS

The system is characterized by the energy storage capacity "E_{ESS}" and the nominal input "N_{in}" and output power "N_{ESS}" of the entire energy storage subsystem. On top of initial costs and other economic considerations, one should take into account some technical parameters of the system, and more specifically, the desired hours of energy autonomy "d_o" of the installation, the maximum permitted depth of discharge "DOD_L," the energy transformation efficiency of the ESS "η_{ESS}" (or "η_{cycle}"), and the power efficiency "η_p" (Kaldellis et al., 2010b). It should be noted that the contribution of the storage

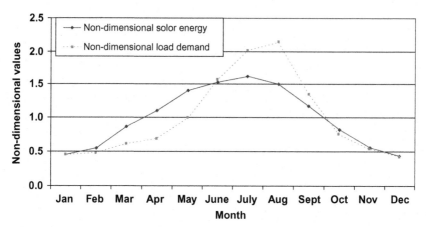

FIGURE 13.8 Relationship between solar potential and load demand distribution for a typical small island case.

system to the total energy consumption "E_{tot}" of the island under investigation may be defined via the parameter "ε"; i.e.:

$$\varepsilon = \frac{E_{stor}}{E_{tot}} = 1 - \frac{E_{dir}}{E_{tot}} \qquad (13.2)$$

where "E_{stor}" (or "E_{out}") is the total energy contribution of the ESS to the annual electricity demand and "E_{dir}" is the energy demand covered directly by the existing power stations, mainly PV generators and complementarily (optionally) by thermal power stations. Obviously, "ε" may take values between zero (no storage system usage) and one (all the energy consumption is covered through the storage system); i.e., $0 \le \varepsilon \le 1.0$. In practice, between these two extreme values, a contribution range determined by the existing power units' principle operating features (including PVs or other RES) dictates the potential use of the ESS on an annual basis.

In order for one to obtain a first estimation of the numerical values of the aforementioned parameters (i.e., DOD_L, η_{ESS}, η_p) along with the service period "n_{ESS}" for every ESS, the data of Table 13.2 can be used, based on the available information in the international literature (see, for example, Sandia National Laboratories, 2003; Nurai, 2004; Gonzalez et al., 2004). In any case, it should be mentioned that real-life values may be quite different from what is currently presented, depending on the technological progress and maturity of the involved technologies in the course of time.

Based on this, the main components of a PV-ESS may be summarized as follows (Kaldellis et al., 2010b):

- One or more PV generators based on the exploitation of the available solar potential. The rated power of the installation is "N_{PV}."
- A number of energy storage devices combined with their corresponding energy production equipment (e.g., charge controllers, inverters, small

TABLE 13.2 Main Characteristics of Several Storage Systems

Storage System	Service Period n_{ESS} (years)	DOD_L (%)	Power Efficiency η_p (%)	Energy Efficiency η_{ESS} (%)
PHS	30÷50	95	85	65÷80
Flywheels	15÷20	75÷80	90÷95	80÷85
CAES	20÷40	55÷70	80÷85	70÷80
Regenesys (flow batteries)	10÷15	100	75÷85	60÷75
FC	10÷20	90	40÷70	30÷45
Lead-acid	5÷8	60÷70	85	75÷80
Na-S	10÷15	60÷80	85÷90	75÷85

Based on Data from Sandia National Laboratories, 2003; Nurai, 2004; Gonzalez et al., 2004

hydroturbines). The energy storage capacity of the installation is equal to "E_{ESS}" and the input and output rated power values are "N_{in}" and "N_{ESS}," respectively. Note that "N_{in}" and "N_{ESS}" are directly dependent on the charging and discharging features of the ESS (e.g., charging and discharging current, operational voltage, charging and discharging rate). The selected ESS should be able to cover the local network electricity requirements for "d_o" typical hours, without the contribution of any other electricity generation device.

- Any existing thermal power units of the already operating autonomous power station (APS), with rated power equal to "N_{APS}," may contribute on covering the local system electricity consumption "δE" under specific circumstances. The main target of the solution is to minimize the contribution of the local APS to the local system electricity consumption ($\delta E \rightarrow 0$).

13.5.2 Dimensions and Characteristics of a PV-ESS

Taking into consideration that the PV energy production is available only during the daytime, one may distinguish the load demand of the island in two parts; see also Equation (13.3). More specifically, one may use the symbols "E_{t1}" and "E_{t2}" to describe the energy consumption during days (sunlight period) and nights, as well as the symbols "N_{p1}" and "N_{p2}" to describe the peak load demand of the local network during the same periods, respectively; see also Equation (13.4) and Figure 13.1.

$$E_{tot} = E_{t1} + E_{t2} \qquad (13.3)$$

$$N_{p\text{-grid}} = \max\{N_{p1}; N_{p2}\} \qquad (13.4)$$

where "$N_{p\text{-grid}}$" is the peak load demand of the electrical network under investigation.

In this context, one may also assume that the total night demand "E_{t2}" is mainly covered by the ESS, while the ESS may also contribute during the daytime (parameter "ξ"), in the case that the PV production is inferior to the load demand (cloudy days; very high load demand). Accordingly, taking into consideration that the PV-based power station should cover the major part of E_{dir} (see also Equation (13.2)) and also provide the necessary energy to the ESS (total energy efficiency η_{ESS}), the corresponding minimum annual energy production "$E_{PV\ min}$" should be expressed as:

$$E_{PV\ min} = (E_{dir} - \delta E) + \frac{E_{stor}}{\eta_{ESS}} = (1 - \varepsilon) \cdot E_{tot} - \delta E + \frac{\varepsilon \cdot E_{tot}}{\eta_{ESS}} \qquad (13.5)$$

At the same time, the energy yield of the PV installation "E_{PV}" is given by the following relation:

$$E_{PV} = CF_{PV} \cdot 8760 \cdot N_{PV} \qquad (13.6)$$

where "CF_{PV}" is the capacity factor of the PV installation on an annual basis (i.e., 8760 h), typically varying between 15% and 25%. Note that by using the "CF_{PV}," the PV generator may be treated as an engine of constant power output equal to "N_{PV}," considering the variability of solar irradiance during a year's period (cloudy days, night periods, etc.), due to which the actual power output of a PV panel does not match "N_{PV}."

Accordingly, defining the capacity factor of the local electrical network "CF_{grid}" using Equation (13.7), i.e.:

$$CF_{grid} = \frac{E_{tot}}{8760 \cdot N_{p\text{-grid}}} \qquad (13.7)$$

one may calculate the required nominal power of the proposed PV-based power station as:

$$N_{PV} = \max\left\{ (1 + SF) \cdot N_{p1}; \ \frac{E_{PV}}{8760 \cdot CF_{PV}} \right\} \Rightarrow$$

$$N_{PV} = N_{p\text{-grid}} \cdot \max\left\{ (1 + SF) \cdot \frac{N_{p1}}{N_{p\text{-grid}}}; \ \frac{CF_p}{CF_{PV}} \cdot \left[(1 - \varepsilon) - \frac{\delta E}{E_{tot}} + \frac{\varepsilon}{\eta_{ESS}} \right] \right\} \qquad (13.8)$$

where "$N_{p\text{-grid}}$" is the peak load demand of the local electricity network and "$SF \geq 0$" is an appropriate safety factor to guarantee that the PV-based power station can meet the local consumption daytime power demand; see also Equation (13.7).

Using the energy balance equation during the daytime, one gets

$$E_{t1} = E_{PV\ dir} + E_{stor1} + \delta E_1 \tag{13.9}$$

where "E_{stor1}" is the contribution of the ESS during the daytime and "δE_1" is the APS participation during the same period. In the same equation, "$E_{PV\ dir}$" is the energy yield of the PV installation absorbed directly by the local network. This parameter may be estimated using the following equation:

$$E_{PV\ dir} = x_3 \cdot E_{t1} \tag{13.10}$$

where "x_3" results from the combination of the solar irradiance and ambient temperature with the corresponding load demand of the local network during the daytime (Kaldellis, 2004; Castronuovo and Lopes, 2004). More specifically, taking into account the distribution of solar irradiance and ambient temperature, along with the operational characteristics of the PV panels employed (current, voltage), one may estimate the detailed energy yield distribution of the PV power plant, determined for a given time step. Comparing the energy yield distribution with the respective daytime energy consumption requirements may lead to the determination of "x_3," providing the proportion of the PV plant energy production used to directly cover the daytime energy demand.

Subsequently, defining the parameter "ξ" describing the contribution of the ESS in the daytime energy demand fulfillment as:

$$\xi = \frac{E_{stor1}}{E_{t1}} \tag{13.11}$$

and using the definitions of Equations (13.10) and (13.11) in Equation (13.9), we get

$$x_3 = (1 - \xi) - \frac{1}{s} \cdot \frac{\delta E_1}{E_{tot}} \leq (1 - \xi) \tag{13.12}$$

Subsequently, the ESS is characterized by the energy storage capacity "E_{ESS}" and the nominal input "N_{in}" and output power "N_{ESS}" of the entire energy storage subsystem. More precisely, the energy storage capacity of ESS may be estimated by the following relation:

$$E_{ESS} = d_o \left(\frac{E_{tot}}{8760} \right) \frac{1}{\eta_{ESS}} \cdot \frac{1}{DOD_L} = d_o \cdot \frac{E_h}{\eta_{ESS}} \cdot \frac{1}{DOD_L} \tag{13.13}$$

where "E_h" is the average hourly load of the electrical network ($E_{tot}/8760$) under investigation.

For the estimation of the ESS contribution to the fulfillment of the local electricity consumption, one may use Equation (13.2) and the corresponding energy storage capacity; hence, one may write

$$\varepsilon = \sum_{j=1}^{j_{max}} \min_{\Delta t_j} \left\{ \frac{E_{PV}(\Delta t_j) - E_{PV\ dir}(\Delta t_j)}{E_{tot}}; \frac{d_o}{8760 \cdot \eta_{ESS}} \cdot \left[\frac{1}{DOD_L} - \frac{1}{DOD(\Delta t_j)} \right] \right\} \tag{13.14}$$

where "j_{max}" is the number of time steps of duration "Δt" in which the entire period (e.g., 1 year) under investigation is divided. According to Equation (13.14), the PV energy production is stored in the ESS, excluding cases where the ESS is almost full.

Equivalently, for the prediction of the "ε" value, one may use the energy balance of the local network for the entire time period examined. Thus, one may write

$$E_{tot} = E_{PV\ dir} + E_{stor} + \delta E \tag{13.15}$$

Introducing Equations (13.2) and (13.10) into Equation (13.15), one gets

$$\varepsilon = (1 - s \cdot x_3) - \frac{\delta E}{E_{tot}} \leq (1 - s \cdot x_3) \tag{13.16}$$

Hence, for a given electricity demand profile and a specific solar potential case, the "ε" distribution varies inversely with the local APS contribution "$\delta E/E_{tot}$." Note that "s" represents the ratio of energy demand during sunlight to energy demand during sunlight absence.

In regard to the nominal output power "N_{ESS}" of the storage unit, it is the power efficiency "η_p" that must be considered as well, i.e.:

$$N_{ESS} = \zeta \cdot \frac{N_{p-grid}}{\eta_p} = \zeta \cdot \frac{E_h}{CF_{grid}} \cdot \frac{1}{\eta_p} \tag{13.17}$$

where "ζ" is the peak power percentage of the local network that the energy storage branch should be able to cover.

Accordingly, the input nominal power "N_{in}" of the ESS depends on the available power excess of the existing PV generators and the corresponding probability distribution, as well as the desired charge time of the installation. For practical cases, and taking into account the limited availability of solar energy defining the charge and the discharge time period of the ESS, finally one may write

$$N_{in} = \lambda \cdot N_{ESS} \leq N_{PV} \tag{13.18}$$

where "λ" depends on the ratio of charge and discharge periods as well as on the efficiency of the energy transformation procedures involved. Generally speaking, for PV applications, "λ" takes values in the range of 1.5-3.0.

13.6 ENERGY STORAGE COSTS

The total initial cost of an ESS primarily comprises two major components: the capital cost of the storage device and the capital cost of the power conversion system's (PCS) nominal power (e.g., inverter, hydroturbine, gas turbine), including the respective capital cost of the various balance-of-system (BOS) components (e.g., pipes, electronic equipment). The specific cost of the storage component is expressed in €/kWh, providing a relationship between the total costs and the storable amount of energy. Accordingly, the specific capacity cost is expressed in €/kW, providing a relationship between the total costs and the nominal power of the PCS. At this point, it is worth mentioning that when comparing different energy storage solutions, the use of capacity costs (in €/kW) is the most common so far. However, this may often be confusing and the results may be just partly meaningful, because PCS capacity determines this parameter exclusively and does not consider the capacity of the energy storage device. Furthermore, for a detailed total investment cost analysis, usually on a life-cycle basis, the fixed and variable maintenance and operation (M&O) costs of the system are also required.

Figure 13.9 presents various typical energy storage solutions as a function of investment costs per unit of energy or unit of power; nevertheless, it should be noted that storage plants' costs may vary substantially across sites due to plant-specific characteristics. For example, PHS plants are characterized by long service period (>50 years) and low M&O costs, while project costs for such systems are very site/project-specific, with some quoted costs in the wide range of 450-2000€/kW and even higher (Deane et al., 2010). In this context, Figure 13.10 presents details about the specific initial cost for some existing and proposed PHS plants in Europe. As shown, most capital costs per kW are found between 500 and 1500€/kW. On the other hand, as far as CAES projects are concerned, the only two currently operational plants, in Bremen and in McIntosh, had the same capital cost of the order of about 400€/kW. Current

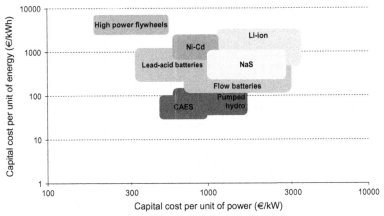

FIGURE 13.9 Various energy storage solutions as a function of investment costs per unit of energy or unit of power. *(Based on data from ESA, 2014.)*

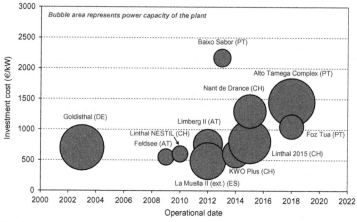

FIGURE 13.10 Specific initial cost of some existing and proposed PHS plants in Europe. *(Based on data from Deane et al., 2010.)*

investment costs are estimated to be between 450 and 1000€/kW, strongly dependent on the structural conditions of the site (e.g., cavern condition; STORE, 2014).

13.6.1 PV-ESS Cost Analysis

The total investment cost "$IC_{PV\text{-}ESS}$" (after "n" years of operation) of a PV-based energy storage installation (Kaldellis, 2008; Kaldellis et al., 2010b; Kaldellis and Zafirakis, 2007) is a combination of the initial installation cost and the corresponding M&O cost, with both quantities expressed in present values. In this context, the initial investment cost takes into account the initial cost of the PV power station "IC_{PV}" and the ESS "IC_{ESS}," as well as the balance of the plant cost, expressed as a function "f" of the initial cost of the PV power station, i.e.:

$$IC_{PV-ESS} = IC_{PV} + f \cdot IC_{PV} + IC_{ESS} \qquad (13.19)$$

According to the available information (Haas, 2002; Kaldellis et al., 2010b), the purchase cost of the PV-based station can be expressed by the following relation:

$$IC_{PV} = Pr \cdot N_{PV} \qquad (13.20)$$

where "Pr" is the specific price (€/kW) of the PV power station.

Accordingly, the initial cost "IC_{ESS}" of an ESS can be expressed (Nurai, 2004; Gonzalez et al., 2004), as previously mentioned, as a function of two coefficients. The first "c_e" (€/kWh) is related to the storage capacity and type of system, and the second "c_p" (€/kW) refers to the nominal power and type of

TABLE 13.3 Main Cost Parameters of Several Storage Systems

Storage System	Specific Energy Cost c_e (€/kWh)	Specific Power Cost c_p (€/kW)	M&O m_{ESS}(%)
PHS	10÷20	500÷1500	0.25÷0.5
Flywheels	250÷350	150÷400	1÷1.5
CAES	3÷5	300÷600	0.3÷1
Regenesys	125÷150	250÷300	0.7÷1.3
FC	2÷15	300÷1000	0.5÷1
Lead-acid	210÷270	140÷200	0.5÷1
Na-S	210÷250	125÷150	0.5÷1

Based on Data from Kaldellis et al., 2009b

storage system in view of Equation (13.18). Hence, one may use the following relation:

$$IC_{ESS} = c_e \cdot E_{ESS} + c_p \cdot N_{ESS} = E_h \cdot \left[\frac{c_e \cdot d_o}{\eta_{ESS} \cdot DOD_L} + \frac{c_p \cdot \zeta}{CF_{grid} \cdot \eta_p} \right] \qquad (13.21)$$

In order to obtain a preliminary idea of the numerical values of the afore-mentioned parameters (i.e., c_e, c_p), one may use the data of Table 13.3, based on the available information found in the international literature (Schoenung and Hassenzahl, 2003; Nurai, 2004; Gonzalez et al., 2004). In the same table, the corresponding annual M&O factor "m_{ESS}" is also included. As is obvious from Table 13.3, a wide range of values has been found for most ESSs under investigation.

In addition to the initial investment cost, one should also take into consideration the M&O cost of the entire installation, including the PV-based power station and the ESS. The M&O cost can be split into the fixed maintenance cost "FC_{PV-ESS}" and the variable cost "VC_{PV-ESS}." Expressing the annual fixed M&O cost as a fraction "m_{PV}" and "m_{ESS}" (see also Table 13.3) of the initial capital invested and assuming a mean annual increase of the cost equal to "g_{PV}" and "g_{ESS}," respectively, the present value of FC_{PV-ESS} is given as:

$$FC_{PV-ESS} = FC_{PV} + FC_{ESS}$$

$$= m_{PV} \cdot IC_{PV} \cdot \sum_{j=1}^{j=n} \left(\frac{(1+g_{PV})}{(1+i)} \right)^j + m_{ESS} \cdot IC_{ESS} \cdot \sum_{j=1}^{j=n} \left(\frac{(1+g_{ESS})}{(1+i)} \right)^j$$

$$(13.22)$$

where "i" is the capital cost of the local market.

Subsequently, the variable M&O cost mainly depends on the replacement of "k_o" and "k_s" major parts of the PV-based power station and the energy storage facility, respectively, which have a shorter lifetime "n_k" or "n_j" compared to the complete installation "$n_{\text{PV-ESS}}$." Using the symbol "r_k" or "r_j" for the replacement cost coefficient of each one of the "k_o" and "k_s" major parts of the entire installation, the "$VC_{\text{PV-ESS}}$" term can be expressed as:

$$
VC_{\text{PV-ESS}} = IC_{\text{PV}} \cdot \sum_{k=1}^{k=k_o} r_k \cdot \left\{ \sum_{l=0}^{l=l_k} \left(\frac{(1+g_k)(1-\rho_k)}{(1+i)} \right)^{l \cdot n_k} \right\}
$$
$$
+ IC_{\text{ESS}} \cdot \sum_{j=1}^{j=k_s} r_j \cdot \left\{ \sum_{l=0}^{l=l_j} \left(\frac{(1+g_j)(1-\rho_j)}{(1+i)} \right)^{l \cdot n_j} \right\}
$$

(13.23)

with "l_k" and "l_j" being the integer part of the following equation, i.e.:

$$
l_k = \left[\frac{n-1}{n_k} \right] \quad \text{and} \quad l_j = \left[\frac{n-1}{n_j} \right]
$$

(13.24)

while "g_k" or "g_j" and "ρ_k" or "ρ_j" describe the mean annual change of the price and the corresponding level of technological improvements for the "kth" major component of the PV-based power station or the "jth" major component of the energy storage installation, respectively.

Recapitulating, the total cost "$C_{\text{PV-ESS}}$" ascribed to the PV-ESS installation after "n" years of operation (in present values) may be estimated using Equation (13.25):

$$
C_{\text{PV-ESS}} = IC_{\text{PV-ESS}} \cdot (1-\gamma) + EC + FC_{\text{PV-ESS}} + VC_{\text{PV-ESS}} - \frac{Y_n}{(1+i)^n} + C_{\text{APS}}
$$

(13.25)

where "γ" is the state subsidization of such kinds of investments ($0\% \leq \gamma \leq 100\%$), "$Y_n$" is the residual value of the installation after n years of operation in current values, and "EC" describes the cost of the input energy "δE" absorbed from the existing thermal power station. For practical applications, this term can be estimated using the following relation; i.e.:

$$
EC = \delta E \cdot c_{\text{tps}} \cdot \sum_{j=1}^{j=n} \left(\frac{(1+w)}{(1+i)} \right)^j
$$

(13.26)

where "c_{tps}" is the specific input energy cost value (current electricity production cost of the existing thermal power stations) and "w" is the mean annual escalation rate of the input energy price. Finally, "C_{APS}" is the cost of keeping the existing thermal power station as a backup station.

Substituting Equations (13.19), (13.22), (13.23), and (13.26) into Equation (13.25), one gets

$$C_{\text{PV-ESS}} = \left[\text{IC}_{\text{PV}} \cdot (1+f) + \text{IC}_{\text{ESS}}\right] \cdot (1-\gamma) + m_{\text{PV}} \cdot \text{IC}_{\text{PV}} \cdot \sum_{j=1}^{j=n}\left(\frac{(1+g_{\text{PV}})}{(1+i)}\right)^{j}$$

$$+ m_{\text{ESS}} \cdot \text{IC}_{\text{ESS}} \cdot \sum_{j=1}^{j=n}\left(\frac{(1+g_{\text{ESS}})}{(1+i)}\right)^{j} + \delta E \cdot c_{\text{w}} \cdot \sum_{j=1}^{j=n}\left(\frac{(1+w)}{(1+i)}\right)^{j} + \text{VC}_{\text{PV-ESS}} - \frac{Y_n}{(1+i)^n} + C_{\text{APS}}$$

$$(13.27)$$

Accordingly, one may express the present value of the electricity generation cost (€/kWh) of the proposed PV-ESS-based installation by dividing the total cost of the installation during the n-year service period with the total energy generation during the same period, taking into consideration the produced electricity price mean annual escalation rate "e." Thus, taking also into account the local market capital cost "i," the electricity generation cost is eventually given as:

$$c_{\text{PV-ESS}} = \frac{C_{\text{PV-ESS}}}{E_{\text{tot}} \cdot \sum_{j=1}^{j=n}\left(\frac{(1+e)}{(1+i)}\right)^{j}} \qquad (13.28)$$

13.7   REPRESENTATIVE CASE STUDY

In this section, the idea of applying a PV-PHS solution is examined for covering the entire electricity demand of the island of Tilos (Aegean Sea, Greece) (Figure 13.11). For this purpose, based on the analysis carried out in Sections 13.5 and 13.6, an integrated computational algorithm is developed (which simulates the hourly operation of the system during an entire year) (see also Kaldellis et al., 2013) and a design case study is then undertaken,

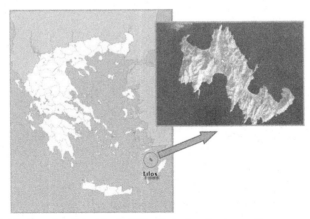

FIGURE 13.11   Location of the island of Tilos in the broader Aegean Sea area.

which is also considered to include a battery storage system that may ensure the local grid stability, providing at the same time a backup solution in case of emergency.

13.7.1 Area of Interest

Tilos is a remote Hellenic island found at the southeast side of the Aegean Sea, belonging to the island complex of Dodecanese (Figure 13.11). The local population of the island is about 600 habitants, although during the summer it may increase to 2500. At the moment, the electrical needs of the islanders (i.e., almost 3.2 GWh per annum, with a peak load demand of approximately 0.92 MW (see Figure 13.12)) are covered exclusively by the operation of the nearby island's (i.e., Kos) autonomous thermal power station, with the required amount of electricity being transferred through an undersea cable of 20 kV. At this point, it should be mentioned that the island suffers from rather frequent "blackouts" (especially during summer, when electricity demand rises sharply), due to the insufficient power generation potential of the Kos power station and several failures of the undersea power transmission cable.

On the other hand, the current use of RES on the island is limited to the use of solar energy for water heating in the residential sector, despite the fact that the entire region appreciates considerable solar potential, in the order of 1760 kWh/m^2.a at the horizontal plane (Figure 13.13), together with a mild climate. At the same time, in the center of the island there exists a water reservoir of 220,000 m^3 (which can be used as a lower reservoir of a PHS facility) that is up to now only partly exploited for irrigation and water supply purposes, and which has, at its proximity, a mountain plateau at an elevation difference of 100 m.

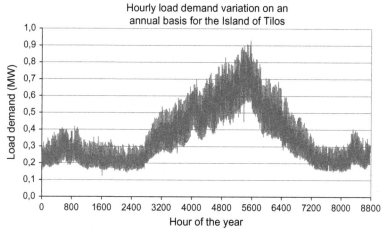

FIGURE 13.12 Load demand variation of Tilos island.

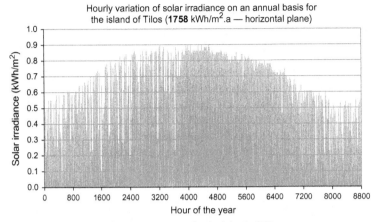

FIGURE 13.13   Solar irradiance measurements for the island of Tilos.

13.7.2   PV-PHS System Components and Operational Modes

The configuration examined comprises the following main components (see also Figure 13.14):

- A PV park (or PV parks) comprising a number of PV panels (determined by their respective panel curves), with total capacity "N_{PV}," installed at a fixed panel-tilt angle of "β" degrees
- A pumping station of total nominal power "N_{pump}" (determined by the maximum appearing energy surplus "$N_{PV} - N_d$" during the daytime, with "N_d" being the mean hourly load demand) and efficiency "η_{pump}," used

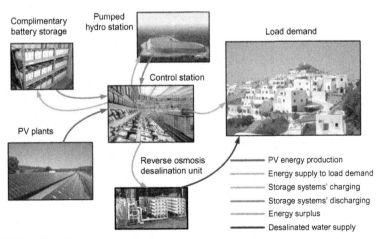

FIGURE 13.14   The examined integrated energy and water solution.

to absorb any energy surplus deriving from the PV plants and pump water to the upper reservoir.

- A hydropower station of total nominal power "N_{hydro}" (determined by the maximum appearing energy deficit or, equivalently, the maximum value of "N_d" during nighttime) and efficiency "η_{hydro}," used to exploit the water stores of the upper reservoir (closed-circuit operation) to cover any energy deficits
- An upper water reservoir of volume "V_{ss}" and maximum depth of discharge "DOD_{PHS}," at a given head "H," combined with a lower reservoir (existing water reservoir of 220,000 m^3) in order to recirculate the required amounts of water
- A piping system comprising two pipelines of absolute roughness "σ," one used to support the pumping stage and the other to support the energy production stage, each one of a certain diameter "D_p" (related to the maximum permitted water speed "U_{w-max}") and certain length "L_p"
- An auxiliary L/A battery storage system of capacity "Q_{bat}," round-trip efficiency "η_{bat}," and maximum depth of discharge "DOD_{bat}," used to support the main PHS unit for a short period of time and also smoothen out any voltage/frequency disturbances of the local grid. Note that to extend its service period, the battery system is assumed to operate on a daily basis under full cycles and minimum depth of discharge, which also requires that any energy surplus is used first to charge the batteries and is accordingly forwarded to feed the pumping station of the PHS plant.
- A central control station used to coordinate operation of the different subsystems, together with any BOS components such as inverters, cabling, and so on.

In this context, the main problem variables currently taken into account for the sizing of the configuration correspond to the PV plant capacity, the panels' tilt angle, and the upper reservoir volume (considering that the lower existing reservoir is of given volume). At the same time, the main problem inputs require detailed solar irradiance and ambient temperature "T_{amb}" measurements along with the hourly electricity load demand of the system under investigation. Furthermore, the technical characteristics of the main system components are also required (see also Table 13.4), while to simulate operation of similar systems, a sizing algorithm (the PV-PHS algorithm) has been developed (see also Figure 13.15 and Kaldellis, 2002; Kaldellis et al., 2013; Kaldellis et al., 2009a for governing system equations).

Recapitulating, the operation scenarios of the proposed configuration include the following:

- In the case that PV energy production is sufficient to cover energy demand, solar energy is fed directly to the local consumption, and any appearing energy surplus is used first to charge the battery bank and then to operate the pumping station. In the case that the available energy surplus cannot be fully absorbed by the storage systems (when the storage devices have

TABLE 13.4 Main System Characteristics

Parameter	Value
Available head "H" (m)	100
Maximum water speed "U_{w-max}" (m/s)	2
Pumping efficiency "η_{pump}"	80%
Hydroturbines' efficiency "η_{hydro}"	85%
Battery storage efficiency "η_{bat}"	75%
PHS maximum depth of discharge "DOD_{PHS}"	95%
Battery maximum depth of discharge "DOD_{bat}"	25%
Water density "ρ_w" (kg/m^3)	1000
Pipeline material absolute roughness "σ" (mm)	0.01

reached their maximum state of charge), a desalination unit (see Kaldellis et al., 2013) may also operate in order to produce fresh water quantities and cover part of the water needs of local islanders.

- In the case that PV energy production is not sufficient to cover the respective load demand, the required amount of water is used in order to operate the hydroturbines, complemented by the battery bank daily discharge performed at a fixed hourly discharge rate from the maximum storage level to the minimum permitted one.

As a result, given a PV capacity value, the hours of load rejection per year are recorded under a fixed upper reservoir storage volume and a fixed panel-tilt angle, while to obtain minimum hours of rejection, the storage capacity is gradually increased within a predefined range of variation. Furthermore, in the case that energy autonomy is not achieved, the PV park capacity is also increased, up to the point that 100% energy autonomy is obtained on the basis of using the PV-PHS solution.

13.7.3   Case-Study Results

Considering the characteristics of the area under investigation, results obtained by the PV-PHS sizing algorithm are presented in the following figures. More precisely, determination of the panel-tilt angle is first undertaken, using the energy yield on a monthly basis. In this context, according to the results of Figure 13.16 for the area of Tilos, the optimum tilt angle is determined in the area of 25-30°, ensuring maximum annual energy yield that also presents its higher values during the summer period (i.e., when the local load demand maximizes).

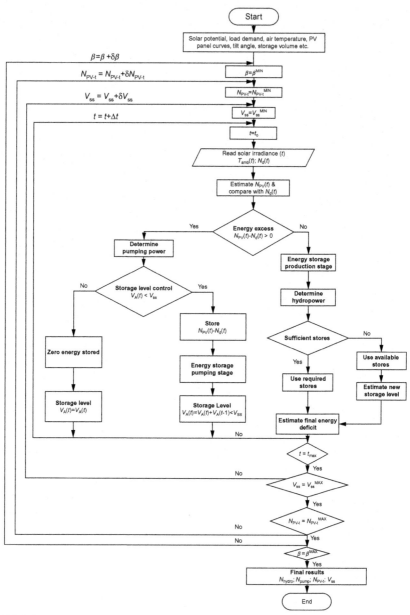

FIGURE 13.15 The PV-PHS sizing algorithm. *(Kaldellis et al., 2009a, 2013.)*

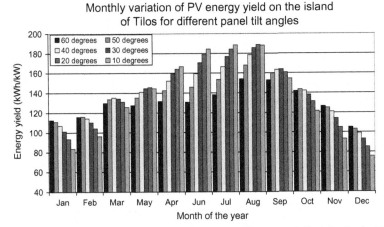

FIGURE 13.16   Monthly PV energy yield variation under different panel-tilt angles for the island of Tilos.

Based on the selected panel-tilt angle, results from the application of the PV-PHS algorithm are then provided in Figure 13.17, where hourly load rejections per year (representative of the energy autonomy levels achieved) are plotted against the upper reservoir storage volume (which is given a high limit that is equal to the available storage volume of the lower reservoir, i.e., 220,000 m^3), examining at the same time different PV capacity ranging from 2 to 4 MW. At this point, it is critical to mention that the respective battery capacity is currently kept constant at 500 kWh, which is equivalent to approximately one hour of autonomy for the entire island (if a deep discharge is taken

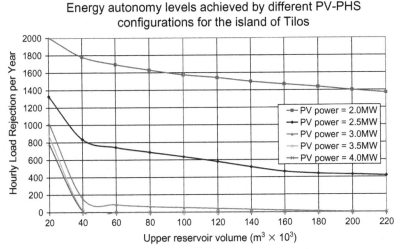

FIGURE 13.17   Energy autonomy levels achieved by different PV-PHS configurations.

into account), considering that this period is adequate in order for a backup plan to be applied (e.g., supply of electricity to the island through the Kos undersea cable) in the case that the PV-PHS solution is unable to cover the forthcoming hours. According to the figure, the PV capacity required to achieve autonomy should be in the order of 3-3.5 MW, combined with an upper storage volume between 50,000 and 150,000 m³.

Subsequently, selection of the system's exact dimensions is undertaken on the basis of the minimum initial cost, in accordance with the results of Figure 13.18, where PV-PHS configurations providing 100% energy autonomy are gathered. As one may obtain from the figure, the minimum initial cost combination suggests employment of 3.1 MW of PV power together with a storage volume of 100,000 m³, which along with the rest of components used leads to an overall cost in the order of 7.5 M€. The resulting cost also takes into account employment of 2.2 MW of pumping power as well as 0.8 MW of hydropower (see also the duration curves of energy surplus and energy deficit from the comparison between the PV plants' production and local load demand in Figure 13.19), combined with pipelines of 650 m long and diameter of 1.1 and 0.8 m, respectively. To this end, the initial cost breakdown of the proposed configuration is obtained from Figure 13.20, with the greater cost attributed as expected to the components of PVs and the upper reservoir.

Finally, in order to demonstrate the energy balance analysis of the proposed configuration, Figure 13.21 is provided for a representative week. As one may see, during the daytime, the available PV power output is used first to cover the local load demand and accordingly support charging of the battery bank and operation of the pumping units to increase water stores in the upper water reservoir. When energy deficit appears the hydroturbines are set to operate,

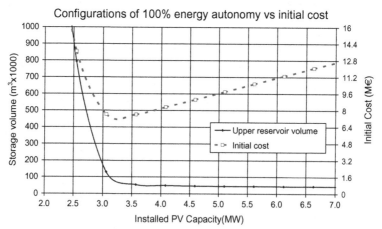

FIGURE 13.18 Selection of the minimum initial cost of the energy autonomous PV-PHS configuration and upper reservoir volume.

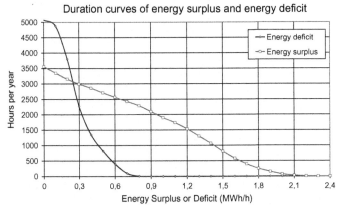

FIGURE 13.19 Duration curves of energy surplus and energy deficit to be absorbed and covered by the PHS system.

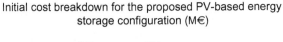
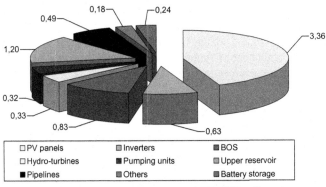

FIGURE 13.20 Initial cost breakdown of the proposed PV-PHS configuration.

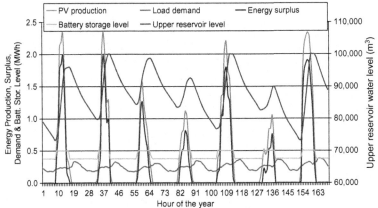

FIGURE 13.21 Energy balance analysis of the proposed system.

drawing water from the upper reservoir, while, as it may be noted, daily cycling of the battery bank is also ensured.

13.8 CONCLUSIONS

The opportunities for increasing RES penetration into remote island networks seem abundant; however, the barriers and challenges are also significant. In many island regions, renewable energy technologies, such as solar and wind power, have vast potential to reduce local dependence on the low quality and expensive oil-fired power generation. However, their intermittent nature has prompted concerns about grid stability and raised questions on how much these resources can penetrate without the contribution of energy storage. On the other hand, up to now most storage technologies are still treated with some skepticism, arising from high initial costs, inherent transformation losses, and certain geographic requirements (especially for large-scale energy storage). In any case, what should be noted is that, although until recently electricity storage was mostly feasible for very large systems (via PHS consuming conventional electricity) or for the remote consumer level (stand-alone systems such as PV-battery), technological advances call for the deployment of energy storage as an essential component of future energy systems that use large amounts of variable renewable resources into remote island networks.

REFERENCES

Anagnostopoulos, J.S., Papantonis, D.E., 2007. Pumping station design for a pumped-storage wind-hydro power plant. Energy Convers. Manag. 48 (11), 3009–3017.

Benitez, L.E., Benitez, P.C., Van Kooten, G.C., 2008. The economics of wind power with energy storage. Energ. Econ. 30 (4), 1973–1989.

Boyes, J.D., 2000. Overview of energy storage applications. In: IEEE Power Engineering Society 2000 Summer Meeting, Seattle, Washington, USA, July 16-20.

Bueno, C., Carta, J.A., 2006. Wind powered pumped hydro storage systems, a means of increasing the penetration of renewable energy in the Canary Islands. Renew. Sustain. Energy Rev. 10, 312–340.

Castronuovo, E.D., Lopes, J.A.P., 2004. Optimal operation and hydro storage sizing of a wind–hydro power plant. Int. J. Electr. Power Energy Syst. 26, 771–778.

Deane, J.P., Ó.Gallachóir, B.P., McKeogh, E.J., 2010. Techno-economic review of existing and new pumped hydro energy storage plant. Renew. Sustain. Energy Rev. 14, 1293–1302.

Denholm, P., Kulcinski, G.L., 2003. Net Energy Balance and Greenhouse Gas Emissions from Renewable Energy Storage Systems. Energy Center of Wisconsin, Wisconsin, USA.

Denholm, P., Ela, E., Kirby, B., Milligan, M., 2010. The role of energy storage with renewable electricity. Technical Report: NREL/TP-6A2-47187.

ESA (Electricity Storage Association), 2009. Technologies. Morgan Hill, California, USA. Available at: http://www.electricitystorage.org/site/technologies/.

ESA (Electricity Storage Association), 2014. Available at: http://www.energystorage.org/.

European Photovoltaic Industry Association (EPIA), 2013. Global Market Outlook for Photovoltaics 2013-2017. Available at: http://www.epia.org/.

Fthenakis, V.M., Nikolakakis, T., 2012. Storage options for photovoltaics. In: van Sark, W.G.J.H.M. (Ed.), Photovoltaic Solar Energy. In: Sayigh, A. (Ed.), Comprehensive Renewable Energy Encyclopedia, vol. 1. Elsevier. ISBN: 978-0-08-087873-7 (chapter 11).

Gonzalez, A., Ó'Gallachóir, B., McKeogh, E., Lynch, K., 2004. Study of Electricity Storage Technologies and Their Potential to Address Wind Energy Intermittency in Ireland. Available at: https://www.seai.ie/uploadedfiles/FundedProgrammes/REHC03001FinalReport.pdf.

Haas, R., 2002. Building PV markets: customers and prices. Renew. Energ. World J. 5, 98–111.

Jaramillo, O.A., Borja, M.A., Huacuz, J.M., 2004. Using hydropower to complement wind energy: a hybrid system to provide firm power. Renew. Energy 29, 1887–1909.

Kaldellis, J.K., 2002. Parametrical investigation of the wind–hydro electricity production solution for Aegean Archipelago. Energy Convers. Manag. 43, 2097–2113.

Kaldellis, J.K., 2004. Optimum techno-economic energy-autonomous photovoltaic solution for remote consumers throughout Greece. Energy Convers. Manag. 45, 2745–2760.

Kaldellis, J.K., 2008. Integrated electrification solution for autonomous electrical networks on the basis of RES and energy storage configurations. Energy Convers. Manag. 49, 3708–3720.

Kaldellis, J.K., Zafirakis, D., 2007. Optimum energy storage techniques for the improvement of renewable energy sources-based electricity generation economic efficiency. Energy J. 32 (12), 2295–2305.

Kaldellis, J.K., Kavadias, K.A., Filios, A.E., Garofallakis, S., 2004. Income loss due to wind energy rejected by the Crete island electrical network—the present situation. Appl. Energy 79, 127–144.

Kaldellis, J.K., Simotas, M., Zafirakis, D., Kondili, E., 2009a. Optimum autonomous photovoltaic solution for the Greek islands on the basis of energy pay-back analysis. J. Cleaner Prod. 17, 1311–1323.

Kaldellis, J.K., Zafirakis, D., Kavadias, K., 2009b. Techno-economic comparison of energy storage systems for island autonomous electrical networks. Renew. Sustain. Energy Rev. 13 (2), 378–392.

Kaldellis, J.K., Kapsali, M., Kavadias, K.A., 2010a. Energy balance analysis of wind-based pumped hydro storage systems in remote islands electrical networks. Appl. Energy 87 (8), 2427–2437.

Kaldellis, J.K., Zafirakis, D., Kondili, E., 2010b. Optimum sizing of photovoltaic-energy storage systems for autonomous small islands. Int. J. Electr. Power Energy Syst. 32, 24–36.

Kaldellis, J.K., Kapsali, M., Tiligadas, D., 2012. Presentation of a stochastic model estimating the wind energy contribution in remote island electrical networks. Appl. Energy 97, 68–76.

Kaldellis, J.K., Kapsali, M., Kondili, E., Zafirakis, D., 2013. Design of an integrated PV-based pumped hydro and battery storage system including desalination aspects for an island of Tilos-Greece. In: International Conference on Clean Electrical Power (ICCEP), Alghero, Sardinia, Italy.

Kapsali, M., Kaldellis, J.K., 2010. Combining hydro and variable wind power generation by means of pumped-storage under economically viable terms. Appl. Energy 87 (11), 3475–3485.

Kapsali, M., Anagnostopoulos, J.S., Kaldellis, J.K., 2012. Wind powered pumped-hydro storage systems for remote islands: a complete sensitivity analysis based on economic perspectives. Appl. Energy 99, 430–444.

Katsaprakakis, D.A., Christakis, D.G., Zervos, E.A., Papantonis, D., Voutsinas, S., 2008. Pumped storage systems introduction in isolated power production systems. Renew. Energy 33 (3), 467–490.

Katsaprakakis, D.A., Christakis, D.G., Pavlopoylos, K., Stamataki, S., Dimitrelou, I., Stefanakis, I., Spanos, P., 2012. Introduction of a wind powered pumped storage system in the isolated insular power system of Karpathos-Kasos. Appl. Energy 97, 38–48.

Makansi, J., Abboud, A., 2002. Energy Storage. The Missing Link in the Electricity Value Chain. Energy Storage Council, Saint Louis, USA.

Mills, A., Wiser, R., 2010. Implications of wide-area geographic diversity for short-term variability of solar power. LBNL-3884E. Ernest Orlando Lawrence Berkeley National Laboratory, Berkeley, CA.

Nurai, A., 2004. Comparison of the Costs of Energy Storage Technologies for T&D Applications. Available at: www.electricitystorage.org.

Papantonis, D., 1995. Hydrodynamic Machines: Pumps-Hydro Turbines, second ed. Symeon, Athens (in Greek).

Sandia National Laboratories, 2003. Long vs. short-term energy storage technologies analysis: a life-cycle cost study. A Study for the DOE Energy Storage Systems Program, SAND2003-2783.

Sauer, D. 2006. The demand for energy storage in regenerative systems. In: 1st International Renewable Energy Storage Conference (IRES I), Science Park Gelsenkirchen, Germany, October 30–31.

Schoenung, S.M., Hassenzahl, W.V., 2003. Long vs. short-term energy storage technologies analysis: a life-cycle cost study. A Study for the DOE Energy Storage Systems Program. Sandia National Laboratories, California. Available at: http://www.sandia.gov/ess/Publications/.

STORE, 2014. Facilitating energy storage to allow high penetration of intermittent renewable energy. Report summarizing the current status, role and costs of energy storage technologies. Available at: http://www.store-project.eu/.

Zafirakis, D., 2010. Overview of energy storage technologies for renewable energy systems. In: Kaldellis, J.K. (Ed.), Stand-Alone and Hybrid Wind Energy Systems: Technology, Energy Storage and Applications. Woodhead Publishing Limited, Cambridge, UK. ISBN: 1 84569 527 5.

Chapter 14

Solar Thermal Energy Storage for Solar Cookers

Ashmore Mawire
Department of Physics and Electronics, Northwest University, Mmabatho, South Africa

Chapter Contents

14.1 INTRODUCTION

Solar thermal energy storage (TES) for solar cookers allows for cooking of food during periods when the sun is not available, thus enhancing their usefulness. The viable options of storing thermal energy for solar cookers are sensible-heat thermal energy storage (SHTES) and latent-heat thermal energy storage (LHTES).

Solar Energy Storage. http://dx.doi.org/10.1016/B978-0-12-409540-3.00014-1

327

In SHTES, heat is stored by heating a material (or extracted by cooling) without any change in its phase. The specific heat capacity of the material and the temperature change during the heating cycle determine the amount of heat that can be stored in a given volume. A variety of materials can be used for such systems, including water, heat-transfer oils, inorganic molten salts, pebbles, and rocks. With solids, the material is often in the porous form, and heat is stored or extracted by the flow of a fluid through the solid pores or the bed voids.

LHTES is based on the heat absorbed or released when a storage material undergoes a phase change. A solid phase change material (PCM) is A material with a high heat of fusion and it is capable of storing large amounts of energy when melting at a certain temperature. This energy is then released when the material solidifies. Because heat is absorbed or released when the material changes phase, PCMs are classified as LHTES units. PCMs can be classified as organic, inorganic, and eutectic. Organic PCMs include fatty acids and paraffins, while inorganic PCMs are usually hydrated salts. A eutectic PCM is a melting composition of two or more components.

In this chapter, a brief general overview of the main types of solar cookers is presented. Basic operating principles of the solar cookers are discussed. Solar cookers with TES are reviewed and discussed to cater to the intermittent behavior of the solar energy resource. Solar cookers using both SHTES and LHTES are presented. Methods of characterization of solar cookers with TES systems are also presented.

14.2 SOLAR COOKING SYSTEMS

A solar cooker is a device that uses energy from the sun to cook (Mawire, 2009). Solar cookers have been in existence for more than a century, with one of the first reported in India by Adams (1878). Essentially, three types of solar cookers exist, and these are classified according to their different designs. The three types of solar cookers are direct-focusing solar cookers, oven solar cookers, and indirect solar cookers.

14.2.1 Direct-Focusing Solar Cookers

Direct-focusing solar cookers, also referred to as concentrating solar cookers, use reflectors to focus and concentrate sunlight directly onto a usually smaller and darker cooking pot compared to the reflector. The pot is either suspended or set on a stand at the focal region. These cookers consist of one or more reflectors and a framework that supports both the reflectors and the pot. Numerous arrangements of these cookers have been devised to allow the reflector to be tilted to always point toward the sun, with the pot remaining at the focal region. The types of reflectors used for these cookers include parabolic dish reflectors, spherical reflectors, plane mirror reflectors, and parabolic trough reflectors. A direct-focusing parabolic dish solar cooker is shown in Figure 14.1.

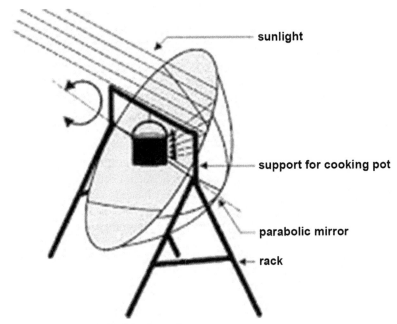

FIGURE 14.1   A parabolic dish solar cooker.

A direct-focusing solar cooker operates by concentrating reflected solar radiation from a larger aperture area onto an absorbing device referred to as a receiver. The receiver absorbs energy and is able to cook food contained in it. The ratio of the larger reflecting aperture area to the smaller absorber area is the geometric concentration ratio given as (Rabl, 1985):

$$C = C_{geo} = \frac{A_{ap}}{A_{abs}} \tag{14.1}$$

where A_{ap} is the aperture area and A_{abs} is the absorber area. The total rate of solar radiation (or power) incident onto the absorber from the reflector aperture is given as:

$$P_i = \eta_o I A_{ap} \tag{14.2}$$

where P_i is the incident power, η_o is the optical efficiency, I is the solar radiation power flux measured in W/m^2, and A_{ap} is the aperture area. Equation (14.2) is rather idealistic because no heat losses have been considered, and it is assumed that the reflector aperture receives all the incident solar radiation and concentrates it all onto the absorber.

If a known amount of water m_w is placed in a pot at the focal region of the solar cooker, the energy absorbed by the water for a time interval Δt or (the output energy) is given as:

$$Q_w = m_w c_w (T_{wf} - T_{wi}) \tag{14.3}$$

where c_w is the specific heat capacity of water, and T_{wf} and T_{wi} are the final and initial water temperatures in the time interval. The instantaneous energy efficiency is thus the ratio of the output energy to the incident absorbed solar radiation, and is expressed as:

$$\eta_e = \frac{m_w c_w (T_{wf} - T_{wi})}{\eta_o I A_{ap} \Delta t} \tag{14.4}$$

Direct-focusing cookers are reasonably priced and cheap to construct and achieve high temperatures in a short interval of time. However, these cookers have a number of disadvantages, some of which may be alleviated by the use of a TES system. Some of the disadvantages are (a) the reflective surface material degrades, (b) the cooking pot is exposed to many hazards, (c) they are less versatile, (d) some designs with long focal points may hurt the eyes, and (e) they perform poorly during cloudy or hazy periods because they utilize direct solar radiation.

Most reflectors for the solar cookers focus the solar radiation from below the receiver, whereas only a few focus the radiation from above the receiver. Spherical reflectors, parabolic reflectors, and Fresnel reflectors are examples of reflectors that focus solar radiation from below the receiver. Different types of direct-focusing solar cookers described by Muthusivagami et al. (2010) are shown in Figure 14.2.

FIGURE 14.2 Direct-focusing types of solar cookers: (a) panel cooker, (b) funnel cooker, (c) spherical reflector, (d) parabolic reflector, (e) Fresnel concentrator, and (f) cylindro-parabolic concentrator.

14.2.2   Oven Solar Cookers

Oven solar cookers are basically insulated boxes with glazed covers that cook food using the greenhouse effect. Variations of these types of solar cookers exist, and these cookers are also referred to as solar box cookers (SBCs). Solar radiation that enters the oven cooker through a glazed window (plastic or glass) gets absorbed inside and heats the darker inside walls and the cooking vessels. Because heat cannot escape through the glazed window, the oven gets hot. Booster mirrors around the window can also be used to direct more solar radiation into the oven. A principal advantage of oven solar cookers over direct-focusing cookers is that they can use both the direct and the diffuse components of solar radiation. A further advantage of these cookers is that no solar tracking is required to focus the solar radiation. Operating temperatures of about 200 °C can be achieved with these solar cookers when booster mirrors are utilized. These temperatures are adequate for cooking most types of food, except for prolonged frying. Heat transfer into and within an SBC occurs by conduction, convection, and radiation.

For a very simple SBC with no reflectors, the energy entering the glazed collecting window is

$$Q_c = A_{ap} \tau_g I_p \qquad (14.5)$$

where A_{ap} is area of the glazed collecting window, τ_g is the transmissivity of the glazing material, and I_p is the global solar radiation that is incident normal to the collecting window. In reality, the apparent area of the window will change with the angle of the sun's rays. This variation is given by

$$A_{app} = A_{per} \cos(\theta) \cos(\varphi) \qquad (14.6)$$

where A_{app} is the apparent area, A_{per} is the perpendicular area, θ is the solar azimuth, and φ represents the difference between the solar elevation angle and the collecting window tilt angle. The azimuth angle is the coplanar angle between a line pointing due south (in the northern hemisphere) and a line pointing toward the sun as seen from a stationary point. The solar elevation angle is the angle of the sun's position relative to a plane tangent to the earth at a point on which the observer is standing (Shaw, 2004). The tilt angle is the angle between the collecting window's normal and a plane tangent to the earth upon which the collecting window is sitting (Shaw, 2004).

The useful energy efficiency of an SBC is defined as the ratio of the energy output to the energy input, and can be expressed as (El-Sebaii and Ibrahim, 2005):

$$\eta_u = \frac{M_f C_f \Delta T_f}{I_{av} A_{ap} \Delta t} \qquad (14.7)$$

where M_f and C_f are the mass and the specific heat capacity of the cooking fluid inside a pot, respectively. ΔT_f is the change in the cooking fluid temperature,

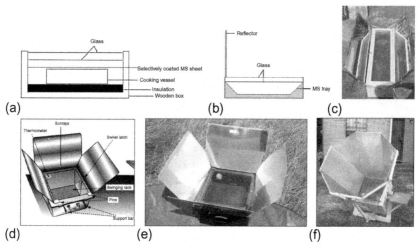

FIGURE 14.3 Oven types of solar cookers: (a) without reflector, (b) with a single reflector, (c) with double reflectors, (d) with three reflectors, (e) with four reflectors, and (f) with eight reflectors.

Δt is the time required to achieve the maximum temperature of the cooking fluid, and I_{av} is the average solar intensity during the time interval Δt with a reference value equal to 900 W/m².

Different designs of oven solar cookers as presented by Muthusivagami et al. (2010) are shown in Figure 14.3 according to the number of reflectors. Although oven solar cookers do possess some advantages over direct-focusing cookers, their main disadvantages are that they have low efficiencies, they require more time to cook food, they have limited capacity dependent on the size of the cooker, they have limited varieties, and they cannot be used for indoor cooking.

14.2.3 Indirect Solar Cookers

Indirect solar cookers are solar cookers constructed such that the solar energy collectors are separated from the cooking vessels. A heat transfer medium is usually required to bring the collected energy into the cooking vessel. The cooking vessel can be placed farther away from the solar energy collector, allowing for an indirect cooking mode (Mawire, 2009). Solar energy collectors can be placed on the roof, while the cooking vessel can be placed indoors. In theory, the distance between the solar energy collector and the cooking vessel can be very large. However, practical challenges such as heat loss and the circulation of the heat transfer medium limit this distance. Close proximity between the cooking vessel and the solar energy collector allows for heat transfer through natural convection. Indirect solar cookers have the advantages of indoor cooking, stability, ease of use, controlled cooking, and easy incorporation into a TES unit.

A heat transfer fluid (HTF) circulates around the solar energy collector to capture the solar radiation for cooking, and the rate of energy absorption by the circulating fluid is given by

$$\dot{Q}_f = \dot{m}_f c_f (T_{fout} - T_{fin}) \tag{14.8}$$

where $\dot{m}_f$ is the mass flow rate, c_f is the specific heat capacity, T_{fout} is the outlet fluid temperature, and T_{fin} is the inlet fluid temperature. The solar collection instantaneous efficiency is defined as the ratio of the rate of heat transfer to the absorbed power, and is expressed as:

$$\eta_C = \frac{\dot{m}_f c_f (T_{fout} - T_{fin})}{IA_C} \tag{14.9}$$

where A_C is the solar collector area. The thermal sensible efficiency (da Silva et al., 2002) is the ratio of the energy used to heat a certain mass of water in the cooking vessel from the ambient temperature to 95 °C to the absorbed solar energy in a time interval dt. This is expressed as:

$$\eta_{95} = \frac{m_w c_w (95 - T_{amb})}{\int_0^t IA_C dt} \tag{14.10}$$

where m_w is the mass of the water in the cooking vessel, c_w is the specific heat capacity of the water, and T_{amb} is the ambient temperature. The denominator represents the absorbed solar energy, which is equivalent to the integral of the absorbed power for a time interval dt. The end temperature of 95 °C is used to avoid the uncertainty in the start of boiling.

The thermal sensible power is the rate of energy used to heat the water in the cooking vessel, and it is given as:

$$\dot{Q}_h = \frac{m_w c_w (95 - T_{amb})}{\Delta t} \tag{14.11}$$

where Δt is the time interval. To estimate the boiling power, the numerator of Equation (14.11) can be replaced by the corresponding latent heat energy expression $m_w h_{fg}$ to become

$$\dot{Q}_h = \frac{m_w h_{fg}}{\Delta t} \tag{14.12}$$

where h_{fg} is the latent heat of vaporization of water. The average latent heat energy efficiency is determined as the ratio of latent heat energy to the absorbed solar energy, and is expressed as:

$$\eta_{boiling} = \frac{m_w h_{fg}}{\int_0^t IA_C dt} \tag{14.13}$$

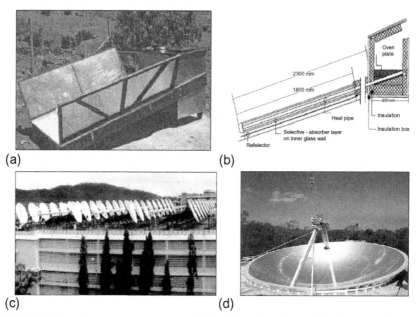

(a)   (b)

(c)   (d)

FIGURE 14.4 Indirect types of solar cookers: (a) with flat plate collector, (b) with evacuated tube collector, (c) with parabolic concentrators at Tirumala Tirupathi Devasthanam, and (d) with spherical reflectors at Auroville.

Indirect solar cookers of different designs are shown in Figure 14.4, as presented by Muthusivagami et al. (2010). A major disadvantage of indirect solar cookers is that they are rather expensive to build and maintain. Another disadvantage is that some of the solar cookers, especially those using solar concentrators, require constant tracking.

14.3 SOLAR COOKERS USING SENSIBLE HEAT THERMAL ENERGY STORAGE (SHTES)*

In this section, different types of solar cookers that have been designed using SHTES are discussed according to the three main types of solar cookers.

14.3.1 Direct-Focusing Solar Cookers Using HCTES

Direct-focusing solar cookers using sensible heat storage are rather rare, and only a few designs using SHTES have been proposed that operate principally in the indirect mode. A portable solar cooker and water heater using a parabolic concentrator, shown in Figure 14.5, was designed by Badran et al. (2010). The device was able to cook food and heat water in the storage tank. An umbrella type of parabolic dish

---

*"Sensible energy" is the popular name sometimes used to describe heat-capacity energy.

FIGURE 14.5   Portable solar cooker and water heater designed by Badran et al. (2010).

concentrator that uses oil TES material was designed by Chandra et al. (2013) to heat the oil that was in thermal contact with the cooking surface. At night, water is poured through a funnel that leads into the oil storage vessel. The water in the pipes gets heated because of the hot oil inside the storage container. The water turns into vapor and comes out through pores that are used to cook rice.

Mawire et al. (2008) proposed a solar cooker using a parabolic dish concentrator that could operate both in the direct and the indirect modes with an oil/pebble bed TES system. Models for the solar energy capture (SEC) system and the TES system were developed to simulate the performance of the TES system using two charging methods. The first method charged the TES system at constant flow rate, while the second method charged the TES system at constant temperature by varying the flow rate. Simulation results showed a greater degree of thermal stratification and energy stored in the TES system for constant temperature charging compared to constant flow-rate charging. Maximum energy efficiencies using both methods were comparable; however, the constant temperature method produced a greater exergy efficiency under high solar radiation conditions. The conceptual diagram of the solar cooker with an oil/pebble bed TES system is shown in Figure 14.6.

The simulation investigation done by Mawire et al. (2009) used an experimentally validated one-dimensional model for SHTES material in the solar cooker of Figure 14.6. The study compared three SHTES materials: fused silica, alumina, and stainless steel. The thermal performance of these materials was evaluated in terms of axial temperature distribution, total energy stored, total exergy stored, and transient charging efficiency. The results indicated that not only was the value of the total amount of energy stored important for the thermal performance of oil/pebble bed TES systems, but also the amount of exergy stored and the degree of thermal stratification should be considered. A high ratio of the total exergy to total energy stored was suggested as a good

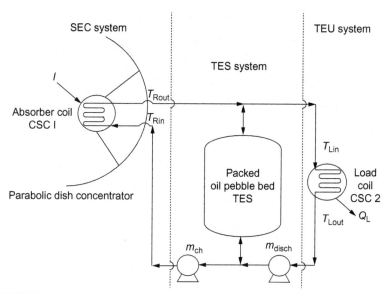

FIGURE 14.6 A conceptual diagram of the solar TES and cooking system proposed by Mawire et al. (2008), showing solar energy capture (SEC), thermal energy storage (TES), and thermal energy utilization (TEU).

measure of the thermal performance of the pebble material. From the simulations, it was concluded that fused silica possesses the best thermal stratification performance, while stainless steel achieved the highest total energy stored at the expense of a greater drop in energy from the peak value as charging progressed. Alumina, on the other hand, was found to have the fastest energy storage rate and had the best exergy-to-energy ratio variation during the charging process, which was comparable to that of fused silica at the end of the charging process.

Discharging simulations of an oil/pebble bed TES system were done by Mawire et al. (2010). Two methods of discharging were compared, which were constant flow-rate discharging and controlled power discharging at a fixed load inlet temperature to maintain a fixed discharging power by varying the discharging flow rate. Results of discharging the TES system at a constant flow rate indicated a higher rate of heat utilization, which was not beneficial to the cooking process, because the maximum cooking temperature was not maintained for the duration of the discharging period. On the other hand, the controlled load power discharging method had a slower initial rate of heat utilization, but the maximum cooking temperature was maintained for most of the discharging process, which was desirable for the cooking process.

An experimental comparison of different storage materials for a small parabolic trough solar cooker was done recently by Mussard and Nydal (2013a). Although the storage tanks stored PCM during the phase change period, the

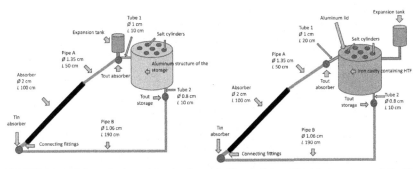

FIGURE 14.7 Schematic view of the aluminum- (left) and oil-based (right) storages coupled with a self-circulated loop. *(Mussard and Nydal, 2013a.)*

storage tanks also stored sensible heat before the phase change process. Oil- and aluminum-based storage tanks were compared experimentally. The loop connecting the collector and the storage was filled with the heat transfer oil, which circulated by self-circulation. The first storage was mainly made of aluminum and salts, while the second was based on oil and salts. Figure 14.7 shows the two experimental configurations that were deemed to be operating in direct mode because the distance between the collector and the storage tank was small. Results showed that the oil-based system reached higher temperatures than the aluminum-based system, and the efficiency of the oil-based system was more than that of the aluminum-based system.

The charging of an oil-based heat storage tank coupled with a low-cost, small-scale, solar parabolic trough for cooking purposes was done by Mussard and Nydal (2013b). Two tests were carried out: one with an uninsulated absorber and the other with an insulated absorber. The results showed that at low temperatures, the absorber without insulation was much more effective. When the storage temperature approached 200 °C, the insulated tube became more effective. An SK-14 direct-focusing solar cooker without heat storage was experimentally compared with a solar parabolic trough solar cooker using a storage unit (Mussard et al., 2013). The SK-14 performed better than the solar cooker with storage due to the nonoptimized design of the cooking surface, which could be improved to match that of an electrical cooker.

14.3.2 Oven Solar Cookers Using SHTES

Designs of oven solar cookers or SBCs using sensible heat storage are also limited, and a few designs will be presented in this section. A hot-box solar cooker that uses engine oil as a storage material was designed, fabricated, and tested so that cooking can be performed at late evening times (Nahar, 2003). A photograph of the hot-box storage solar cooker is shown in Figure 14.8 (Muthusivagami et al., 2010). The device is a double-walled solar hot box.

FIGURE 14.8 A photograph of the hot-box storage cooker designed by Nahar (2003). *(Muthusivagami et al., 2010.).*

The outer box is made of a galvanized steel sheet, and the inner box is made of a double-walled aluminum sheet tray. The space between the inner trays is filled with 5.0 kg of used engine oil, and it is completely sealed. The space between the outer tray and the outer box is filled with glass wool insulation and separated by a wooden frame. The inner tray is painted black with blackboard paint. Nahar (2003) found out that the maximum stagnation temperature achieved inside the cooking chambers of the hot-box solar cooker with storage material was the same as that of the hot-box solar cooker without storage during the daytime, but it was 23 °C more in the storage solar cooker from 17:00 to 24:00 hours. The efficiency of the hot-box storage solar cooker was found to be 27.5%. Cooking trials were also conducted with rice and green vegetables using the hot-box storage cooker and with a hot-box solar cooker without storage from 17:30 hours. The food inside the hot-box storage cooker was cooked perfectly by 20:00 hours, while the food inside the hot box cooker without storage was not cooked at all.

Ramadan et al. (1988) developed a hot-box solar cooker using sand as the storage medium, as shown in Figure 14.9. It consisted mainly of a wooden box with one opening. A copper absorbing flat plate painted black with two glass covers (3 mm thick and 25 mm apart) was placed on top of the box. A copper cylinder cooking pot was used. The pot cover was welded to the absorbing plate to obtain maximum possible thermal conduction. Four square reflecting plane mirrors were attached to the sides of the box to concentrate the solar insolation according to the angle of incidence. Six hours/day of cooking time were recorded, and approximately 3 h/day of indoor cooking were achieved due to the storage material. An overall efficiency of 28.4% was reported by the authors.

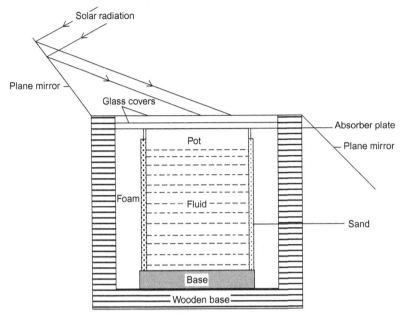

FIGURE 14.9   A schematic diagram of the solar cooker developed by Ramadan et al. (1988).

The experimental thermal performance evaluation of a box-type solar cooker using stone pebbles as TES material was done in Nepal (Shrestha and Byanjankar, 2007). For comparison purposes, a cooker without stone pebbles was tested with a solar cooker with stone pebbles. The experimental results of both no-load testing and load testing showed that with stone pebbles inside the cooker, the time for cooking food could be delayed by a considerable amount of time of about 2 h after noon, thus making cooker suitable for evening meals due to the stored heat. Alozie et al. (2010) compared the performance of three solar hot boxes: (a) a solar hot-box cooker without collectors, (b) a solar hot-box cooker with collectors, and (c) a solar hot-box cooker with heat storage stone pebbles. Results indicated that the storage type of cooker could keep temperatures high enough at around 90 °C by 18:00 hours for the possibility of evening cooking.

Oven solar cookers with SHTES have the disadvantages of achieving low temperatures due to low efficiencies, slow cooking speeds, and limited capacity depending on the size of the cooker.

14.3.3   Indirect Solar Cookers Using SHTES

The most popular designs of solar cookers with SHTES are indirect solar cookers; substantial research has been done on these kinds of cookers. Flat-plate indirect solar cookers, evacuated tube indirect solar cookers, and concentrating indirect solar cookers using SHTES have been designed, and a few of these designs are presented in this section.

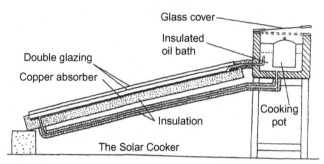

FIGURE 14.10 Natural convection flat-plate collector solar cooker using coconut oil TES, designed by Haraksingh et al. (1996).

A flat-plate collector natural convection solar cooker with short-term coconut oil TES, shown in Figure 14.10, was designed by Haraksingh et al. (1996). A double-glazed flat-plate collector covered with a selective surface was used as the power source for the solar cooker. Coconut oil was used as the HTF, and at the highest part of the thermosiphon loop there was an oil bath in which two cooking pots were immersed to facilitate good heat transfer between the working fluid and the cooking pots. Temperatures of approximately 150 °C could be achieved between 10:00 and 14:00 hours under high solar radiation conditions. A flat-plate collector indirect solar cooker using a vegetable oil as the HTF and an oil/pebble bed TES system, which also uses the thermosiphon principle, is shown in the photograph of Figure 14.11 (Schwarzer and da Silva, 2003). The oil is heated up in the collector with reflectors and moves by a natural flow to the cooking unit. Manually controlled valves guide the oil flow rate either to the pots or to the storage tank. This type of solar cooker can be incorporated into a kitchen. The major advantages of this solar cooker are the possibility of indoor cooking, the use of a thermal storage tank to keep the food warm for longer

FIGURE 14.11 Indirect flat-plate collector solar cooker with TES, designed by Schwarzer and da Silva (2003).

periods of time for night cooking, and the high temperatures of the working fluid reached in a short period of time, allowing fast cooking as well as frying and roasting.

Solar cookers based on conventional flat-plate solar collectors suffer from the drawback of the performance deteriorating due to the reversed cycle during night and cloudy periods of the day. Further disadvantages are that they are expensive to construct, and the nonremovable pots make cleaning and dishing of food difficult.

Evacuated tube solar collectors (ETSCs) have a number of advantages over other types of solar collectors. These advantages include the following: the need for solar tracking is removed because they operate with direct as well as diffuse solar radiation, high temperatures can be achieved, cooking can take place in the shade or inside a building because of the spatial separation of collecting part and oven unit, their thermal conductance is extremely high, and the heat transfer between the evaporator and the condenser section is nearly isothermal.

Kumar et al. (2001) designed the community-type solar pressure cooker based on an ETSC, as shown in Figure 14.12. It consists of an evacuated tubular solar collector and a pressure cooker that acts both as a cooking unit and a TES unit. Both units are coupled together by a heat exchanger. The incident solar irradiance falls onto the collector and heats up the working fluid inside the tubes. The vaporized fluid rises upward to the heat exchanger and transfers energy by condensation to the water flowing in the secondary loop of the heat exchanger. The condensed fluid then returns to the collector tubes, and the process of heat transfer continues. Batch-type cooking was suggested by the

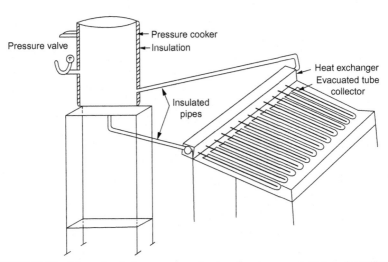

FIGURE 14.12 Schematic of solar pressure cooker based on an evacuated tube solar collector. *(Kumar et al., 2001.)*

experimental results. In Australia, Morrison et al. (1993) developed an indoor type of solar cooker using evacuated heat pipes with a pressurized water heat storage tank, with an appeal similar to a normal electrical hot-plate cooker. The cooker used a sophisticated system whereby steam generated in an evacuated tubular absorber was transferred via a long pipe system into a storage vessel connected to the cooking plates.

Balzar et al. (1996) developed a solar cooking system consisting of a vacuum-tube collector with integrated long heat pipes directly leading to the oven plate. The cooker was tested during several clear days. Detailed temperature distributions and their time dependences were measured. The maximum temperature obtained in a pot containing 51 L of edible oil was 252 °C. The design developed by Esen (2004) of an ETSC using different heat transfer refrigerants in the heat pipes with an oil TES system is shown in Figure 14.13 The oil reservoir of a capacity of 9 L was used for heat storage, allowing the cooker to be preheated and the foods to be kept warm after cooking. The maximum temperature obtained in a pot containing 7 L of edible oil was 175 °C. The cooker was successfully used to cook several foods, and cooking processes were performed with the cooker in 27-70 min periods.

Evacuated tube indirect solar cookers with TES are rather complex and expensive to fabricate. Added to this disadvantage, the tubes tend to deteriorate with time, thus reducing their overall performance.

Concentrating solar collectors can achieve higher temperatures than the other types of solar collectors; hence, it is possible to perform high-temperature

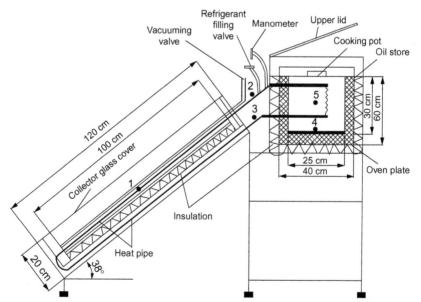

FIGURE 14.13 Cross-sectional view of the evacuated collector tube with integrated heat pipes and oil thermal storage, developed by Esen (2004).

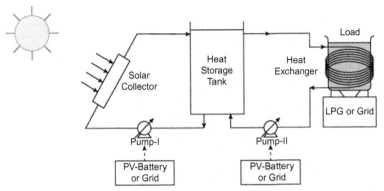

FIGURE 14.14   Block diagram of hybrid solar cooking system. *(Prasanna and Umanand, 2011.)*

cooking applications like baking and frying. Two designs of indirect solar cookers that can also be used in the direct mode using parabolic concentrators have been discussed previously (Mawire et al., 2008; Mussard and Nydal, 2013a). A hybrid indirect solar cooker with an oil TES system and a parabolic dish concentrator was designed by Prasanna and Umanand (2011). A schematic diagram showing the hybrid solar cooking system is shown in Figure 14.14. The energy source is a combination of solar thermal energy and liquefied petroleum gas (LPG). Solar thermal energy is transferred to the kitchen by means of a circulating fluid. The transfer of solar heat is a twofold process, whereby the energy from the collector is transferred first to an intermediate oil storage tank, and this energy is subsequently transferred from the storage tank to the cooking load. During periods when the sun is not available, stored heat, LPG, or electricity can be used to cook foods such that cooking can be carried out at any time during the day.

An innovative design of a Hawkeye solar cooker has been made by students at the University of Iowa and the University of California, Berkeley (Hawkeye solar cooker, 2011), using sand as a SHTES unit. A compound parabolic reflector was used to concentrate the solar radiation onto an absorbing box. The cooker was able to cook food and store heat. A photograph of the innovative Hawkeye solar cooker is shown in Figure 14.15.

Instead of using a HTF in indirect concentrating solar cookers, a secondary reflector can be used to focus the solar radiation onto a cooking device or onto a cooking device with a TES unit, as shown in Figure 14.16. Nyahoro et al. (1997) performed charging and discharging simulations using cast iron and granite charging blocks, and the results showed that cast iron had more energy stored and less energy lost during a charging and discharging sequence. The results also indicated that the height of the storage block should be at least one-fifth of the diameter of the block after different heights and diameters were simulated.

FIGURE 14.15 Hawkeye solar cooker.

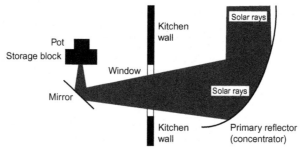

FIGURE 14.16 Layout of components of an indoor focusing hybrid cooker, similar to the design by Oelher and Scheffler (1994), but including a TES block. *(Nyahoro et al., 1997.)*

14.4 SOLAR COOKERS USING LHTES

LHTES based on PCMs has the advantages of a higher energy storage density and an isothermal behavior during phase change, compared to SHTES material. This means that there is a significant decrease in the storage volume when using LHTES material, as compared to SHTES material. In the sections that follow, solar cookers using LHTES are discussed and reviewed.

14.4.1 Direct-Focusing Solar Cookers Using LHTES

Direct-focusing solar cookers using LHTES are an emerging technology; there are only very recent research studies done on these types of solar cookers. An experimental investigation of a solar cooker based on a parabolic dish collector with a PCM storage unit for Indian climatic conditions was performed by

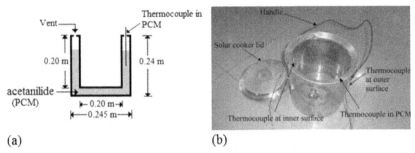

(a)   (b)

FIGURE 14.17   (a) Schematic diagram; (b) Photograph of the solar cooker receiver designed by Chaudhary et al. (2013).

Chaudhary et al. (2013). Figure 14.17 shows a schematic diagram and a photograph of the receiver of the solar cooker. The solar cooker with the PCM TES unit was kept on the absorber plate of a parabolic dish concentrator. During the daytime, acetanilide PCM stored heat, and during the evening, the solar cooker was kept in an insulator box and the PCM delivered heat to the food. To enhance the performance of the solar cooker, three cases were considered: an ordinary solar cooker, a solar cooker with the outer surface painted black, and a solar cooker with the outer surface painted black along with glazing. Results indicated that the solar cooker with the outer surface painted black along with glazing performed better compared to the other two cases.

A portable solar cooker of a standard concentrating parabolic type that incorporated a daily PCM TES unit was evaluated by Lecuona et al. (2013a, b). The storage unit was made by using two conventional coaxial cylindrical cooking pots: an internal one and a larger external one. The space between the two coaxial pots was filled with PCM, forming an intermediate jacket. Figure 14.18 shows photographs of the solar cooker in operation, with the sun focused on the cooking pot, and the inner pot view with water. A model was developed to evaluate the thermal performance of the cooker, which was validated with experimental results. Two types of PCMs were evaluated: technical-grade paraffin and erythritol. Results indicated that cooking lunch for a family was possible with the simultaneous storage of heat during the day. Keeping the utensil afterward inside an insulating box indoors allowed for cooking dinner with the retained heat, and also allowed for using the heat for breakfast the next day.

Other types of direct-focusing solar cookers using PCM storage have also been reported (Foong et al., 2011; Arunasalam et al., 2012; Abinaya and Rajakumar, 2013). Direct-focusing solar cookers with PCM storage have the disadvantages of the PCM being relatively expensive compared to SHTES, thermal degradation of the PCM after numerous charging and discharging cycles, poor heat transfer due to the low thermal conductivity of the PCM, and the need of a solar tracking mechanism.

FIGURE 14.18 (a) Solar cooker in operation, (b) focused sun on utensil, (c) inner pot view of water. *(Lecuona et al., 2013a,b.)*

14.4.2 Oven Solar Cookers Using LHTES

When compared to direct-focusing solar cookers, oven solar cookers using LHTES have been in existence for a longer time and research work done on them is well documented. Some types of oven solar cookers using PCM storage will be presented in this section. The different types of oven solar cookers using PCM TES, as presented by Muthusivagami et al. (2010), are shown in Figure 14.19.

Buddhi and Sahoo (1997) designed and tested the solar cooker shown in Figure 14.19a with LHTES for cooking food late in the evening. In their design,

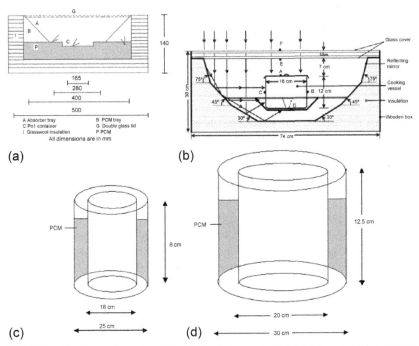

FIGURE 14.19   Oven solar cookers with latent heat storage: (a) Buddhi et al. model, (b) Domanski et al. model, (c) Sharma et al. model, and (d) Buddhi et al. model.

the PCM was filled below the absorbing plate. Commercial-grade stearic acid was used as the PCM. In this design, the rate of heat transfer from the PCM to the cooking pot during the discharging mode of the PCM was slow, and more time was required for cooking food in the evening. Figure 14.19b shows the design by Domanski et al. (1995) of a solar hot-box cooker with LHTES. The possibility of cooking during nonsunshine hours using PCMs as storage media was investigated. Two concentric cylindrical vessels made from aluminum were connected together at their tops, using four screws to form a double-walled vessel with a gap between the outer and inner walls. The gap between the outer and the inner vessels was filled with 1.1 kg of stearic acid (melting temperature 69.8 °C) or 2 kg of magnesium nitrate hexahydrate (melting temperature 89.8 °C), which left sufficient space for expansion of the PCMs during melting. The cooker performance was evaluated in terms of charging and discharging times of the PCMs under different conditions. Results indicated that the performance depended on the solar irradiance, mass of the cooking medium, and the thermophysical properties of the PCM. The overall efficiency of the cooker during discharging of the PCM was found to be three to four times greater than that for steam and heat-pipe solar cookers, which can be used for indoor cooking.

Sharma et al. (2000) designed and developed a cylindrical PCM storage unit for a box-type solar cooker to cook food late in the evening, as shown in Figure 14.19c. The PCM surrounded the cooking vessel, hence the rate of heat transfer between the PCM and the food was high. The designed PCM container had two hollow concentric aluminum cylinders, and the space between the cylinders was filled with acetamide (melting point 82.8 °C, latent heat of fusion 263 kJ/kg) as the PCM. To enhance the rate of heat transfer between the PCM and the inner wall of the PCM container, eight fins were welded at the inner wall of the PCM container. Results obtained from the experimental tests showed that by using 2 kg of acetamide as the PCM, a second batch of food could be cooked if it was loaded before 3:30 p.m. during the winter season. The researchers also recommended that the melting temperature of the PCM be between 105 and 110 °C for evening cooking, and thus there was a need to identify a storage material with an appropriate melting point and quantity that could be used to cook food late in the evening. Buddhi et al. (2003) developed a latent heat storage unit, shown in Figure 14.19d, for a box type of solar cooker with three reflectors. They used acetanilide (melting point 118.9 °C, latent heat of fusion 222 kJ/kg) as a PCM for night cooking. From the experimental results, the authors concluded that cooking experiments were successfully conducted for evening time cooking up to 20:00 hours, with 4.0 kg of PCM used in the storage unit.

Yuksel et al. (2012) experimentally investigated the potential use and effectiveness of paraffin wax in a solar hot-box cooker during daylight and late evening hours. A paraffin wax was used as the PCM, and metal shavings were used in conjunction with the PCM to enhance heat transfer. The effect of the reflector angle on the thermal efficiency of the cooker was tested with different solar insolation conditions on different days. It was concluded that the designed cooker could be used effectively at an angle of 30°. The maximum temperature of the paraffin achieved during the experiments was in the range of 75.1-80.5 ° C. The rectangular solar cooker filled with the paraffin wax was found to have a high thermal performance, which was indicated by high temperatures and decreased cooking times for the given design conditions.

The major drawbacks of oven solar cookers with PCM storage units are low heat transfer rates and low operating temperatures; thus, different heat transfer enhancement mechanisms have to be employed to improve their efficiencies.

14.4.4 Indirect Solar Cookers Using LHTES

Indirect solar cookers utilizing LHTES are a relatively new technology with a limited literature base, compared to oven cookers with LHTES. A few examples of different designs will be presented in this section.

Hussein et al. (2008) developed a novel indirect solar cooker, shown in Figure 14.20, with outdoor elliptical cross-section wickless heat pipes coupled to a flat-plate solar collector with an integrated indoor PCM TES and cooking

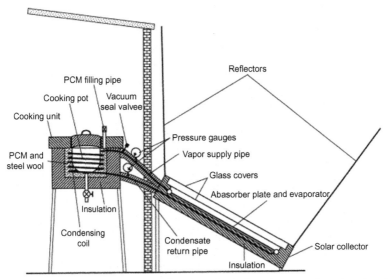

FIGURE 14.20   Latent-heat storage type flat-plate solar cooker using magnesium nitrate hexahydrate as PCM, developed by Hussein et al. (2008).

unit. Two plane reflectors were used to enhance the solar radiation incident onto the collector, while magnesium nitrate hexahydrate (melting temperature 89 °C and latent heat of fusion 134 kJ/kg) was used as the PCM inside the indoor cooking unit of the cooker. Different experiments were performed with the solar cooker without loading and with different loads at different loading times to study the benefit of the elliptical cross-section wickless heat pipes and PCM in the indirect solar cooker. The PCM was evaluated in terms of cooking food at noon, cooking food in the evening, and keeping food warm at night and early in the morning. The experimental results indicated that the solar cooker could be used to successfully cook different kinds of meals at noon, afternoon, and evening times. The cooker could also be used for heating or keeping meals hot at night and early in the morning.

An indirect solar cooker based on an ETSC with a PCM storage unit was developed by Sharma et al. (2005). A schematic diagram of the indirect solar cooker is shown in Figure 14.21. The cooker consists of an ETSC, a closed loop pumping line-containing water as the HTF, a PCM storage unit, a cooking unit, a pump, a relief valve, a flow meter, and a stainless steel tubular heat exchanger. The PCM storage unit has two hollow concentric aluminum cylinders, and the space between the cylinders is filled with 45 kg erythritol (melting point 118 °C, latent heat of fusion 339.8 kJ/kg). A pump circulates the heated water (HTF) from the ETSC through the insulated pipes to the PCM storage unit by using a stainless steel tubular heat exchanger that is wrapped around the cooking unit. During sunshine hours, heated water transfers its heat to the PCM and

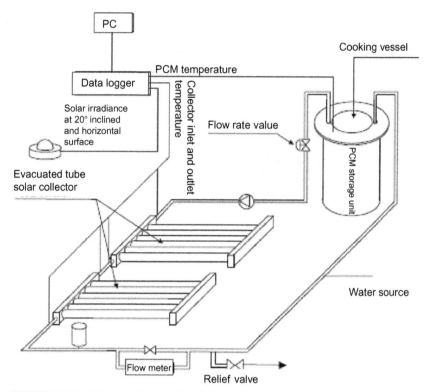

FIGURE 14.21 Schematic diagram of the indirect solar cooker, based on evacuated tube solar collector with a PCM storage unit designed by Sharma et al. (2005).

stores it in the form of latent heat through the stainless steel tubular heat exchanger. The stored heat is utilized to cook food in the evening or when sun intensity is not sufficient to cook food. Results of the experimental tests concluded that evening and noontime cooking were possible. Evening time cooking was also found to be faster than noontime cooking. Experimental results also indicated that this solar cooker yielded satisfactory performance despite the low heat transfer. A modified design of the heat exchanger in the TES unit was suggested to enhance the rate of heat transfer in that solar cooker.

Besides using flat-plate collectors or ETSCs, concentrating collectors may be employed with an LHTES unit for indirect solar cooking applications. One such design has been discussed previously (Mussard and Nydal, 2013a,b). Murty et al. (2007) designed and developed an inclined heat exchanger unit for an SK-14 parabolic solar concentrator (PSC) for off-place cooking, shown in Figure 14.22. The principal objective of this study was to use an inclined HTF column as heat exchanger unit and to evaluate the thermal performance of a PSC assisted with an inclined cylindrical heat exchanger unit for off-place cooking with and without PCM. Experiments were conducted for cooking foods

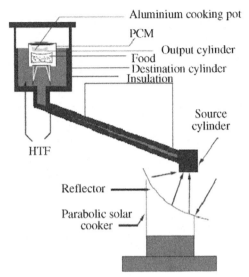

FIGURE 14.22   Schematic of the inclined heat exchanger unit assisted SK-14 PSC. *(Murty et al., 2007.)*

on a normal day, with commercial-grade sodium acetate (melting point is 104 ° C and latent heat of fusion is 230 kJ/kg) and acetanilide (melting point is 115.42 °C and latent heat of fusion is 189.4 kJ/kg) as the LHTES materials. The cooking experiments were conducted with the PCMs as TES media during charging and discharging of the PCMs. It was observed that the cooking time was less during discharging of the PCMs.

Complex-type indirect solar cookers using PCM storage for the future have also been proposed (Parida et al., 2013; Kenesarin and Mahkamov, 2007). Indirect solar cookers using PCM have the major drawbacks of poor heat transfer, complexity in design, and the expense in construction.

14.5   CHARACTERIZATION OF SOLAR COOKERS WITH TES

There are no general methods for characterizing solar cookers with TES, because different designs exist and general thermal performance measures only give rough comparative estimates. Most general methods or standards reported in the literature are applicable to the solar cooking device. This section presents existing solar cooking performance standards, and also presents new proposed figures of merit applicable to solar cookers with TES.

14.5.1   Existing Solar Cooking Standards

The thermal performance of solar cooking systems is determined with useful energy evaluations, and there are currently three well-developed testing standards (Shaw, 2004).

The first standard is the American Society of Agricultural Engineers (ASAE) standard, ASAE S580, which was developed by Funk (2000). This standard is simple and produces a useful and meaningful measure of solar cooker performance based on an energy evaluation. ASAE S580 monitors the average temperature inside a pot of water during operation of the cooker under guidelines given in the standard for tracking and thermal loading. Temperature measurements of the water are taken and averaged every 10 min. The ambient temperature as well as the normal irradiance (solar energy flux per area) are also measured and recorded. The tests are not conducted during windy conditions, low direct solar radiation conditions, and low ambient temperature. The primary figure of merit used by ASAE S580 is the cooking power, P, calculated as:

$$P = \frac{m_w c_w (T_2 - T_1)}{600} \tag{14.14}$$

where c_w is 4.186 kJ/kg/K, T_2 is the final temperature, and T_1 is the initial temperature. 600 is the number of seconds in the 10-min interval. P is normalized to 700 W/m^2 to yield.

$$P_S = P\left(\frac{700}{I}\right) \tag{14.15}$$

where P_S is the standardized cooking power and I is the interval average solar radiation. P_S is plotted against ΔT and a linear regression is performed. For standard reporting procedures, a temperature difference of 50 °C is used (i.e., $T_{water} - T_{ambient} = 50$ °C), and the corresponding value of P_S is reported as the "Cooking Power." The performance evaluation only examines thermal performance, without looking at qualitative issues such as heat losses, safety, ease of use, and affordability.

The second testing standard is referred to as the Indian standard (Mullick et al., 1987). This standard is more technical than ASAE S580, and it provides two figures of merit. These figures are calculated to be as independent of environmental conditions as possible. The first figure of merit is

$$F_1 = \frac{T_P - T_A}{H_S} \tag{14.16}$$

where T_p is the stagnation temperature of the absorber plate, T_A is the ambient air temperature, H_S is the solar radiation on a horizontal surface (taken at a time of stagnation), and the second is

$$F_2 = \frac{F_1 m_w c_w}{A_{ap} t} \ln \left[\frac{1 - \dfrac{F_1(T_{w1} - T_A)}{H_{av}}}{1 - \dfrac{F_1(T_{w2} - T_A)}{H_{av}}} \right] \tag{14.17}$$

where H_{av} is the average horizontal solar radiation, T_{w1} is the initial water temperature, and T_{w2} is the final water temperature. By setting a reference temperature and solving Equation (14.17) for t, a characteristic curve can be developed that describes how long the cooker will take to reach the reference temperature. This method, however, does not take into account qualitative factors like cost and safety.

The third standard was proposed by the European Committee of Solar Cooking Research (ECSCR) and covers a wider scope than the other two standards (ECSCR, 1995). The ECSCR standard devotes more attention to safety factors and uses an exhaustive testing methodology. The evaluation process is driven by several detailed data sheets, which are filled out by the tester. The benefit of this procedure is that there is no need to measure normal incidence solar radiation, which usually requires a tracking pyrheliometer. The disadvantages of this method are errors encountered due to the rigorous testing method, the fact that the method is not suitable for multiple tests, and that it takes a long time (3 days) to complete a basic test.

To cater to the drawbacks of these three standards, a fourth and new standard for the testing of solar cookers that has flexible environmental testing conditions and also addresses qualitative factors like safety was proposed by Shaw (2004). Four thermal performance figures of merit were proposed: standard stagnation temperature, standard cooking power, standard sensible heating time, and unattended cooking time. The author also proposed that the tests be carried out between 10:00 and 17:00 hours, that the cooking vessels be filled with 5 kg of water per square meter of aperture area, that the cooker be tracked according to the manufacturer's specifications, that tests be conducted with pots from the manufacturer, and that an electronic data logger be used. For the tests to be valid, it was also proposed that the wind speed be less than 2.5 m/s for a 10-min data-logging interval, that the ambient temperature be 20-35 °C during the tests, that the solar radiation be 500-1000 W/m^2 for the 10-min interval, and that any presence of precipitation would make the affected results invalid. Though this standard is thorough, it has not been widely adopted by researchers because it is a recent standard.

Besides the generalized standards usually used for direct-focusing cookers and SBCs, methods of characterizing solar cookers, including the indirect types of cookers, have been proposed by Schwarzer and da Silva (2008). The cookers were classified into four types: Type A, flat-plate collector with direct use; Type B, flat-plate collector with indirect use; Type C, parabolic cooker with direct use; and Type D, parabolic cooker with indirect use. The characterization methods suggested that the cooking pots be filled with 5 kg of water or oil per 1 m^2 of collector surface, and that the average solar radiation be higher than 800 W/m^2. The authors also suggested that water be stirred to avoid stratification. For temperatures above 100 °C, oil was to be used with the caution of observing the self-ignition temperature of the oil. They proposed that four thermal performance parameters be evaluated, similar to those of Shaw (2004).

Another recent unified method of characterizing and comparing solar cookers of different geometries based on exergy was proposed and implemented by Kumar et al. (2012). Four exergy-based thermal performance parameters were proposed for solar cookers of different topological design. The four thermal performance parameters were peak exergy, quality factor, exergy temperature difference gap product, and heat loss coefficient. The exergy output power and temperature difference were plotted, and they resembled a parabolic curve for each design. The peak exergy (vertex of the parabola) was deemed acceptable as a measure of the device's fuel ratings. The ratio of the peak exergy power gained to the exergy power lost at that instant of time was considered the quality factor of the solar cooker. The exergy power was found to vary linearly with temperature difference, irrespective of the topology of the device and the slope of the straight line obtained through curve fitting, representing the heat loss coefficient of the cooker.

14.5.2 New Solar Cooking Figures of Merit for Solar Cookers with TES

The thermal performance standards described in the previous section only consider the cooking device, and shed no insight into the storage capability of a solar cooker with TES. New thermal performance figures of merit to cater to a direct-focusing type of solar cooker and a box type of solar cooker with PCM TES were proposed by Lecuona et al. (2013a,b). The nominal heat stored in a cooking utensil can be expressed as

$$Q_s = M_s[L + c_s(T_0 - 70\ °\text{C})] + M_p c_p(T_0 - 70\ °\text{C}) + M_w c_w(T_0 - 70\ °\text{C})$$

$$(14.18)$$

where M_s is the mass of the PCM, L is the latent heat of fusion, and c_s is the specific heat capacity of the PCM, considered to the same for the solid and the liquid phases. The temperature range is defined for $T_0 > T_m > 70\ °\text{C}$, where T_m is the melting temperature for the PCM and 70 °C is the minimum temperature for sterilization. $M_p c_p$ is the heat capacity of the pot and $M_w c_w$ is the heat capacity of the walls. This heat is available for increasing the temperature of the food above the ambient conditions and for keeping the temperature against losses either (i) with no solar input (in the shade) or (ii) under the sun, depending whether it is cooking lunch, breakfast, or dinner. The time for cooking is very dependent on the nature of the food, so that establishing a standard time for keeping the temperature above heat losses is difficult.

A figure of merit can be evaluated considering the theoretical heat storage capacity compared with heat desired for cooking the nominal liquid mass load $M_{l,nom}$ whose specific heat capacity is c_l. Choosing the worst case scenario described previously (i), and equating Q_s to the heat required to increase the temperature of the nominal water load $M_{1,nom}$ to an equilibrium temperature T_{eq}, leads to the limiting temperature that is expressed as:

$$T_{eq} - T_a = \underbrace{\frac{M_s}{M_{1,nom}} \frac{L + c_s(T_0 - T_{eq})}{c_1}}_{\Delta T_s} + \underbrace{\frac{M_p}{M_{1,nom}} \frac{c_p(T_0 - T_{eq})}{c_1}}_{\Delta T_p}$$

$$+ \underbrace{\frac{M_w}{M_{1,nom}} \frac{c_w(T_0 - T_{eq})}{c_1}}_{\Delta T_w}; T_0 \geq T_m \geq T_{eq} \tag{14.19}$$

where the first term represents ΔT_s, the second term ΔT_p, and the last term ΔT_w. In heat storing solar cookers, generally ΔT_p, $\Delta T_w \ll \Delta T_s$, owing to the high melting heat L, $M_p \sim M_s$ and generally c_p, $c_w < c_s$.

As with the stagnation test, if the resulting temperature is higher than 100 °C, oil must be selected for the liquid. ΔT_s in Equation (14.19) varies between ≈ 50 °C for paraffins up to ≈ 110 °C for erythritol, so that with no losses, boiling is possible with a unity ratio of PCM mass to water mass, using the reported values. Another parameter also evaluated by Lecuona et al. (2013a,b) using detailed modeling for solar cookers with PCM TES storage is the useful heat power.

The figures of merit presented by Lecuona et al. (2013a,b) require detailed modeling, and these parameters are not valid for solar cookers with SHTES systems. Mawire et al. (2009) presented a very simple parameter, referred to as the exergy factor, which evaluates the ratio of the exergy stored to the energy stored during charging and discharging of a TES system, and is useful for both LHTES and SHTES systems. The exergy factor is simply a measure of the quality of energy obtained from a given quantity of energy. A high exergy factor indicates that high quality energy is available in a given quantity of energy. The exergy factor can also be used to evaluate the thermal energy performance of TES systems for indirect cookers.

14.6 CONCLUSION

An overview of the three main types of solar cookers has been presented, and the basic operating principles of direct-focusing, oven, and indirect solar cookers have been outlined. These three types of cookers have been also reviewed and discussed when they are used in conjunction with solar TES units to enhance their usefulness during periods when solar radiation is not available. Solar cookers using both SHTES and LHTES have been reviewed and discussed. Advantages and disadvantages of the different types of solar cookers with TES have also been highlighted. Methods of characterizing solar cookers have been presented, and it was concluded that more research efforts must be carried out to develop methods of characterizing solar cookers with TES systems.

The most viable options for solar cookers with TES for developing countries are oven solar cookers and direct-focusing solar cookers, because they are relatively cheap to fabricate and maintain. On the other hand, when issues of efficiency and safety are concerned, indirect solar cookers with TES are more viable; these can be implemented for community-scale cooking because they are relatively expensive to construct. Solar cookers with TES offer an alternative to polluting fossils and LPG in rural areas of developing countries.

REFERENCES

Abinaya, P., Rajakumar, S., 2013. Performance of PCM and cooking vessel in solar cooking system. Int. J. Eng. Invent. 2, 83–89.

Adams, W., 1878. Cooking by solar heat. Sci. Am. 38, 376–377.

Alozie, G., Mejeha1, I., Ogungbenro, O., Nwandikom, G., Akujor, C., 2010. Design and construction of a solar box cooker as an alternative in Nigerian kitchens. ISESCO Sci. Technol. Vis. 6 (9), 57–62.

Arunasalam, A., Ravi, A., Srivatsa, B., Senthil, R., 2012. Thermal performance analysis on solar integrated collector storage. Int. J. Instrum. Control Autom. 1, 68–71.

Badran, A., Yousef, I., Joudeh, N., Al Hamad, R., Halawa, H., Hassouneh, H., 2010. Portable solar cooker and water heater. Energy Convers. Manag. 51, 1605–1609.

Balzar, A., Stumpf, P., Eckhoff, S., Ackermann, H., Grupp, M., 1996. A solar cooker using vacuum-tube collectors with integrated heat pipes. Sol. Energy 58, 63–68.

Buddhi, D., Sahoo, L., 1997. Solar cooker with latent heat storage: design and experimental testing. Energy Convers. Manag. 38, 493–498.

Buddhi, D., Sharma, S., Sharma, A., 2003. Thermal performance evaluation of a latent heat storage unit for late evening cooking in a solar cooker having three reflectors. Energy Convers. Manag. 44, 809–817.

Chandra, J., Pradyumna, K., Teja, A., Badu, A., 2013. Design of a solar cooker. Design project. Gitam University, Rushikonda, Vizag, India.

Chaudhary, A., Kumar, K., Yadav, A., 2013. Experimental investigation of a solar cooker based on parabolic dish collector with phase change thermal storage unit in Indian climatic conditions. J. Renew. Sustain. Energy 5, 023107 (1-13).

da Silva, M., Schwarzer, K., Medeiros, M., 2002. Experimental results of a solar cooker with heat storage. In RIO 02 -World Climate and Energy Conference, Rio DeJanairo, Brazil, 89–93.

Domanski, R., El-Sebaii, A., Jaworski, M., 1995. Cooking during off sunshine hours using PCMs as storage media. Energy 20, 607–616.

ECSCR, 1995. Second international solar cooker test: summary of results. Technical report, European Committee of Solar Energy Research, Germany.

El-Sebaii, A., Ibrahim, A., 2005. Experimental testing of a box-type solar cooker using the standard procedure of cooking power. Renew. Energy 30, 1861–1871.

Esen, M., 2004. Thermal performance of a solar cooker integrated vacuum-tube collector with heat pipes containing different refrigerants. Sol. Energy 76, 751–757.

Foong, C., Nydal, O., Løvseth, J., 2011. Investigation of a small scale double-reflector solar concentrating system with high temperature heat storage. Appl. Therm. Eng. 31, 1807–1815.

Funk, P., 2000. Evaluating the international standard procedure for testing solar cookers and reporting performance. Sol. Energy 68, 1–7.

Haraksingh, I., Mcdoom, M., Headley, O., 1996. A natural convection flat-plate solar cooker with short term storage. In: WREC 1996 conference, Denver, USA, 729-732.

Hawkeye solar cooker, 2011. Mechanical Engineering Senior Design Project Final Report. https://www.engineeringforchange.org/news/2012/02/04/ten_solar_cookers_that_work_at_night.html(accessed January 2014).

Hussein, H., El-Ghetany, H., Nada, S., 2008. Experimental investigation of novel indirect solar cooker with indoor PCM thermal storage and cooking unit. Energy Convers. Manag. 49, 2237–2246.

Kenesarin, M., Mahkamov, K., 2007. Solar energy storage using phase change materials. Renew. Sustain. Energy Rev. 11, 1913–1965.

Kumar, R., Adhikari, R., Garg, H., Kumar, A., 2001. Thermal performance of a solar pressure cooker based on an evacuated tube solar collector. Appl. Therm. Eng. 21, 1699–1706.

Kumar, N., Vishwanath, G., Gupta, A., 2012. An exergy based unified test protocol for solar cookers of different geometries. Renew. Energy 44, 457–462.

Lecuona, A., Nogueira, J., Ventas, R., Rodríguez-Hidalgo, M., Legrand, M., 2013a. Solar cooker of the portable parabolic type incorporating heat storage based on PCM. Appl. Energy 111, 1136–1146.

Lecuona, A., Nogueira, J., Vereda, C., Ventas, R., 2013b. Solar cooking figures of merit. Extension to heat storage. In: Mendez-Vilas, A. (Ed.), In: Materials and Processes for Energy: Communicating Current Research and Technological Developments. Vol 1. Formatex Publishing, Spain, pp. 134–141.

Mawire, A., 2009. Characterisation of a thermal energy storage system developed for indirect solar cooking. PhD thesis. North West University, Mafikeng, South Africa.

Mawire, A., McPherson, M., van den Heetkamp, R., 2008. Simulated energy and exergy analyses of the charging of an oil–pebble bed thermal energy storage system for a solar cooker. Sol. Energy Mater. Sol. Cells 92, 1668–1676.

Mawire, A., McPherson, M., van den Heetkamp, R., Mlatho, J., 2009. Simulated performance of storage materials for pebble bed thermal energy storage (TES) systems. Appl. Energy 86, 1246–1252.

Mawire, A., McPherson, M., van den Heetkamp, R., 2010. Discharging simulations of a thermal energy storage (TES) system for an indirect solar cooker. Sol. Energy Mater. Sol. Cells 94, 1100–1106.

Morrison, G., Di, J., Mills, D., 1993. Development of a solar thermal cooking system. Technical report, School of Physics, University of Sydney, Sydney, Australia.

Mullick, S., Kumar, S., Kandpal, T., 1987. Thermal test procedure for box-type solar cookers. Sol. Energy 39, 353–360.

Murty, V., Gupta, A., Patel, K., Patel, N., Shukla, A., 2007. Design, development and thermal performance evaluation of an inclined heat exchanger unit assisted SK-14 parabolic solar cooker for off-place cooking with and without phase change material. In: 3rd International Conference on Solar Radiation and Day Lighting (SOLARIS 2007), Delhi, India, vol. II, pp. 8–15.

Mussard, M., Nydal, O., 2013a. Comparison of oil and aluminum-based heat storage charged with a small-scale solar parabolic trough. Appl. Therm. Eng. 58, 146–154.

Mussard, M., Nydal, O., 2013b. Charging of a heat storage coupled with a low-cost small-scale solar parabolic trough for cooking purposes. Sol. Energy 95, 144–154.

Mussard, M., Gueno, A., Nydal, O., 2013. Experimental study of solar cooking using heat storage in comparison with direct heating. Sol. Energy 98, 375–383.

Muthusivagami, R.M., Velraj, R., Sethumadhavan, R., 2010. Solar cookers with and without thermal storage—a review. Renew. Sustain. Energy Rev. 14, 691–701.

Nahar, M., 2003. Performance and testing of a hot box storage solar cooker. Energy Convers. Manag. 44, 1323–1331.

Nyahoro, P., Johnson, R., Edwards, J., 1997. Simulated performance of thermal storage in a solar cooker. Sol. Energy 59, 11–17.

Oelher, U., Scheffler, W., 1994. The use of indigenous materials for solar conversion. Sol. Energy Mater. Sol. Cells 33, 379–387.

Parida, O., Tripathy, S., Dash, J., 2013. Night cooking solar cooker using molten sodium chloride as phase change material. Int. Innov. Technol. Res. 1, 201–206.

Prasanna, U., Umanand, R., 2011. Modeling and design of a solar thermal system for hybrid cooking application. Appl. Energy 88, 1740–1755.

Rabl, A., 1985. Active Solar Collectors and Their Applications. Oxford University Press, New York.

Ramadan, M., Aboul-Enein, S., El-Sebaii, A., 1988. A model for an improved low cost-indoor-solar cooker in Tanta. Sol. Wind Technol. 5, 387–393.

Schwarzer, K., da Silva, M., 2003. Solar cooking system with or without heat storage for families and institutions. Sol. Energy 75, 35–41.

Schwarzer, K., da Silva, M., 2008. Characterisation and design methods of solar cookers. Sol. Energy 82, 157–163.

Sharma, S., Buddhi, D., Sawhney, R.L., Sharma, A., 2000. Design, development and performance evaluation of a latent heat storage unit for evening cooking in a solar cooker. Energy Convers. Manag. 41, 1497–1508.

Sharma, D., Iwata, T., Kitano, H., Sagara, K., 2005. Thermal performance of a solar cooker based on an evacuated tube solar collector with a PCM storage unit. Sol. Energy 78, 416–426.

Shaw, S., 2004. Development of a comparative framework for evaluating the performance of solar cooking devices. Master's thesis. Rensselaer Polytechnic Institute, New York, USA.

Shrestha, J., Byanjankar, M., 2007. Thermal performance evaluation of box type solar cooker using stone pebbles for thermal energy storage. Int. J.Renew. Energy 2, 11–21.

Yuksel, N., Arabacgl, B., Avc, A., 2012. The thermal analysis of paraffin wax in a box-type solar cooker. J. Renew. Sustain. Energy 4, 063126(1–9).

Chapter 15

Isolated and Mini-Grid Solar PV Systems: An Alternative Solution for Providing Electricity Access in Remote Areas (Case Study from Nepal)

B. Mainali[1] and R. Dhital[2]

[1]*Energy and Climate Studies, Royal Institute of Technology, KTH, Stockholm, Sweden*
[2]*Alternative Energy Promotion Center, Ministry of Science, Technology and Environment, Lalitpur, Nepal*

Chapter Outline

15.1 INTRODUCTION

Access to electricity is indispensable for the development of any society. The role of electricity is important for productive and economic activities, as well as

Solar Energy Storage. http://dx.doi.org/10.1016/B978-0-12-409540-3.00015-3

359

for the overall health and well-being of communities (Chaurey and Kandpal, 2010). However, most of the rural areas in Nepal are sparsely populated, isolated, and remote in terms of accessibility. The electricity market in poor rural areas is characterized by a low access rate and low load factors (Haanyika, 2008; Mainali, 2011). This rural characteristic increases the unit delivery cost of electricity (Banerjee, 2006).

In the past, electrification had been perceived as the responsibility of a government utility (Ilskog and Kjellström, 2008), and grid extension has remained one of the most common pathways for rural electrification. However, we have learned from our experience that the extension of the grid line may not meet the immediate needs of electricity access for billions of rural people in developing regions. There is a need for innovation in off-grid renewable energy technologies that can be disseminated in rural areas in a cost-effective way, attracting new investment from the private sector and deregulation of energy markets (IEA, 2008).

Off-grid and on-grid options are normally promoted in parallel when pursuing rural electrification in Nepal. Proper resource assessment and analysis of the cost-efficiency of the options at hand is important for making the right choice.

The Alternative Energy Promotion Center (AEPC), an apex government body for the promotion of renewable energy, used to execute various donor-supported, off-grid, rural electrification programs in Nepal. Recently, all those programs have been merged under a single program: National Rural and Renewable Energy Programme (NRREP). Micro hydro and solar technologies are the current choices in Nepal because of the provision of a government subsidy and a well-defined market structure (Mainali and Silveira, 2012). On-grid electrification is developed by the Nepal Electricity Authority (NEA), the national utility accountable for generation, transmission, and distribution of electricity. The Community Rural Electrification Department (CRED) has been established under NEA to perform community-based rural electrification. In addition, there are some independent power producers (IPPs) involved in power generation that is then sold to NEA.

This study presents solar photovoltaics (PV) as an alternative for rural electrification, considering off-grid solar PV for individual households and solar mini-grids for electrifying the rural community, and comparing them with grid extension and with conventional diesel generator for a specific case study in the Kyangshing village of the Sindhupalchowk district in Nepal. Levelized cost of electricity (LCOE) production with these various alternatives is compared, and the most cost-effective option is suggested for meeting all the electricity demand in the village. The business model and operational and management model for supplying electricity in the village are discussed in this chapter.

Following this introduction, Section 15.2 presents the socioeconomic information of the Kyangshing village. Section 15.3 highlights existing energy consumption patterns and the potential electricity demand of the village. The methodology adopted in this study and the data source are presented in Section 15.4. Various technological options for providing electricity access and their specific component sizing are discussed in Section 15.5. Section 15.6 presents the LCOE of various technological options, along with the sensitivity analysis for some of the crucial input assumptions. The possible business model and operation and management model for supplying electricity access in the village is discussed in Sections 15.7 and 15.8. Final conclusions are drawn in Section 15.9. This study serves as a feasibility study and helps the community and government agencies like AEPC to better appreciate the costs behind different technological options and to choose the appropriate technological option for the village. The study also provides the basic business model and operational and management model to implement such a project in a sustainable way.

15.2   SITE DESCRIPTION

The case study taken is from Kyangshing village, situated in Gumba VDC of the Sindhupalchowk district. It takes around 8 h by walking from the nearest road head, Kartike. The village has 48 households, with one commune hall, a Buddhist monastery, and a primary school. The main occupation of this village is agriculture. The total population of Kyangshing village is about 235, with an almost equal share of male (49%) and female (51%). The place is sunny every month, except for a few days of winter. The village is located at the south face, giving longer sunny days throughout the year. The annual average solar insolation in Nepal varies from 3.5–7.0 kWh/m^2/day (Schillings et al., 2004), and the estimated number of days with sunshine in a year is 300 days (WECS, 2010).

The village is about 30 km from the national grid and the extension in the near future. So, a solar PV-based system is one of the potential options for rural electrification. The study team has interacted with the local people to identify existing energy use patterns, future demands, and trends.

Literacy in the village is poor. A large segment of the population (i.e., 177 out of 235 people) is illiterate. None of the population has attended university. Agriculture is the main occupation of the surveyed area. Around 4% of the population has gone to the gulf countries for employment, and the rest of the people are involved in agricultural farming and livestock rearing as their main occupation. This village only has access to primary school, with no direct access to education for secondary school and university, no post office for communication, and no proper market for shopping for daily needs and agricultural products (Figure 15.1).

FIGURE 15.1   Glimpse of Kyangshing village. *(SETM/AEPC (2012).)*

15.3 EXISTING ENERGY CONSUMPTION PATTERNS AND POTENTIAL ELECTRICITY DEMAND

The choice of energy largely depends on the available energy resources, disposable income, education level, type/size of household, and so on (Mainali et al., 2012). Firewood, which is normally collected from the community forest, is the most widely used energy resource in the village for cooking and heating purposes. The use of modern energy sources like kerosene, liquefied petroleum gas (LPG), and biogas are negligible.

For lighting purposes, most families in the village use "Tuki" powered by dry cell batteries. About 15% of the families use solar home systems (SHS) with 100% financial support from the local government (village development committee). However, most of these systems are not performing well, mainly due to poor maintenance.

It has been observed that families with kids in school have more lighting demands. Based on the survey data, demands for lighting, various appliances, and other potential end uses are estimated. The probable and potential demands of various appliances like radios, TVs, mobile chargers, and others as well as productive end uses like a grinder and cyber center have been investigated for the village.

For the lighting load, it is estimated that 48 households will use an average of four lamps in each household, with an average power of 7 W. Only about 60% of the households are assumed to have TV at their houses in the near future; hence, the demand of 80 W TV each is estimated for 30 HH. Four vaccine refrigerators of 70 W each are expected to run for 24 h a day for the existing health post. The electricity demand of the school and monastery is also included. Income-generating activities like running a grinder and cyber center are also included in the design so that the project not only meets basic needs, but

TABLE 15.1 Estimated Load Demands for Kyangshing Village

Load Type	Watt Peak	Hours of Operation	Energy Required (Wh/day)
Lighting load	1344	6	8064
TV load	2400	6	14,400
Mobile charging	144	5	720
ICT (Cyber center)	374	7	2618
Huller	4000	3	12,000
Grinder	4000	3	12,000
Gumba/Monastery	110	8	880
School (Lighting, Computers)	90	4	360
Health Post (Lighting, TV)	90	5	450
Vaccine refrigerator	300	12	3600
	12,852		**55,092**

also supports productive end uses. Table 15.1 shows the lighting demand, various appliances, and productive end uses expected to run, along with the hours of operation and energy required.

15.4 METHODS AND DATA SOURCE

This chapter presents different possible options for providing electricity access using solar technology (i.e., off-grid solar PV for individual households and solar mini-grids for electrifying the rural community) and compares them with grid extension. The system sizing and cost estimations for the mini-grid option are made to meet two demand conditions: household-only electricity demand, and household demand along with productive end use demand of Kyangshing village.

Data are sourced from the field survey conducted by the AEPC. The cost parameters are based on the recent market price.

LCOE is used to quantify and compare the monetary value of electricity produced from various generation technical alternatives irrespective of type, scale of production operation, investment, or life span (Mainali and Silveira, 2013). Levelized cost is the discounted average cost per kWh of useful electrical energy produced by the system over the life of the technology, which can be expressed as

$$\text{LCOE} = \frac{C_c + C_{om} + C_r + C_f}{E_i}$$

Life-cycle costs are compared instead of simple capital costs. Such lifetime costs are basically the discounted costs of the project incurred each year and summed over the life span of the project. These costs comprise capital cost (C_c), operation and maintenance cost (C_{om}), replacement cost (C_r), and fuel cost (C_f). The procedures for estimating such costs have been discussed in Mainali and Silveira (2013) and Nguyen (2007). The input assumptions highly influence the estimated LCOE. Thus, it is important to have a clear understanding and transparent assumptions for accuracy. We further carry out some sensitivity analysis to reflect the uncertainty associated with the various parameters, namely, capital cost and life span of the energy storage system (in the case of a solar PV system), and rise in fuel price (in the case of fossil-based technology).

15.5 TECHNOLOGY SELECTION AND COMPONENT SIZING

In this section, different technological options for providing electricity access and their specific component sizing are discussed. The households can be supplied with individual SHS, with an average size of 40 Wp. SHS basically consist of solar PV modules, a bank of battery, and charge controller, and deliver direct current (DC) output. So, household loads (i.e., compact fluorescent lamps (CFLs), TVs, and small radios) can be served by such systems.

Households in the whole village can also be served with a solar PV minigrid. Appropriate component sizing for solar PV mini-grid systems is important to ensure reliable operation.

Power generation may vary depending on the size of PV modules. The capacity/size of PV modules should be sufficient to meet daily energy demand as well as system losses that may arise due to different efficiencies of the various components. First, the total wattage hours (WHs) that need to be served per day is estimated depending on the load demand. Then, to estimate the total WH to be supplied by the panel, this WH is increased by 30% to capture various system losses. To determine the sizing of the PV modules, the total peak watts required to be produced should be known. The peak watt (Wp) produced depends on watt hours/day to be supplied and local climatic conditions (mainly determined by sunshine hours and solar irradiation). This can be considered taking the "panel generation factor," which is different for each location. We do not have a site-specific factor, so the panel generation factor for Nepal (4.15) has been used in this case study.

A deep-cycle battery bank has been recommended for the solar PV system, as these batteries are capable of discharging up to low energy level and rapid recharging frequently for years. When selecting the size of the battery, the capacity should be large enough to store sufficient energy to operate the load demands at night and on cloudy days. The battery backup system should be able to store the required energy in such cases. For reliable and consistent supply of energy during rainy or cloudy days, the battery system should be designed with some autonomy. Battery loss of 85%, depth of discharge (DoD) of 0.6, and two days of autonomy have been assumed in these estimations. Use of deep-cycle solar tubular batteries is suggested for added compatibility with the solar PV and electrical load system.

Inverters are used for generating alternating current (AC) outputs. The size of the inverter should be sufficiently higher than the total watts of appliances, and should consider the inverter efficiency. Specifically, the inverter must be large enough to handle the total amount of watts that will be used at one time (total peak load) in the system when it is operated in stand-alone mode. The inverter must have the same nominal voltage as the battery. The inverter size has been chosen 30% larger than the total watts of the appliance. If the appliance is a motor or compressor, then the size of the inverter should be as high as 3 times the capacity of those appliances. It has been assumed that the huller and grinder will be operated in the daytime, when there is no other major load in the system. For end uses with critical health and communication devices, pure sine wave solar inverters are preferred. The sizing of the solar charge controller is determined by the number of solar PV modules connected in parallel in the array and the short circuit current of those PV modules. The system voltage rating shall be matched with the battery bank voltage in the case of the pulse-width modulated (PWM) charge controller. However, if the selected charge controller is of maximum power point tracking (MPPT) type, the solar PV array voltage will be adjusted by the MPPT controller itself to synchronize the voltage.

In this study, three different situations have been analyzed: (i) household lighting load only (supplying with SHS), (ii) household lighting and other appliances, and (iii) household and other productive end use in the village. Mini-grid solar PV has been proposed for meeting the demand of the latter two cases. For all three cases, total capacity of the proposed mini-grid system and daily energy demand have been estimated.

   i. ***Only household lighting—23.18 kWh/day
  ii. Household lighting and other appliances—31.09 kWh/day
 iii. All loads with productive end use—55.09 kWh/day

Similarly, optimal size of solar PV modules, storage capacity of the battery, and capacity of the charge controller and inverter were identified as per the demand forecasted in various cases. The technical specifications and component sizing for the isolated PV system and mini-grid solar systems, considering different load demands, are tabulated in Table 15.2.

15.6 LEVELIZED COST OF ELECTRICITY (LCOE)

15.6.1 Solar PV System

After discussing the design aspect of the solar PV systems, in this section, LCOE has been estimated for (i) SHS supplying only lighting loads and the solar mini-grid system for (ii) household energy demand and (iii) household

TABLE 15.2 Design Parameters of Solar PV for Different Cases

Description	Isolated Solar Home System	Solar Mini-Grid System	
Load type	Household lighting load only	Household energy demand only	Household and other productive endues
PV module size	40 Wp*47 Nos (1 in each household)	250 Wp*30 Nos	250 Wp*70 Nos
PV module type	Mono or polycrystalline	Mono or polycrystalline	Mono or polycrystalline
Inverter	NA	5 KVA; 48 volt; pure sine wave inverter	13 KVA; 48 volt; pure sine wave inverter
Battery size	60 AH, 12 V	2500 AH, 48 V	6000 AH, 48 V
Battery type	Ordinary deep-cycle lead-acid battery	2 volt type, tubular gel maintenance-free valve- regulated lead-acid (VRLA) battery	2 volt type, tubular gel maintenance-free valve-regulated lead-acid (VRLA) battery
Solar charge controller	Maximum power point tracking (MPPT)	Maximum power point tracking (MPPT)	Maximum power point tracking (MPPT)

energy demand and other productive end uses in the village. In this study, we have adopted a discount rate of 10% and the general escalation factor of 5%. Technical and economical parameters for solar PV systems in various cases are presented in Table 15.3.

Analysis has shown that LCOE is 0.645 USD/kWh if the system has to supply only DC power to household lighting loads with individual SHS of 40-watt peak. If the system is designed to meet household lighting loads together with some other appliances (e.g., TV, mobile charger, fan etc.) with a mini-grid system (with AC), then the LCOE can be as high as 0.820 USD/kWh. If the system is designed to meet the entire village load (i.e., household lighting loads and other household appliances, electricity demand of schools, health post, monastery and other productive end uses like a huller, grinder, and others, as discussed in Section 15.3), the estimated LCOE is 0.750 USD/kWh.

Sensitivity analysis was performed to see the variation in the levelized cost due to uncertainty in different input parameters. It has been found that the LCOE is more sensitive in the variation in energy storage cost (i.e., cost of the battery and its life span).

TABLE 15.3 Technical and economical parameters for solar PV systems in various cases

Description	Isolated Solar Home System	Solar Mini-Grid System	
Load type	Household lighting load only	Lighting load and other household appliances	Household lighting, appliances, and other productive endues in the village
PV module cost in USD[a]	140	7970	18,600
PV module life span in years	20	20	20
Inverter cost in USD	-	1875	4690
Inverter life span in years	-	10	10
Battery cost in USD	56	24,500	59,000
Battery life span in years	3	3	3
Charge controller cost in USD	14	1355	1355
Charge controller life span in years	10	10	10
Overall system life span in years	20	20	20
Other accessories' cost in USD	25	3770	4180
Distribution cost in USD	-	3000	3825
Transportation cost in USD	10	1060	2350
Installation cost in USD	25	3590	4650
Levelized cost in USD/ kWh	**0.645**	**0.820**	**0.750**

[a] 1 USD = 98 Nepalese rupees.

The LCOE from the mini-grid system is more sensitive with the cost of the energy storage system (batteries). With the 25% change in the capital cost of the energy storage system (battery cost), the variation in the LCOE is around 8% for the solar mini-grid system supplying lighting loads, household appliances, and other productive end uses in the village. Variation in the LCOE is only 3% in the case of individual SHS supplying household lighting load only (Figure 15.2). Life spans of the batteries also have some impact on the LCOE, which is shown in Figure 15.3.

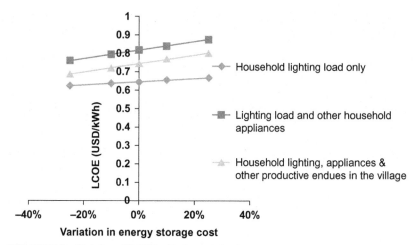

FIGURE 15.2 Variation of LCOE with the variation in energy storage cost.

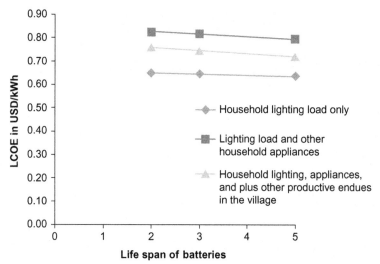

FIGURE 15.3 Variation of LCOE with variation in the life span of the energy storage system (batteries).

15.6.2 Comparison with Diesel Generator Supply Option

Diesel generators (DGs) can also be an option to supply power in the mini-grid approach to meet the demand of rural areas, especially because the capital investment of such generators is lower compared to other energy technologies. We have also explored the electrification option with a DG in the village for the purpose of cost comparison. A brushless, three-phase, 400 V DG with power factor of 0.8 with revolving field and directly coupled was used for the cost

TABLE 15.4 Technical and Cost Parameters for DG Sets

Description	Values
Diesel generator capacity in kW	15 kW
System life span in hours	20,000
Diesel generator and its accessories' installation cost in USD	10,500
Cabling distribution cost in USD (with 200 HH/km^2 and 200 HH and 4 km/km^2)	2090
Annual operation and maintenance cost in USD (5% of the total DG installation cost)	660
Fuel tank cost in USD	600
Fuel tank life span in years	3
Fuel cost at the local market USD/ltr	1.3
Average diesel price escalation in %	4
Levelized cost (USD/kWh)	**0.869**

analysis. A generator of minimum 15 kW is needed to meet the local household lighting and other appliance demands, plus the other end uses in the village.

The technical and cost information of DG sets is tabulated in Table 15.4.

The estimated LCOE from the DG set at this village is 0.869 USD/kWh. When estimating LCOE, a fuel price escalation factor of 4% has been considered. However, the fossil-fuel cost in the international market is volatile, which also impacts the diesel price in the local market. Therefore, to understand the sensitivity of the levelized cost with the fuel price escalation factor, a sensitivity analysis has been performed. This has significant impact on the levelized cost, which has been shown in the figure. With a fuel escalation factor of 15%, the LCOE can be increased by 40% (Figure 15.4).

15.6.3   Comparison with Grid Line Supply Option

As mentioned previously, this village is about 30 km away from the national grid line. In this study, we have also explored the option of supplying electricity in the village with the national grid line extension. Apart from the distance from the grid line, the grid line extension cost was also subjected to average household energy demand, load density (i.e., number of households per km^2 periphery of service area), and total number of households to be supplied. The technical and cost parameters considered for grid line extension in the village are tabulated in Table 15.5.

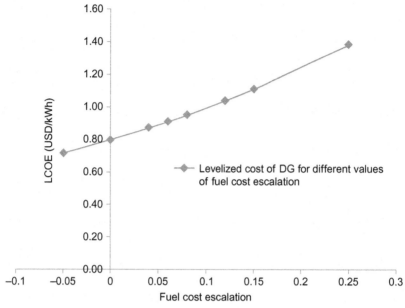

FIGURE 15.4 Variation of LCOE with variation in the fuel cost escalation factor.

The analysis has been carried out with the load density of 75 households per km^2 (hilly regions) for supplying 50 households. Average incremental cost (AIC) has been used as a proxy for marginal costing in electricity system analysis (Mainali and Silveira, 2013). This AIC has been projected to be 5.83 cents/kWh for the year 2014 in Nepal (ADB, 2004). The same value had been adopted for this case as well. The unit cost model has been used to estimate the transmission and distribution cost per km, referring to the current market price. The estimated LCOE of grid line extension is 0.867 USD/kWh. So among various alternative explored, mini grid solar PV is the most cost effective option f to meet both the household and other productive end use demand of the village.

15.7 BUSINESS MODEL FOR MINI-GRID SOLAR PV SYSTEM

Most often, government investments and public budgets are insufficient to expand modern energy access in rural areas in a sustainable manner. Therefore, mobilizing all available financial resources and creating innovative business models is crucial for the expansion of local energy access in the developing world (Sovacool, 2013a). Realizing the importance of the market mechanism and the success of privatization efforts in various countries has suddenly increased interest in the public-private partnership (PPP) concept (Jamali, 2004). Public–private partnerships have been widely used as a mechanism where unique characteristics of both private and public sectors are utilized

TABLE 15.5 The Technical and Cost Parameters for the LCOE of Grid Line Extension

Description/Specification	Quantity
Average household electricity demand (kWh/month/HH)[a]	34
AIC of electricity supply (USD/kWh)[b]	0.583
Transmission line (11 kV)	
Line extension (USD/km)	3975
Annual maintenance and operation (M&O) cost of lines and transformer (USD/km/yr)	205
Transformer 11 kV[c]	2850
Distribution line (440 V)	
Distribution line length per km^2 load area (km/km^2)	4
Distribution line extension cost (USD/km)	2075
Annual M&O cost of distribution (USD/km/yr)	105
Service wire connection and house wiring (USD/HH)	85
Overall transmission/distribution line loss	0.1
Life span in years	40
LCOE (USD/kWh)	**0.867**

[a]*Demand includes productive end uses in the village.*
[b]*ADB (2004).*
[c]*Mainali and Silveira (2013).*

for efficient service delivery and for developing the infrastructure projects. In such partnerships, responsibilities, investments, risks, as well as rewards are shared among partners (Sovacool, 2013b).

Doing business with the poor can involve substantial business risk. To address the problems associated with doing business with the bottom section of the economic pyramid, an innovative PPP approach has been emerging in recent years, engaging governments as well as private companies, microfinance institutions, multilateral development banks/agencies, and nonprofit organizations (including nongovernmental organizations (NGOs)) in expanding access to energy services (Felsinger, 2010). A pro-poor PPP, denoted by the abbreviation "5P," has evolved to explicitly target the provision of services to poor rural communities. The main feature of the 5P business model is that it just does not view the poor as consumers who receive benefits, but also considers the poor as partners in business ventures. Pro-poor PPPs are one of the best mechanisms to supplement and overcome government budgetary constraints for widening access to energy services, especially to the poor, as this can allocate project risks between the public and private sector (Sovacool, 2013a). This model can be used for the implementation of a solar mini-grid.

Proposed partnership modality for the 5P business model:

- A special purpose vehicle (SPV) company shall be formed to own the projects, with joint ownership by a private company, the local community, and an NGO. This will allow the SPV to make highly leveraged or speculative investments, without exposing the entire stakeholder (i.e., private company, community, and the NGO) to any financial risk.
- The NGO shall be in the ownership structure of the 5P project, with major roles of social mobilization and facilitation between the local community and the private company.
- The SPV should also be eligible for the government grant. However, the government grant shall not exceed 50% of the total project cost.
- The bank finance should cover at least 15% of the project installation cost as a loan. The bank loan shall be secured against the projected cash flow of the project and project assets, and there should be buyback guarantee from the equipment supplier in case of project failure.
- The project partners are expected to continue their collaboration throughout the estimated project life span.

15.8 OPERATIONAL AND MANAGEMENT MODEL FOR THE SOLAR MINI-GRID SYSTEM

Though the operation and management aspects of a project are important, they are often given less attention. Formation of a project management committee is the most essential part for the sustainability of such a project. Active participation of local community members on such a management committee is a must. A well-managed project with a strong institutional aspect can overcome any challenges with much less difficulty compared to those with a weak institutional aspect. Components like regular monitoring, transparency, systematic record keeping, logbook maintenance, and effective community mobilization are crucial for sustainable management. At least one operator and one project manager/account keeper have been proposed for the smooth operation of the power plant. The operator needs to be given some specific training on plant operation and basic repair and maintenance, while the manager/account keeper should be trained for plant management, tariff collection and follow-up, customer relations, and more.

15.9 CONCLUSION

This study specifically presents solar PV as an alternative for rural electrification, considering off-grid solar PV for individual households and solar mini-grids for electrifying rural communities, and compares them with grid extension and supplying through a DG for a specific case study in the Kyangshing village of the Sindhupalchowk district. Potential load has been estimated based on survey data obtained from AEPC Nepal. LCOE has been estimated to compare the cost-effectiveness of these options.

The analysis has shown that SHS (individual household system) have the lowest LCOE (0.645 USD/kWh) among various options considered; however, under such cases, the uses of electricity are limited to lighting and some other small applications. For meeting the different productive end uses, the solar mini-grid option looks most cost-effective (LCOE: 0.750 USD/kWh). The LCOE can be as high as 0.867 USD/kWh for grid line extension and 0.869 USD/kWh for the DG option. Sensitivity analysis has also shown that the assumed cost parameters and life span of the energy storage system (battery system) are crucial in determining the LCOE for the solar PV mini-grid system.

In contrast to a normal rural setup, the settlements in this particular village are dense, with short distance among the households. Also, almost all of these households are south facing in the orientation favoring solar mini-grid electrification. Community property like the roof of the school building, which is centrally located in the village, can be used for solar panel installation. A small separate room can be constructed for the energy storage system (battery bank and other equipment). The micro-hydro option has not been taken into consideration, as there is no technically feasible site nearby to fulfill the village demand. Grid extension and a DG were not the most cost-effective options compared to the solar mini-grid. Therefore, the solar mini-grid option seems the most cost-effective, meeting all kinds of demand in the village. Financing such schemes is a real challenge, especially in the rural parts of the country where most of the population is financially poor. A special purpose vehicle (SPV) company under the 5P business model has been proposed for this project, so as to reach the poorer section of the people. Likewise, an effective operational and maintenance model is necessary to ensure sustainable operation of the project. The operator needs to be given some specific training on plant operation and basic repair and maintenance training, and account keeping and project management training for the plant manager will help the sustainable operation and management of the project.

REFERENCE

ADB, 2004. Nepal: Power Sector Reforms in Nepal. Technical Assistance Consultant's Report. International Resources Group, Washington D.C, USA.

Banerjee, R., 2006. Comparison of option for distributed generation in India. Energy Policy 34, 101–111.

Chaurey, A., Kandpal, T.C., 2010. Assessment and evaluation of PV based decentralized rural electrification: an overview. Renew. Sust. Energ. Rev. 14, 2266–2278.

Felsinger, K., 2010. The Public Private Partnership Handbook. Asian Development Bank, Manila.

Haanyika, C.M., 2008. Rural electrification in Zambia: a policy and institutional analysis. Energy Policy 36, 1044–1058.

IEA, 2008. Electricity Access Database in 2008. International Energy Agency, Paris. Available at: http://www.iea.org/weo/database_electricity/electricity_access_database.htm.

Ilskog, E., Kjellström, B., 2008. And then they lived sustainably ever after? Assessment of rural electrification cases by means of indicators. Energy Policy 36, 2674–2684.

Jamali, D., 2004. Success and failure mechanisms of public private partnerships (PPPs) in developing countries Insights from the Lebanese context. Int. J. Public Sect. Manage. 17, 414–430.

Mainali, B., 2011. Renewable energy market for rural electrification in developing countries: country case Nepal. Licentiate thesis, Division of Energy and Climate Studies, KTH.

Mainali, B., Silveira, S., 2012. Renewable energy market in rural electrification: country Case Nepal. Energy Sustainable Dev. 16, 168–178.

Mainali, B., Silveira, S., 2013. Alternative pathways for providing access to electricity in developing countries. Renew. Energy 57, 299–310.

Mainali, B., Shonali, P., Nagai, Y., 2012. Analyzing cooking fuel and stove choices in China till 2030. J. Renewable and Sustainable Energy 4, 0318051-14.

Nguyen, K.Q., 2007. Alternatives to grid extension for rural electrification: decentralized renewable energy technologies in Vietnam. Energy Policy 35, 2579–2589.

Schillings, C., Meyer, R., Trieb, F., 2004. High resolution solar radiation assessment for Nepal. Country paper prepared for UNEP/GEF.

SETM/AEPC, 2012. Feasibility Study of Community Solar Electrification (Solar Mini Grid) in Selected Site. Sustainable Energy and Technology Management and Alternative Energy Promotion Centre, Nepal.

Sovacool, B.K., 2013a. Expanding renewable energy access with pro-poor public private partnerships in the developing world. Energy Strategy Rev. 1, 181–192.

Sovacool, B.K., 2013b. Energy Access and Energy Security in Asia and the Pacific, ADB Economics Working Paper, Series No. 383

WECS, 2010. Energy synopsis report: Nepal. Water and Energy Commission Secretariat, Government of Nepal.

Index

Note: Page numbers followed by *f* indicate figures and *t* indicate tables.

Printed in the United States
By Bookmasters